Lecture Notes in Control and Information Sciences

Volume 455

For further volumes:
http://www.springer.com/series/642

About this Series

This series aims to report new developments in the fields of control and information sciences—quickly, informally and at a high level. The type of material considered for publication includes:

1. Preliminary drafts of monographs and advanced textbooks
2. Lectures on a new field, or presenting a new angle on a classical field
3. Research reports
4. Reports of meetings, provided they are
   (a) of exceptional interest and
   (b) devoted to a specific topic. The timeliness of subject material is very important.

Harald Waschl · Ilya Kolmanovsky
Maarten Steinbuch · Luigi del Re
Editors

Optimization and Optimal Control in Automotive Systems

Springer

Editors
Harald Waschl
Luigi del Re
Institute for Design and Control
of Mechatronical Systems
Johannes Kepler University Linz
Linz
Austria

Maarten Steinbuch
Department of Mechanical Engineering
Eindhoven University of Technology
Eindhoven
The Netherlands

Ilya Kolmanovsky
Department of Aerospace Engineering
University of Michigan
Ann Arbor, MI
USA

ISSN 0170-8643 ISSN 1610-7411 (electronic)
ISBN 978-3-319-05370-7 ISBN 978-3-319-05371-4 (eBook)
DOI 10.1007/978-3-319-05371-4
Springer Cham Heidelberg New York Dordrecht London

Library of Congress Control Number: 2014933667

Printed on acid-free paper

Springer is part of Springer Science+Business Media (www.springer.com)

Preface

The use of optimization techniques is becoming essential to address rapidly increasing stringency of requirements for automotive systems. In particular, there is a growing interest in systematic optimization approaches that can be exploited for automotive vehicle development at various levels. Against this background, a workshop organized by the Austrian Center of Competence in Mechatronics (ACCM) was held at the Johannes Kepler University in Linz from July 15 to 16, 2013. This workshop aimed to bring together specialists in optimization theory and methods on one hand and practitioners from inside the automotive community. The purpose was to foster an exchange of information and an open discussion between representatives of different areas and approaches about the problems, methods, tools and about applications of optimization in general, and dynamic optimization, in particular.

The contents of this book are peer reviewed versions of selected workshop contributions and are structured into four parts, starting with a survey on optimization fundamentals and tailored methods, then followed by three parts addressing different types of automotive optimization problems, at vehicle-to-vehicle and inter-vehicle applications level, at single vehicle powertrain optimization level and finally at purely engine-related level.

Neither the workshop nor this collection of contributions would have been possible without the support of several people (in particular Daniela Hummer and Sandra Pfistermüller). Thanks are due also to the reviewers of the single chapters who have done an important and essential work.

Organization

Steering Organization

Austrian Center of Competence in Mechatronics, Linz, Austria

Hosting Organization

Johannes Kepler University Linz, Austria

Programme Committee

Luigi del Re	Johannes Kepler University Linz, Austria
Ilya Kolmanovsky	University of Michigan, USA
Maarten Steinbuch	Eindhoven University of Technology, The Netherlands
Harald Waschl	Johannes Kepler University Linz, Austria

Organizing Committee

Daniela Hummer	Johannes Kepler University Linz, Austria
Harald Waschl	Johannes Kepler University Linz, Austria

Referees

I. Besselink
P. Colaneri
A. G. De Jager
S. Di Cairano
M. C. F. Donckers
L. Eriksson
L. F. P. Etman
D. Filev
D. J. Guerriero Tome Antunes
R. Gupta
I. Haskara
E. Hellström
H. Hjalmarsson
T. Hofman
M. Huang
M. Jankovic
N. Killingsworth
M. Lazar
C. Manzie
K. McDonough
S. Onori
J. P. Pauwelussen
J. Ploeg
A. Y. Pogromski
A. Rao
A. Saccon
M. Sassano
A. Schilling
A. Sciarretta
R. Shorten
T. Stanger
M. Steinbuch
G. Steinmaurer
S. Szwabowski
M. Tanelli
P. Tunestal
T. A. C. Van Keulen
Y. Wang
F. Willems
K. Zaseck

Contents

Introduction

Luigi del Re, Ilya Kolmanovsky, Maarten Steinbuch and Harald Waschl

Most user requirements for technical products can be formulated in terms of a constrained optimization problem. For instance one may wish to develop an engine with the highest fuel efficiency under the constraints of providing the desired torque characteristics and limiting noise, emissions, weight, size, cost, etc., given by the legislators or set by the customer expectations.

In the case of automotive systems, most requirements must be met not only in special, well-defined driving scenarios but also in a wide variety of environmental and road conditions under the action of an unknown driver. Furthermore, automotive systems exhibit significant production and lifetime deviations, which makes the 'real' parameters frequently unknown. All this means that optimization in the automotive context must include in some way the uncertainty–roughly speaking, we are dealing with the optimization of partly unknown systems under greatly unknown conditions.

Traditionally, automotive systems have been optimized rather heuristically, and the impressive results are more the consequence of an enormous effort than of a systematic optimal design. While commercial aspects–first of all the production scale and the possibly catastrophic consequences of errors–have made this approach viable and maybe even necessary, lately the interest in more systematic approaches has been increasing driven by the need to reduce development time and cost. This has led to two main directions: on one hand, tools for calibration (tuning) support have been developed, based usually on local data-based models and optimizations; on the other hand the academic work in the field of model-based control has been gaining additional momentum and has found some applications in production.

Between 15 and 16 July 2013, a workshop on the subjects of optimization in automotive systems took place in Linz to assess the status and discuss ways to improve the availability of optimization methods in the industrial practice. Since optimization can be exploited at different design stages, starting from the system design level and the selection of an optimal topology, to the operation level, and the development of an optimal control strategy, many relevant topics were addressed. Most attention, nevertheless, was given to the control-related optimization as it and the coupled problem of control and design optimization are comparatively less studied and understood.

The book begins with a discussion of dynamic optimization techniques and methodologies for automotive applications, and includes a basic introduction into dynamic and trajectory optimization techniques (Chap. 1), and three more specific methods which are extremum seeking (Chap. 2), Model Predictive Control (Chap. 3) and an optimal nonlinear feedback regulation technique based on the Hamilton–Jacobi–Bellman equation (Chap. 4). The latter approach is an example of a theoretically advanced methodology that has been shown to be effective for automotive systems once combined with a suitable system identification method.

The book then proceeds to discuss automotive applications of dynamic optimization and optimal control at various levels. At a very high level, significant benefits are expected from its use to control the traffic flows and enforce their fluidity as discussed in Chap. 5. Much can also be gained at the level of single vehicle operation and active safety as discussed in Chaps. 6–8. While these three contributions are mainly centered on handling and safety, two other contributions (Chaps. 9 and 10) examine applications of dynamic optimization in assistant systems for vehicle speed control.

Demonstrating optimization potential at powertrain level is pursued next. This topic addressed especially in the context of hybrid electrical vehicles (HEVs), in terms of topology optimization in Chap. 11, optimal energy management (Chaps. 12 and 13) where the latter contribution includes battery ageing which is a significant consideration in view of battery ageing and cost.

Another application of dynamic optimization, now to coordinated control of Diesel engine aftertreatment, is the topic of the contribution of Chap. 14. While the use of dynamic optimization for HEV energy management is relatively well understood, there is now a growing interest in the use of dynamic optimization to achieve effective control of aftertreatment systems and emission reductions.

Finally the optimization of engine operation, a topic that has received many theoretical but also industrial contributions is addressed. Specifically, Chap. 15 addresses the problem of optimal calibration of engine maps by learning methods from an industrial point of view, and Chap. 16 presents development aspects of a commercially available optimization tool for a similar task. Chapter 17 tackles the problem of optimal control of homogeneous charge compression ignited (HCCI) engines, while Chap. 18 in some sense closes the gap between the vehicle and engine control addressing the optimal operation at vehicle level while taking into account the engine operation.

Some general conclusions can be drawn from the discussion. Briefly, there is a gap between academic and industrial communities. This gap is due to legacy issues which make it difficult for industrial users to employ or test new optimization and optimal control methods and the need of the academic community to address more complex questions that incorporate more realistic models and system engineering aspects and requirements.

More in detail:

Scope — Optimization is important in many different fields, but it must be seen as a part of the development process, i.e. it must take into account system engineering aspects and support a sequential process of design, modelling and optimization.

Models — Not surprisingly, different kinds of models greatly affect the complexity of the optimization task. As a general trend, simple first principle-based models initially used for design, are increasingly complex and accurate data-based models that tend to deliver better results for optimization replaced by more. Still, first principle models tend to be used also for optimization and are in fact preferred by some legacy users.

Industrial priorities — Optimal control and dynamic optimization are becoming more important in industry, yet so are related tasks, like optimal calibration and tuning, that are frequently overlooked by the academic community. Addressing these related tasks can lead to immediate and direct impact on industrial development processes. On the other hand, the academic work is concentrating more on the algorithmic solutions of the standard problems and these solutions are of growing interest in new applications.

Need of hand-on proof — For the final user, it is not easy to choose between different approaches and to see the benefits of a specific optimization method, especially because many algorithms are developed under simplified conditions which would not be realistic in the normal product operation. There is a need to establish evaluation methods that take into account implementation elements, robustness and certifiability.

Adaptive methods — Adaptive optimal methods are potentially an excellent approach to improve the operation under real, changing conditions, but methods are required to guarantee and/or test their performance/convergence/safety under all conditions. This is a critical point also in view of certification.

Human in the loop Human interaction is an important factor for the operation of optimal control-based systems, metrics to evaluate subjective criteria are needed. Human behaviour and reactions can also be embedded in the optimization chain.

Acknowledgments

The authors gratefully acknowledge the sponsoring of this work by the COMET K2 Center "Austrian Center of Competence in Mechatronics (ACCM)". The COMET Program is funded by the Austrian federal government, the Federal State Upper Austria and the Scientific Partners of ACCM.

Part I
Optimization Methods

Chapter 1
Trajectory Optimization: A Survey

Anil V. Rao

Abstract A survey of numerical methods for trajectory optimization. The goal of this survey is to describe typical methods that have been developed over the years for optimal trajectory generation. In addition, this survey describes modern software tools that have been developed for solving trajectory optimization problems. Finally, a discussion is given on how to choose a method.

1.1 Introduction

Trajectory optimization is a process where it is desired to determine the path and the corresponding input (control) to a dynamical system that meet specified constraints on the system while optimizing a specified performance index. Typically, optimal trajectory generation is performed *off-line*, that is, such problems are not solved in real time nor in a closed-loop manner. Because of the complexity of most applications, optimal trajectories are typically generated using numerical methods. Numerical for trajectory optimization date back nearly five decades to the 1950s with the work of Bellman [5–10]. Because complexity of modern applications has has increased tremendously as compared to applications of the past, methods for trajectory optimization continue to evolve and the discipline is becoming increasingly relevant in a wide range of subject including virtually all branches of engineering, economics, and medicine.

Numerical methods for trajectory optimization are divided into two major classes: indirect methods and direct methods. In an indirect method, the first-order optimality conditions from variational calculus are employed. The trajectory optimization problem is then converted into a multiple-point Hamiltonian boundary-value problem. The HBVP is then solved numerically to determine candidate optimal trajectories

A. V. Rao (✉)
University of Florida, Gainesville, FL, USA
e-mail: anilvrao@ufl.edu

H. Waschl et al. (eds.), *Optimization and Optimal Control in Automotive Systems*,
Lecture Notes in Control and Information Sciences 455,
DOI: 10.1007/978-3-319-05371-4_1,

called *extremals*. Each extremal solution of the HBVP is then examined to see if it is a local minimum, maximum, or a saddle point, and the extremal with the lowest cost is chosen. In a direct method, the state and/or control of the original trajectory optimization problem is approximated by parameterizing the state and/or the control and the trajectory optimization problem is transcribed to a finite-dimensional nonlinear programming problem (NLP). The NLP is then solved using well known optimization techniques.

It is seen that indirect methods and direct method emanate from two different philosophies. On the one hand, the indirect approach solves the problem indirectly (thus the name, *indirect*) by converting the trajectory optimization problem to a boundary-value problem. As a result, in an indirect method the optimal solution is found by solving a system of differential equations that satisfies endpoint and/or interior point conditions. On the other hand, in a direct method the optimal solution is found by transcribing an infinite-dimensional optimization problem to a finite-dimensional optimization problem.

The two different philosophies of indirect and direct methods have led to a dichotomy in the trajectory optimization community. Researchers who focus on indirect methods are interested largely in the numerical solution of differential equations, while researchers who focus on direct methods are interested primarily in the numerical solution of optimization problems. While at first glance these two approaches may seem completely unrelated, they have a great deal in common. As will be described in the survey, recent years researchers have delved quite deeply into the connections between the indirect and direct forms. This research has uncovered that the optimality conditions from many direct methods have a well-defined meaningful relationship. Thus, these two classes of methods are merging as time goes by.

1.2 Trajectory Optimization Problem

A fairly general trajectory optimization problem is posed formally as follows. Typically, the problem is divided into P *phases* [15] and the phases are connected in some meaningful way. A multiple-phase trajectory optimization problem is posed as follows. Optimize the cost functional

$$J = \sum_{k=1}^{P} \left[\Phi^{(k)} \left[\mathbf{y}^{(k)}(t_0), t_0, \mathbf{y}^{(k)}\left(t_f\right), t_f; \mathbf{s} \right] + \int_{t_0^{(k)}}^{t_f^{(k)}} \mathscr{L}\left[\mathbf{y}^{(k)}(t), \mathbf{u}^{(k)}(t), t; \mathbf{s}^{(k)} \right] dt \right] \tag{1.1}$$

subject to the dynamic constraints

$$\dot{\mathbf{y}}^{(k)}(t) = \mathbf{f}\left(\mathbf{y}^{(k)}(t), \mathbf{u}^{(k)}(t), t; \mathbf{s}^{(k)} \right), \tag{1.2}$$

the *boundary conditions*,

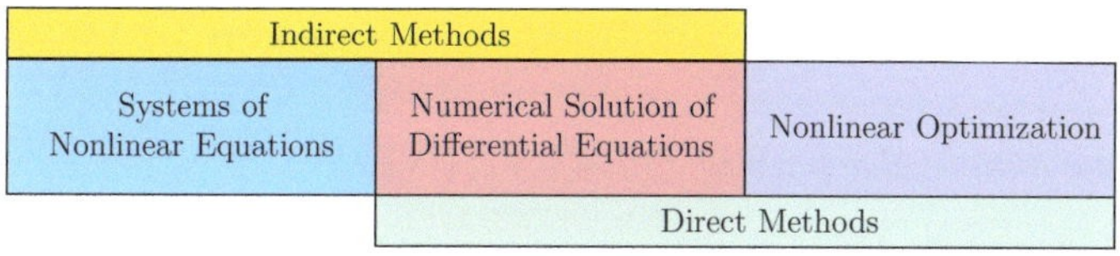

Fig. 1.1 The three major components of trajectory optimization and the class of methods that uses each component

$$\phi_{\min}^{(k)} \leq \phi^{(k)}\left(\mathbf{y}^{(k)}(t_0^{(k)}), t_0^{(k)}, \mathbf{y}^{(k)}(t_f^{(k)}), \mathbf{s}^{(k)}, t_f^{(k)}\right) \leq \phi_{\max}^{(k)}, \tag{1.3}$$

the algebraic *path constraints*

$$\mathbf{c}_{\min}^{(k)} \leq \mathbf{c}^{(k)}\left(\mathbf{y}^{(k)}(t), \mathbf{u}^{(k)}(t), \mathbf{s}^{(k)}, t\right) \leq \mathbf{c}_{\max}^{(k)} \tag{1.4}$$

and the *linkage constraints* (also known as *phase continuity constraints*)

$$\mathbf{L}_{\min}^{(s)} \leq \mathbf{L}\left(\mathbf{y}^{(l_s)}\left(t_f^{(l_s)}\right), \mathbf{u}^{(l_s)}\left(t_f^{(l_s)}\right), \mathbf{s}^{(l_s)}, t_f^{(l_s)}, \mathbf{y}^{(r_s)}(t_f^{(r_s)}), \mathbf{u}^{(r_s)}\left(t_f^{(r_s)}\right), \mathbf{s}^{(r_s)}, t_f^{(r_s)}\right) \leq \mathbf{L}_{\max}^{(s)}, \tag{1.5}$$

where $s \in [1, \ldots, S]$ and S is the number of pairs of phases that are being linked. In Eq. (1.5) the parameter S is the number of pairs of phases to be linked, $r_s \in [1, \ldots, S]$ and $l_s \in [1, \ldots, S]$ are the *right phases* and *left phases*, respectively, of the linkage pairs, $r_s \neq l_s$ (implying that a phase cannot be linked to itself), and $s \in [1, \ldots, S]$.

1.3 Numerical Methods for Trajectory Optimization

At the heart of a well-founded method for solving trajectory optimization problems are the following three fundamental components: (1) a method for solving differential equations and integrating functions; (2) a method for solving a system of nonlinear algebraic equations; and (3) a method for solving a nonlinear optimization problem. Methods for solving differential equations and integrating functions are required for all numerical methods in trajectory optimization optimal control. In an indirect method, the numerical solution of differential equations is combined with the numerical solution of systems of nonlinear equations while in a direct method the numerical solution of differential equations is combined with nonlinear optimization. A schematic with the breakdown of the components used by each class of optimal control methods is shown in Fig. 1.1.

1.4 Numerical Solution of Differential Equations

Consider the *initial-value problem* [30, 42, 92] (IVP)

$$\dot{\mathbf{y}} = \mathbf{f}(\mathbf{y}(t), t), \qquad \mathbf{y}(t_0) = \mathbf{y}_0. \tag{1.6}$$

Next, let $[t_i, t_{i+1}]$ be a time interval over which the solution to Eq. (1.6) is desired. Integrating Eq. (1.6), we can write

$$\mathbf{y}_{i+1} = \mathbf{y}_i + \int_{t_i}^{t_{i+1}} \dot{\mathbf{x}}(s)ds = \mathbf{y}_i + \int_{t_i}^{t_{i+1}} \mathbf{f}(\mathbf{y}(s), s)ds. \tag{1.7}$$

The two most common approaches for solving differential equations are time marching and collocation. In a time-marching method, the solution of the differential equation at each time step t_k is obtained sequentially using current and/or previous information about the solution. In a *multiple-step time marching method*, the solution at time t_{k+1} is obtained from a defined set of previous values $t_{k-j}, \ldots, t_k$ where j is the number of steps. The simplest multiple-step method is a *single-step method* (where $j = 1$). The most common single-step methods are *Euler methods*, while most commonly used multiple-step methods are the *Adams-Bashforth* and *Adams-Moulton* multiple-step methods [30]. Euler backward and Crank-Nicolson are examples of *implicit methods* whereas Euler forward is an example of an *explicit method*. When employing implicit method, the solution at t_{k+1} is obtained using a *predictor-corrector* where the predictor is typically an explicit method (that is, Euler-forward) while the corrector is the implicit formula. Implicit methods methods are more stable than explicit methods [42], but an implicit method requires more computation at each step due to the need to implement a predictor-corrector.

An alternative to a multiple-step time marching method is a *multiple-stage* method. In a multiple-stage method, the interval $[t_i, t_{i+1}]$ into K subintervals $[\tau_j, \tau_{j+1}]$ where

$$\tau_j = t_i + h_i\alpha_j, \quad (j = 1, \ldots, K), \quad h_i = t_{i+1} - t_i, \tag{1.8}$$

and $0 \le \alpha_j \le 1, \ (j = 1, \ldots, K)$. Each value τ_j is called a *stage* . The integral from t_i to t_{i+1} can be approximated via *quadrature* as

$$\int_{t_i}^{t_{i+1}} \mathbf{f}(\mathbf{y}(s), s)ds \approx h_i \sum_{j=1}^{K} \beta_j \mathbf{f}(\mathbf{y}_j, \tau_j) \tag{1.9}$$

where $\mathbf{y}_j \equiv \mathbf{y}(\tau_j)$. It is seen in Eq. (1.9) that the values of the state at each stage are required in order to evaluate the quadrature approximation. These intermediate values are obtained as

$$\mathbf{y}(\tau_j) - \mathbf{y}(t_i) = \int_{t_i}^{\tau_j} \mathbf{f}(\mathbf{y}(s), s)ds \approx h_i \sum_{l=1}^{K} \gamma_{jl}\mathbf{f}(\mathbf{y}_l, \tau_l) \tag{1.10}$$

The combination of Eqs. (1.9) and (1.10) leads to the family of $K-$stage *Runge-Kutta* methods [15, 24, 25, 30, 50, 51, 92]. A Runge-Kutta method is called *explicit* if $\gamma_{jl} = 0$ for all $l \geq j$ and is called *implicit* otherwise. In an explicit Runge-Kutta method, the approximation at t_{k+1} is computed using information prior to t_{k+1} whereas in an implicit Runge-Kutta method $\mathbf{y}(t_{k+1})$ is required in order to determine the solution at t_{k+1}. In the latter case, the solution is updated using a predictor-corrector approach.

1.4.1 Collocation

Another way to solve differential equations is as follows. Suppose over a subinterval $[t_i, t_{i+1}]$ we choose to approximate the state using the following Kth-degree piecewise polynomial:

$$\mathbf{Y}(t) \approx \sum_{k=0}^{K} \mathbf{a}_i (t - t_i)^k, \quad t \in [t_i, t_{i+1}]. \tag{1.11}$$

Suppose further that the coefficients $(\mathbf{a}_0, \ldots, \mathbf{a}_K)$ of the piecewise polynomial are chosen to match the value of the function at the beginning of the step, that is,

$$\mathbf{Y}(t_i) = \mathbf{y}_i. \tag{1.12}$$

Finally, suppose we choose to match the derivative of the state at the points defined in Eq. (1.8), that is,

$$\dot{\mathbf{y}}(\tau_j) = \mathbf{f}(\mathbf{y}(\tau_j), \tau_j), \quad (j = 1, \ldots, K). \tag{1.13}$$

Equation (1.13) is called a *collocation condition* because the approximation to the derivative is set equal to the right-hand side of the differential equation evaluated at each of the intermediate points $(\tau_1, \ldots, \tau_K)$. Collocation methods fall into three general categories [15]: Gauss methods, Radau methods, and Lobatto methods. In a *Gauss method*, neither of the endpoints t_k or t_{k+1} are collocation points. In a *Radau method*, at most one of the endpoints t_k or t_{k+1} is a collocation point. In a *Lobatto method*, both of the endpoints t_k and t_{k+1} are collocation points.

As it turns out, Euler and Runge-Kutta methods can be thought of equivalently as either time-marching or collocation methods. When an Euler or a Runge-Kutta method is employed in the form of collocation, the differential equation is said to be solved *simultaneously* because all of the unknown parameters are determined at the same time. Furthermore, collocation methods are said to simulate the dynamics of the system *implicitly* because the values of the state at each collocation point are obtained at the same time (as opposed to solving for the state sequentially as in a time-marching method). In order to implement simultaneous simulation, the discretized dynamics are written as *defect constraints* of the form

$$\boldsymbol{\zeta}_j = \dot{\mathbf{y}}(\tau_j) - \mathbf{f}(\mathbf{y}(\tau_j), \tau_j). \tag{1.14}$$

As an example, the defect constraints for the Crank-Nicolson method are given as

$$\boldsymbol{\zeta}_k = \mathbf{y}_{k+1} - \mathbf{y}_k - \frac{h_k}{2}\left(\mathbf{f}_k + \mathbf{f}_{k+1}\right). \tag{1.15}$$

In collocation (that is, implicit simulation) it is desired to find a solution such that all of the defect constraints are zero. Finally, one of the key differences between collocation and time-marching is that in collocation it is not necessary to use a predictor-corrector because the values of the state at each discretization point are being solve for simultaneously.

1.4.2 Integration of Functions

Because the objective is to solve a trajectory optimization problem, it is necessary to approximate the cost function of Eq. (1.1). Typically, the cost is approximated using a quadrature that is consistent with the numerical method for solving the differential equation (for example, if one is using an Euler-forward rule for solving the differential equation, the cost would also be approximated using Euler-forward integration). The requirement for consistency in the approximation of the differential equations and the cost can be thought of in another manner. Consider a one-phase trajectory optimization problem. The cost functional

$$J = \Phi(\mathbf{y}(t_0), t_0, \mathbf{y}(t_f), t_f) + \int_{t_0}^{t_f} \mathscr{L}[\mathbf{y}(t), \mathbf{u}(t), t; \mathbf{s}]dt \tag{1.16}$$

can be converted to a Mayer problem by adding a state y_{n+1} and adding the differential equation

$$\dot{y}_{n+1} = g[\mathbf{y}(t), \mathbf{u}(t), t; \mathbf{s}] \tag{1.17}$$

with the initial condition

$$y_{n+1}(t_0) = 0. \tag{1.18}$$

The cost functional of Eq. (1.16) would be given as

$$J = \Phi[\mathbf{y}(t_0), t_0, \mathbf{y}(t_f), t_f; \mathbf{s}] + y_{n+1}(t_f) \tag{1.19}$$

and the resulting augmented system of differential equations would then be written as

$$\begin{aligned}\dot{\mathbf{y}}(t) &= \mathbf{f}[\mathbf{y}(t), \mathbf{u}(t), t; \mathbf{s}],\\ \dot{y}_{n+1} &= \mathscr{L}[\mathbf{x}(t), \mathbf{u}(t), t; \mathbf{s}].\end{aligned} \tag{1.20}$$

Equation (1.20) could then be solved using any well established numerical integration method. Using this approach, is it seen that the method used to integrate Eq. (1.17) must be the same method that is used to integrate

$$\mathscr{L}[\mathbf{y}(t), \mathbf{u}(t), t; \mathbf{s}].$$

1.5 Nonlinear Optimization

A key ingredient to solving trajectory optimization problems is the ability to solve *nonlinear optimization* or *nonlinear programming problems* [4, 11, 21, 44] (NLPs). An NLP takes the following general mathematical form. Determine the decision vector $\mathbf{z} \in \mathbb{R}^n$ that minimizes the cost function

$$f(\mathbf{z}) \tag{1.21}$$

subject to the algebraic constraints

$$\mathbf{g}(\mathbf{z}) = \mathbf{0}, \tag{1.22}$$

$$\mathbf{h}(\mathbf{z}) \leq \mathbf{0}, \tag{1.23}$$

where $\mathbf{g}(\mathbf{z}) \in \mathbb{R}^m$ and $\mathbf{h}(\mathbf{z}) \in \mathbb{R}^p$. The NLP may either be *dense* (that is, a large percentage of the derivatives of the objective function and the constraint functions with respect to the components of $\mathbf{z}$ are nonzero) or may be *sparse* (that is, a large percentage of the derivatives of the objective function and the constraint functions with respect to the components of $\mathbf{z}$ are zero). Dense NLPs typically are small (consisting of at most a few hundred variables and constraints) while sparse NLPs are often extremely large (ranging anywhere from thousands of variables and constraints to millions of variables and constraints).

1.6 Methods for Solving Trajectory Optimization Problems

With the exception of simple problems, trajectory optimization problems must be solved numerically. The need for solving optimal control problems numerically has given rise to a wide range of numerical approaches. These numerical approaches are divided into two broad categories: (1) indirect methods and (2) direct methods. The major methods that fall into each of these two broad categories are described in the next two sections.

1.6.1 Indirect Methods

In an indirect method, the *calculus of variations* [3, 20, 23, 40, 56, 62, 67, 68, 91, 94, 97] is used to determine the first-order optimality conditions of the trajectory optimization problem given in Eqs. (1.1)–(1.5). Unlike ordinary calculus (where the objective is to determine points that optimize a function), the calculus of variations is the subject of determining *functions* that optimize a *function of a function* (also known as *functional optimization*). Applying the calculus of variations to the functional optimization problem given in Eqs. (1.1)–(1.5) leads to the *first-order necessary conditions* for an extremal trajectory. The first-order optimality conditions for a single-phase continuous-time trajectory optimization problem with no static parameters are given as

$$\dot{\mathbf{y}} = \mathscr{H}_{\boldsymbol{\lambda}}^{\mathsf{T}}, \quad \dot{\boldsymbol{\lambda}} = -\mathscr{H}_{\mathbf{y}}^{\mathsf{T}}, \tag{1.24}$$

$$\mathbf{u}^* = \arg\min_{\mathbf{u}\in\mathscr{U}} \mathscr{H}, \tag{1.25}$$

$$\boldsymbol{\phi}(\mathbf{y}(t_0), t_0, \mathbf{y}(t_f), t_f) = \mathbf{0}, \tag{1.26}$$

$$\boldsymbol{\lambda}(t_0) = -\Phi_{\mathbf{y}}(t_0) + \boldsymbol{\nu}^{\mathsf{T}}\boldsymbol{\phi}_{\mathbf{y}(t_0)}, \quad \boldsymbol{\lambda}(t_f) = \Phi_{\mathbf{y}}(t_f) - \boldsymbol{\nu}^{\mathsf{T}}\boldsymbol{\phi}_{\mathbf{y}(t_f)}, \tag{1.27}$$

$$\mathscr{H}(t_0) = \Phi_{t_0} - \boldsymbol{\nu}^{\mathsf{T}}\boldsymbol{\phi}_{t_0}, \quad \mathscr{H}(t_f) = -\Phi_{t_f} + \boldsymbol{\nu}^{\mathsf{T}}\boldsymbol{\phi}_{t_f}, \tag{1.28}$$

$$\begin{aligned} \mu_j(t) &= 0, \ \text{when} \ \ C_j(\mathbf{x}, \mathbf{u}, t) < 0, \quad j = 1, \ldots, c, \\ \mu_j(t) &\le 0, \ \text{when} \ \ C_j(\mathbf{x}, \mathbf{u}, t) = 0, \quad j = 1, \ldots, c, \end{aligned} \tag{1.29}$$

where $\mathscr{H} = \mathscr{L} + \boldsymbol{\lambda}^{\mathsf{T}}\mathbf{f} - \boldsymbol{\mu}^{\mathsf{T}}\mathbf{C}$ is the augmented Hamiltonian, $\mathscr{U}$ is the feasible control set and $\boldsymbol{\nu} \in \mathbb{R}^q$ is the Lagrange multiplier associated with the boundary condition $\boldsymbol{\phi}$. Finally, it is noted that the solution to the optimal control problem may lie along a singular arc [23] where the control cannot be determined from the first-order optimality conditions. If a singular arc is a possibility, additional conditions must be derived to determine the control along the singular arc.

Because the dynamics of Eq. (1.24) arise from differentiation a Hamiltonian, Eq. (1.24) is called a *Hamiltonian system* [3, 66, 67]. Furthermore, Eq. (1.25) is known as *Pontryagin's Minimum Principle* [75] (PMP) and is a classical result to determine theoptimal control. Finally, the conditions on the initial and final costate given in Eq. (1.27) are called *transversality conditions* [3, 23, 40, 62, 66, 68, 91, 93, 94] while the conditions on the Lagrange multipliers of the path constraints given in Eq. (1.29) are called *complementary slackness conditions* [4, 11, 21]. The Hamiltonian system, together with the boundary conditions, transversality conditions, and complementary slackness conditions, is called a *Hamiltonian boundary-value problem* (HBVP) [2, 3, 66]. Any solution $(\mathbf{y}(t), \mathbf{u}(t), \lambda(t), \mu(t), \nu)$ is called an *extremal solution* and consists of the state, costate, and any Lagrange multipliers that satisfy the boundary conditions and any interior-point constraints on the state and costate. In an indirect method extremal trajectories (that is, solutions of the HBVP) are determined numerically. Because an indirect method requires solving a multiple-point boundary-value problem, the original trajectory optimization problem is turned into the problem of solving a system of nonlinear equations of the form

$$\begin{aligned} \mathbf{f}(\mathbf{z}) &= \mathbf{0}, \\ \mathbf{g}_{\min} \leq \mathbf{g}&(\mathbf{z}) \leq \mathbf{g}_{\max}. \end{aligned} \tag{1.30}$$

The three two most common indirect methods are the shooting method, the multiple-shooting method, and collocation methods. Each of these approaches is now described.

1.6.1.1 Indirect Shooting Method

Perhaps the most basic indirect method is the *shooting method* [65]. In a typical shooting method, an initial guess is made of the unknown boundary conditions at one end of the interval. Using this guess, together with the known initial conditions, the Hamiltonian system Eq. (1.24) is integrated to the other end (that is, either forward from t_0 to t_f or backward from t_f to t_0). Upon reaching t_f, the terminal conditions obtained from the numerical integration are compared to the known terminal conditions given in Eqs. (1.26) and (1.27). If the integrated terminal conditions differ from the known terminal conditions by more than a specified tolerance ε, the unknown initial conditions are adjusted and the process is repeated until the difference between the integrated terminal conditions and the required terminal conditions is less than some specified threshold.

1.6.1.2 Indirect Multiple-Shooting Method

While a simple shooting method is appealing due to its simplicity, it presents significant numerical difficulties due to ill-conditioning of the Hamiltonian dynamics. The reason for this ill-conditioning is that Hamiltonian systems have the property that the divergence of the flow of trajectories must be zero, that is

$$\sum_{i=1}^{n}\left[\frac{\partial}{\partial x_i}\left(\frac{\partial \mathcal{H}}{\partial \lambda_i}\right)+\frac{\partial}{\partial \lambda_i}\left(-\frac{\partial \mathcal{H}}{\partial x_i}\right)\right]\equiv 0. \tag{1.31}$$

Equation (1.31) implies that, in a neighborhood of the optimal solution, there exist an equal number of directions along which the solution will contract and expand and this expansion and contraction takes place at the same rate (the simultaneous expanding and contracting behavior is due to the fact that many Hamiltonian systems admit an exponential *dichotomy* [2]). As a result, errors made in the unknown boundary conditions will amplify as the dynamics are integrated in either direction of time. The shooting method poses particularly poor characteristics when the trajectory optimization problem is *hyper-sensitive* [76, 78, 79, 81, 82] (that is, when time interval of interest is long in comparison with the time-scales of the Hamiltonian system in a neighborhood of the optimal solution).

In order to overcome the numerical difficulties of the simple shooting method, a modified method, called the *multiple-shooting method* [92], has been developed. In a multiple-shooting method, the time interval $[t_0, t_f]$ is divided into $M+1$ subintervals. The shooting method is then applied over each subinterval $[t_i, t_{i+1}]$ with the initial values of the state and adjoint of the interior intervals being the unknowns that need to be determined. In order to enforce continuity, the following conditions are enforced at the interface of each subinterval:

$$\mathbf{p}(t_i^-) = \mathbf{p}(t_i^+) \Longleftrightarrow \mathbf{p}(t_i^-) - \mathbf{p}(t_i^+) = \mathbf{0}, \tag{1.32}$$

where $\mathbf{p}(t)$ is the combined state-costate vector, that is,

$$\mathbf{p}(t) = \begin{bmatrix} \mathbf{x}(t) \\ \lambda(t) \end{bmatrix}.$$

The continuity conditions of Eq. (1.32) result in vector root-finding problem where it is desired to drive the values of the difference between $\mathbf{p}(t_i^-) - \mathbf{p}(t_i^+)$ to zero. It is seen that the multiple-shooting method requires extra variables be introduced into the problem (that is, the values of the state and adjoint at the interface points). Despite the increased size of the problem due to these extra variables, the multiple-shooting method is an improvement over the shooting method because the sensitivity to errors in the unknown initial conditions is reduced by integrating over subintervals of the original time domain $t \in [t_0, t_f]$. Nevertheless, even multiple-shooting can present issues if a sufficiently good guess of the costate is not used [48].

1.6.1.3 Indirect Collocation Methods

In an indirect collocation method, the state and costate are parameterized using piecewise polynomials as described in Sect. 1.4.1. The collocation procedure leads to a root-finding problem where the vector of unknown coefficients $\mathbf{z}$ consists of the coefficients of the piecewise polynomial. This system of nonlinear equations is then solved using a root-finding technique (for example, Newton's method).

1.6.2 Direct Methods

Direct methods are fundamentally different from indirect methods. In a direct method, the state and/or control of the original optimal control problem are approximated in some appropriate manner. In the case where only the control is approximated, the method is called a *control parameterization method* [46]. When both the state and control are approximated the method is called a *state and control parameterization method*. In either a control parameterization method or a state and control parameterization method, the optimal control problem is transcribed to a *nonlinear optimization problem* or *nonlinear programming problem* [4, 11, 15, 21, 44] (NLP).

1.6.2.1 Direct Shooting Method

The most basic direct method for solving trajectory optimization problems is the *direct shooting method*. The direct shooting method is a control parameterization method where the control is parameterized using a specified functional form. For example, the control could be parameterized as

$$\mathbf{u}(t) \approx \sum_{i=1}^{m} \mathbf{a}_i \psi_i(t), \tag{1.33}$$

where $\psi_i(t), \ (i = 1, \ldots, m)$ are known functions and $\mathbf{a}_i, \ (i = 1, \ldots, m)$ are the parameters to be determined from the optimization. The dynamics are then satisfied by integrating the differential equations using a time-marching algorithm. Similarly, the cost function of Eq. (1.1) is determined using a quadrature approximation that is consistent with the numerical integrator used to solve the differential equations. The NLP that arises from direct shooting then minimizes the cost subject to any path and interior-point constraints.

1.6.2.2 Direct Multiple-Shooting Method

In a manner similar to that for indirect methods, in a *direct multiple-shooting method*, the time interval $[t_0, t_f]$ is divided into $M+1$ subintervals. The aforementioned direct shooting method is then used over each subinterval $[t_i, t_{i+1}]$ with the values of the state at the beginning of each subinterval and the unknown coefficients in the control parameterization being unknowns in the optimization. In order to enforce continuity, the following conditions are enforced at the interface of each subinterval:

$$\mathbf{x}(t_i^-) = \mathbf{x}(t_i^+) \Longleftrightarrow \mathbf{x}(t_i^-) - \mathbf{x}(t_i^+) = \mathbf{0}. \tag{1.34}$$

The continuity conditions of Eq. (1.34) result in vector root-finding problem where it is desired to drive the values of the difference between $\mathbf{x}(t_i^-) - \mathbf{x}(t_i^+)$ to zero. It is seen that the direct multiple-shooting method increases the size of the optimization problem because the values of the state at the beginning of each subinterval are parameters in the optimization. As with indirect multiple-shooting, the direct multiple-shooting method is an improvement over the direct shooting method because the sensitivity to errors in the unknown initial conditions is reduced by integrating over subintervals of the original time domain $t \in [t_0, t_f]$.

1.6.2.3 Direct Collocation Methods

Arguably the most powerful methods for solving general trajectory optimization problems are *direct collocation methods*. A direct collocation method is a state and control parameterization method where the state and control are approximated using a specified functional form. The two most common forms of collocation are *local collocation* and *global collocation*. A *local collocation method* follows a procedure similar to that of Sect. 1.4.1 in that the time interval $[t_0, t_f]$ is divided into S subintervals $[t_{s-1}, t_s]$, $(s = 1, \ldots, S)$ where $t_S = t_f$. In order to ensure continuity in the state across subintervals, the following *compatibility constraint* is enforced at the interface of each subinterval:

$$\mathbf{y}(t_i^-) = \mathbf{y}(t_i^+), \quad (s = 2, \ldots, S-1). \tag{1.35}$$

In the context of trajectory optimization, local collocation has been employed using one of two categories of discretization: Runge-Kutta methods and orthogonal collocation methods. Nearly all Runge-Kutta methods used are *implicit* [31–36, 49, 69, 88, 89] because the stability properties of implicit Runge-Kutta methods are better than those of explicit methods. The seminal work on orthogonal collocation methods in trajectory optimization is due to Reddien [84], where Legendre-Gauss points were used together with cubic splines. Following on Reddien's work, Cuthrell and Biegler used LG points together with Lagrange polynomials [28, 29]. Interestingly, Cuthrell [29] showed mathematically that the indirect transcription using LG points was equivalent to the KKT conditions obtained from the NLP of the direct formulation. In the 1990s, orthogonal collocation methods were developed using higher-order Gauss-Lobatto collocation methods [38, 39, 54, 55]. Finally, the convergence rates of an orthogonal collocation method using Legendre-Gauss-Radau (LGR) points was studied [64].

Generally, employing direct local collocation leads to a *large sparse* NLP, where the NLP contains potentially thousands to hundreds of thousands of variables and constraints. Moreover, such large NLPs arise from trajectory optimization problems that consist of hundreds of of states and controls. Because the NLP is sparse, however, many of the derivatives of the constraint Jacobian are zero. This feature of local direct collocation makes it possible to solve such problem efficiently using appropriate NLP solvers such as *SNOPT* [43, 45], *SPRNLP* [17], and *KNITRO* [26].

1.7 Software for Solving Trajectory Optimization Problems

A wide variety of software tools have been developed for solving trajectory optimization problems. Most of these software programs use direct methods. One well known software program employing indirect methods is *BNDSCO* [71] which employs a multiple-shooting method. Perhaps the oldest software tool that employs direct methods is the *Program to Simulate and Optimize Trajectories* [22] (*POST*). *POST* was originally developed to solve problems in launch vehicle trajectory optimization and it still in use today for such applications.

The late 1980s saw a transformation in the available tools for solving trajectory optimization problems. This transformation was coincident with the observation of the power of direct collocation methods. The first well-known direct collocation software was *Optimal Trajectories by Implicit Simulation* [95] (*OTIS*). *OTIS* is a FORTRAN software that has general-purpose capabilities for problems in aeronautics and astronautics. OTIS has been used widely in the aerospace and defense industries and its theoretical foundations are found in Ref. [52]. Following shortly after the development of *OTIS* is the program *Sparse Optimal Control Software* [18] (*SOCS*). *SOCS* is a highly powerful FORTRAN software that is capable of solving many highly challenging trajectory optimization problems (see Ref. [15] for highly complex optimal control problems solved with *SOCS*). Some of the applications solved using *SOCS* are found in Refs. [12–14, 16, 77]. Finally, three other direct collocation FORTRAN programs are *MISER* [47], *Direct Collocation* [96] (*DIRCOL*), *Graphical Environment for Simulation and Optimization* [1] (*GESOP*), and *Nonlinear Trajectory Generation* [70] (*NTG*). Like *OTIS* and *SOCS*, *DIRCOL* and *GESOP* use local direct collocation techniques while *NTG* is designed for rapid trajectory generation of differentially flat systems.

In recent years, interest in the particular application of optimal control to space flight has led to the development of several useful programs. One such program is *Mission Design and Analysis Software* [87] (*MIDAS*) which is designed to solve complex ballistic heliocentric transfer trajectories for interplanetary space flight missions. Another tool that has been recently developed is the NASA *Generalized Mission Analysis Tool* [61] (*GMAT*). Another tool that has been widely used in the last several years is *COPERNICUS* [72, 73]. Both *GMAT* and *COPERNICUS* are designed to solve trajectory optimization problems where the maneuvers can be treated as either impulsive or finite-thrust burns.

While earlier software programs used compiled languages such as FORTRAN, in recent years, MATLAB® has become increasingly popular for solving optimization problems. The increased appeal for MATLAB emanates from the fact that MATLAB is an extremely easy environment in which to program along with the fact that many of today's most powerful NLP solvers are now available for use in MATLAB® (for example, standalone MATLAB mex versions are now available for the NLP solvers *SNOPT* [43, 45] and *KNITRO* [26]). In addition, the TOMLAB [19, 37, 57–60] package has facilitated additional solvers for use in MATLAB. In addition, because of major computational improvements, the computational efficiency between MATLAB and compiled languages is growing ever closer. Examples of MATLAB-based trajectory optimization software programs include RIOTS_95 [90], *DIDO* [85], *DIRECT* [98], *PROPT* [86], *OPTCONTROLCENTRE* [63], *GPOPS* [80], and *GPOPS-II* [74].

It is important to note that all of the trajectory optimization software programs described above incorporate gradient methods for solving the NLP. In a less formal manner, heuristic methods have also been used to solve trajectory optimization problems. For example, interplanetary trajectory optimization problems using a genetic algorithm have been considered in Refs. [41, 53] while low-thrust orbit transfers using a genetic algorithm have been studied in Refs. [27] and [83]. In addition, a calculus of variations technique has been used together with a genetic algorithm to optimize low-thrust Mars-to-Earth trajectories for the Mars Sample Return Mission [99]. Thus, while gradient methods are somewhat the de facto standard for trajectory optimization, the aforementioned research demonstrates that genetic algorithms may be well-suited for some applications.

1.8 Choosing a Method

Choosing a method for solving a trajectory optimization problem is based largely on the type of problem to be solved and the amount of time that can be invested in coding. An indirect shooting method has the advantage that it is simple to understand and produces highly accurate solutions when it converges. Unfortunately, indirect shooting is extremely sensitive to the unknown boundary conditions. In addition, indirect shooting requires the derivation of the first-order optimality conditions of the trajectory optimization problem [see Eqs. (1.24)–(1.29]! While for simple problems it may be possible to derive the first-order optimality conditions, deriving such conditions for complex optimal control problems is tedious, error-prone, and sometimes impossible (for example, problem with table lookups). Furthermore, the need to derive the optimality conditions makes implementing indirect shooting difficult in a general-purpose software program. For example, if it was required to derive first-order optimality conditions, a program such as *POST* would become nearly impossible to use because every new problem would require the derivation of these conditions! A multiple-shooting method overcomes some of the numerical difficulties of standard shooting, but does not avoid the issue of having to derive the optimality conditions.

The accuracy and robustness of a direct method is highly dependent upon the form of direct method used. Direct shooting methods are very good for problems where the control can be parameterized in a simple manner (for example, piecewise linear functions of time) and the problem can be characterized accurately using a small number optimization parameters. Software programs such as *POST* perform well on launch vehicle ascent trajectories because these problems can be approximated accurately using simple control parameterizations. As the complexity of the problem increases, it becomes more and more apparent that the workhorse for solving trajectory optimization problems is the direct collocation method. The two main reasons that direct collocation methods work so well is because highly complex problems can be formulated and solved with today's NLP solvers. The reason that the NLP solvers can handle such complex problems is because they are designed to converge with poor initial guesses (for example, straight line guesses in the state and control) and are extremely computationally efficient because they exploit the sparsity of the derivatives in the constraints and objective function.

In many cases, the solution of a trajectory optimization problem is a means to an end, that is, the user does not want to know all of the details about a method, but simply wants to use a software program to provide results so that a particular problem of interest can be solved. If one does not wish to become an expert in the technologies associated with trajectory optimization, it is advisable to obtain a canned software package that allows a user to input the problem in an intuitive manner. Then the software can simply be run on the problem of interest. It is always important to understand, however, that canned software can have its issues when things go wrong because the user may often not understand why.

1.9 Applications to Automotive Systems

The numerical methods provided in this survey are designed to generate reference trajectories and corresponding reference controls for systems that have well developed deterministic models. In the context of automotive systems, the methods described in this paper would be of relevance to optimal control in systems where performance is important. For example, state-of-the-art direct collocation software such as GPOPS-II or SOCS could be employed to generate highly accurate trajectories to determine the minimum lap time required in a in high-speed race car problem (for example, Formula One racing). In addition, the indirect methods described in this paper could be the starting point for developing near-optimal feedback controllers for use in engine design or in autonomous ground vehicles. Thus, the numerical methods described in this survey could be used to generate solutions to a wide variety of problems in automotive systems, and the particular numerical method employed would depend upon the intended use of the solution.

1.10 Conclusions

A survey of numerical methods for solving trajectory optimization problems has been given. The problem of solving optimal control problems has been decomposed into the three key components of solving differential equations and integrating functions, solving nonlinear optimization problems, and solving systems of nonlinear algebraic equations. Using these components, the two classes of indirect and direct methods for solving optimal control problems have been described. Subsequently, important computational issues have been discussed and several different software tools for solving optimal control problems have been described. Finally, a brief discussion has been given on how to choose a method.

References

1. GESOP & ASTOS (2003) The new generation of optimization software. Institute of Flight Mechanics and Control of Stuttgart University
2. Ascher UM, Mattheij RM, Russell RD (1996) Numerical solution of boundary-value problems in ordinary differential equations. SIAM Press, Philadelphia
3. Athans MA, Falb PL (2006) Optimal control: an introduction to the theory and its applications. Dover Publications, Mineola, New York
4. Bazaraa MS, Sherali HD, Shetty CM (2006) Nonlinear programming: theory and algorithms, 3 edn. Wiley-Interscience, New Jersey
5. Bellman R (1957) Dynamic programming. Princeton University Press, Princeton
6. Bellman R (1962) Dynamic programming treatment of the travelling salesman problem. J Assoc Comput Mach 9(1):61–63
7. Bellman R (1966) Dynamic programming. Science 1, 153(3731):34–37
8. Bellman R, Dreyfus S (1959) Functional approximations and dynamic programming. Math Tables Other Aids Comput 13(68):247–251
9. Bellman R, Kalaba R, Kotkin B (1963) Polynomial approximation—a new computational technique in dynamic programming: allocation processes. Math Comput 17(82):155–161
10. Bellman RE, Dreyfus SE (1971) Applied dynamic programming. Princeton University Press, Princeton
11. Bertsekas D (2004) Nonlinear programming. Athena Scientific Publishers, Belmont, Massachusetts
12. Betts JT (1993) Using sparse nonlinear programming to compute low thrust orbit transfers. J Astronaut Sci 41:349–371
13. Betts JT (1994) Optimal interplanetary orbit transfers by direct transcription. J Astronaut Sci 42:247–268
14. Betts JT (2000) Very low thrust trajectory optimization using a direct sqp method. J Comput Appl Math 120:27–40
15. Betts JT (2001) Practical methods for optimal control using nonlinear programming. SIAM Press, Philadelphia
16. Betts JT (2007) Optimal lunar swingby trajectories. J Astronaut Sci 55:349–371
17. Betts JT, Huffman WP (1994) A sparse nonlinear optimization algorithm. J Optim Theory Appl 82(3):519–541
18. Betts JT, Huffman WP (1997) Sparse optimal control software—socs. Technical report MEA-LR-085, Boeing information and support services, Seattle, Washington, July 1997
19. Björkman M, Holmström K (1999) Global optimization with the direct user interface for nonlinear programming. Adv Model Simul 2:17–37
20. Bliss GA (1946) Lectures on the calculus of variations. University of Chicago Press, Chicago, IL
21. Boyd S, Vandenberghe L (2004) Convex optimization. Cambridge University Press, Cambridge
22. Brauer GL, Cornick DE, Stevenson R (1977) Capabilities and applications of the program to optimize and simulate trajectories. Technical report NASA-CR-2770, National Aeronautics and Space Administration

23. Bryson AE, Ho Y-C (1975) Applied optimal control. Hemisphere Publishing, New York
24. Butcher JC (1964) Implicit runge-kutta processes. Math Comput 18(85):50–64
25. Butcher JC (2008) Numerical methods for ordinary differential equations. Wiley, New York
26. Byrd RH, Nocedal J, Waltz RA (2006) Knitro: an integrated package for nonlinear optimization. In: Large scale nonlinear optimization. Springer, Berlin, pp 35–59
27. Coverstone-Carroll VL, Hartmann JW, Mason WJ (2000) Optimal multi-objective low-thrust spacecraft trajectories. Comput Methods Appl Mech Eng 186(2–4):387–402
28. Cuthrell JE, Biegler LT (1987) On the optimization of differential-algebraic processes. AIChe J 33(8):1257–1270
29. Cuthrell JE, Biegler LT (1989) Simultaneous optimization and solution methods for batch reactor control profiles. Comput Chem Eng 13(1/2):49–62
30. Dahlquist G, Björck A (2003) Numerical methods. Dover Publications, Mineola, New York
31. Dontchev AL, Hager WW (1998) Lipschitzian stability for state constrained nonlinear optimal control. SIAM J Control Optim 36:696–718
32. Dontchev AL, Hager WW (1998) A new approach to lipschitz continuity in state constrained optimal control. Syst Control Lett 35:137–143
33. Dontchev AL, Hager WW (2001) The euler approximation in state constrained optimal control. Math Comput 70:173–203
34. Dontchev AL, Hager WW, Malanowski K (2000) Error bounds for the euler approximation and control constrained optimal control problem. Numer Funct Anal Appl 21:653–682
35. Dontchev AL, Hager WW, Veliov VM (2000) Second-order runge-kutta approximations in constrained optimal control. SIAM J Numer Anal 38:202–226
36. Dontchev AL, Hager WW, Veliov VM (2000) Uniform convergence and mesh independence of newton's method for discretized variational problems. SIAM J Control Optim 39:961–980
37. Dotzauer E, Holmström K (1999) The tomlab graphical user interface for nonlinear programming. Adv Model Simul 2:9–16
38. Enright PJ (1991) Optimal finite—thrust spacecraft trajectories using direct transcription and nonlinear programming. PhD thesis, Department of Aerospace Engineering, University of Illinois at Urbana-Champaign
39. Enright PJ, Conway BA (1996) Discrete approximations to optimal trajectories using direct transcription and nonlinear programming. J Guidance Control Dyn 19(4):994–1002, Jul–Aug 1996
40. Fleming WH, Rishel RW (1982) Deterministic and stochastic optimal control. Springer, Heidelberg
41. Gage PJ, Braun RD, Kroo IM (1995) Interplanetary trajectory optimization using a genetic algorithm. J Astronaut Sci 43(1):59–75
42. Gear WC (1971) Numerical initial-value problems in ordinary differential equations. Prentice-Hall, Englewood Cliffs, New Jersey
43. Gill PE, Murray W, Saunders MA (2002) Snopt: an sqp algorithm for large-scale constrained optimization. SIAM Rev 47(1):99–131
44. Gill PE, Murray W, Wright MH (1981) Practical optimization. Academic Press, London
45. Gill PE, Murray W, Saunders MA (2006) User's guide for SNOPT version 7: software for large scale nonlinear programming, Feb 2006
46. Goh CJ, Teo KL (1988) Control parameterization: a unified approach to optimal control problems with general constraints. Automatica 24(1):3–18
47. Goh CJ, Teo KL (1988) Miser: a fortran program for solving optimal control problems. Adv Eng Softw 10(2):90–99
48. Grimm W, Markl A (1997) Adjoint estimation from a multiple-shooting method. J Optim Theory Appl 92(2):263–283
49. Hager WW (2000) Runge-kutta methods in optimal control and the transformed adjoint system. Numerische Mathematik 87:247–282
50. Hairer E, Norsett SP, Wanner G (1993) Solving ordinary differential equations I: nonstiff problems. Springer, New York

51. Hairer E, Wanner G (1996) Solving ordinary differential equations II: stiff differential-algebraic problems. Springer, New York
52. Hargraves CR, Paris SW (1987) Direct trajectory optimization using nonlinear programming techniques. J Guidance Control Dyn 10(4):338–342
53. Hartmann JW, Coverstone-Carroll VL (1998) Optimal interplanetary spacecraft trajectories via a pareto genetic algorithm. J Astronaut Sci 46(3):267–282
54. Herman AL (1995) Improved collocation methods with application to direct trajectory optimization. PhD thesis, Department of Aerospace Engineering, University of Illinois at Urbana-Champaign
55. Herman AL, Conway BA (1996) Direct optimization using collocation based on high-order gauss-lobatto quadrature rules. J Guidance Control Dyn 19(3):592–599
56. Hildebrand FB (1992) Methods of applied mathematics. Dover Publications, Mineola, New York
57. Holmström K (1999) New optimization algorithms and software. Theor Stoch Process 1–2:55–63
58. Holmström K (1999) The tomlab optimization environment in matlab. Adv Model Simul 1:47–69
59. Holmström K, Björkman M (1999) The tomlab nlplib toolbox for nonlinear programming. Adv Model Simul 1:70–86
60. Holmström K, Björkman M, Dotzauer E (1999) The tomlab opera toolbox for linear and discrete optimization. Adv Model Simul 2:1–8
61. Hughes S (2008) Gmat—generalized mission analysis tool. Technical report, NASA Goddard Space Flight Center, http://sourceforge.net/projects/gmat
62. Hull DG (2003) Optimal control theory for applications. Springer, New York
63. Jockenhovel T (2002) Optcontrolcentre, software package for dynamic optimization. http://OptControlCentre.com/
64. Kameswaran S, Biegler LT (2008) Convergence rates for direct transcription of optimal control problems using collocation at radau points. Comput Optim Appl 41(1):81–126
65. Keller HB (1976) Numerical solution of two point boundary value problems. SIAM, Philadelphia
66. Kirk DE (2004) Optimal control theory: an introduction. Dover Publications, Mineola, New York
67. Leitmann G (1981) The calculus of variations and optimal control. Springer, New York
68. Lewis FL, Syrmos VL (1995) Optimal control, 2nd edn. Wiley, New York
69. Logsdon JS, Biegler LT (1989) Accurate solution of differential-algebraic optimization problems. Ind Eng Chem Res 28:1628–1639
70. Milam MB (2003) Real-time optimal trajectory generation for constrained dynamical systems. PhD thesis, California Institute of Technology, Pasadena, California, May 2003
71. Oberle HJ, Grimm W (1990) Bndsco: a program for the numerical solution of optimal control problems. Technical report, Institute of Flight Systems Dynamics, German Aerospace Research Establishment DLR, IB 515–89/22, Oberpfaffenhofen, Germany
72. Ocampo C (2003) An architecture for a generalized spacecraft trajectory design and optimization system. In: Proceedings of the international conference on libration point missions and applications, Aug 2003
73. Ocampo C (2004) Finite burn maneuver modeling for a generalized spacecraft trajectory design and optimization system. Ann NY Acad Sci 1017:210–233
74. Patterson MA, Rao AV (2013) GPOPS-II, a matlab software for solving multiple-phase optimal control problems *hp*—adaptive gaussian quadrature collocation methods and sparse nonlinear programming. ACM Trans Math Softw(in Revision) Sep 2013
75. Pontryagin LS (1962) Mathematical theory of optimal processes. Wiley, New York
76. Rao AV (1996) Extension of the computational singular perturbation method to optimal control. PhD thesis, Princeton University
77. Rao AV, Tang S, Hallman WP (2002) Numerical optimization study of multiple-pass aeroassisted orbital transfer. Optim Control Appl Methods 23(4):215–238

78. Rao AV (2000) Application of a dichotomic basis method to performance optimization of supersonic aircraft. J Guidance Control Dyn 23(3):570–573
79. Rao AV (2003) Riccati dichotomic basis method for solving hyper-sensitive optimal control problems. J Guidance Control Dyn 26(1):185–189
80. Rao AV, Benson DA, Darby CL, Francolin C, Patterson MA, Sanders I, Huntington GT (2010) Algorithm 902: GPOPS, a MATLAB software for solving multiple-phase optimal control problems using the gauss pseudospectral method. ACM Trans Math Softw 37(2, Article 22):39 p
81. Rao AV, Mease KD (1999) Dichotomic basis approach to solving hyper-sensitive optimal control problems. Automatica 35(4):633–642
82. Rao AV, Mease KD (2000) Eigenvector approximate dichotomic basis method for solving hyper-sensitive optimal control problems. Optim Control Appl Methods 21(1):1–19
83. Rauwolf GA, Coverstone-Carroll VL (1996) Near-optimal low-thrust orbit transfers generated by a genetic algorithm. J Spacecraft Rockets 33(6):859–862
84. Reddien GW (1979) Collocation at gauss points as a discretization in optimal control. SIAM J Control Optim 17(2):298–306
85. Ross IM, Fahroo F (2001) User's manual for DIDO 2001 α: a MATLAB application for solving optimal control problems. Technical Report AAS-01-03, Department of Aeronautics and Astronautics, Naval Postgraduate School, Monterey, California
86. Rutquist P, Edvall M (2008) PROPT: MATLAB optimal control software. Tomlab Optimization Inc, Pullman, Washington
87. Sauer CG (1989) Midas—mission design and analysis software for the optimization of ballistic interplanetary trajectories. J Astronaut Sci 37(3):251–259
88. Schwartz A (1996) Theory and implementation of numerical methods based on runge-kutta integration for solving optimal control problems. PhD thesis, Department of Electrical Engineering, University of California, Berkeley
89. Schwartz A, Polak E (1996) Consistent approximations for optimal control problems based on runge-kutta integration. SIAM J Control Optim 34(4):1235–1269
90. Schwartz A, Polak E, Chen Y (1997) Recursive integration optimal trajectory solver (RIOTS_95)
91. Stengel RF (1994) Optimal control and estimation. Dover Publications, Mineola, New York
92. Stoer J, Bulirsch R (2002) Introduction to numerical analysis. Springer, Berlin
93. Vinh N-X (1981) Optimal trajectories in atmospheric flight. Elsevier Science, New York
94. Vintner R (2000) Optimal control (systems and control: foundations and applications). Birkhäuser, Boston
95. Vlases WG, Paris SW, Lajoie RM, Martens MJ, Hargraves CR (1990) Optimal trajectories by implicit simulation. Technical report WRDC-TR-90-3056, Boeing Aerospace and Electronics, Wright-Patterson Air Force Base, Ohio
96. von Stryk O (1999) User's guide for DIRCOL version 2.1: a direct collocation method for the numerical solution of optimal control problems. Technische Universitat Darmstadt, Darmstadt, Germany
97. Weinstock R (1974) Calculus of variations. Dover Publications, Mineola, New York
98. Williams P (2008) User's guide for DIRECT 2.0. Royal Melbourne Institute of Technology, Melbourne, Australia
99. Wuerl A, Crain T, Braden E (2003) Genetic algorithm and calculus of variations-based trajectory optimization technique. J Spacecraft Rockets 40(6):882–888

Chapter 2
Extremum Seeking Methods for Online Automotive Calibration

Chris Manzie, Will Moase, Rohan Shekhar, Alireza Mohammadi, Dragan Nesic and Ying Tan

Abstract The automotive calibration process is becoming increasingly difficult as the degrees of freedom in modern engines rises with the number of actuators. This is coupled with the desire to utilise alternative fuels to gasoline and diesel for the promise of lower CO_2 levels in transportation. However, the range of fuel blends also leads to variability in the combustion properties, requiring additional sensing and calibration effort for the engine control unit (ECU). Shifting some of the calibration effort online whereby the engine controller adjusts its operation to account for the current operating conditions may be an effective alternative if the performance of the controller can be guaranteed within some performance characteristics. This tutorial chapter summarises recent developments in extremum seeking control, and investigates the potential of these methods to address some of the complexity in developing fuel-flexible controllers for automotive powertrains.

2.1 Introduction

Reciprocating engines are used in transportation and stationary power generation, with diesel and gasoline representing the vast majority of the fuels used. Their environmental impact is observable in the fact that the Australian transport sector contributes approximately 15 % of national CO_2 equivalent emissions [20], while this ratio is slightly higher in the EU and US at 17.5 and 22 % respectively. Meanwhile, over the period 1990–2007 the relative cost of oil has also risen nearly 300 % [33].

C. Manzie (✉) · W. Moase · R. Shekhar · A. Mohammadi
Department of Mechanical Engineering, The University of Melbourne, Victoria, Australia
e-mail: manziec@unimelb.edu.au

D. Nesic · Y. Tan
Department of Electrical and Electronic Engineering, The University of Melbourne, Victoria, Australia

H. Waschl et al. (eds.), *Optimization and Optimal Control in Automotive Systems*, Lecture Notes in Control and Information Sciences 455,
DOI: 10.1007/978-3-319-05371-4_2,

The environmental impact of petroleum fuels has led to substantial consideration and investigation of lower CO_2 emitting alternatives. Regional dependencies dictate that the best option from a fuel security and cost perspective are not unique with the possible options including liquified petroleum gas, compressed natural gas and various levels of ethanol blending with gasoline.

While the concern over the environmental ramifications of exhaust emissions has been well publicised, the public health implications of emissions are only just coming to light. A study by the Californian Air Resources Board [5] estimated 9,000 premature deaths during 2007 in California alone are attributable to particulate matter of diameter less that 2.5 microns. By way of comparison, natural gas has less than half the particulate matter of diesel [23], and liquefied petroleum gas (LPG) has 70 % fewer particulate emissions than gasoline [4].

The downside of both these and other alternative fuels, however, is that in unrefined form their composition is variable. Consequently, subsequent operation can vary markedly, for example a 20 % change in fuel consumption was observed across a range of typical CNG blends in [10], while LPG can vary from propane-butane ratios of 25:75–100:1, with emissions performance significantly impacted [24].

From the engine control systems perspective, the challenge of changing composition is reflected in Fig. 2.1. In this figure, the torque produced from an internal combustion engine for spark sweeps on two CNG blends at a close to idle operating condition are presented. Here, an incorrect (fixed) assumption on fuel composition would lead to an incorrect MBT estimate used by the engine controller—leading to efficiency degradation and possible increased coefficient of variation of indicated mean effective pressure. The same scenario of lost optimality applies for other engine inputs.

While this approach focuses on the immediate penalty associated with changing fuel composition, the presence of hierarchical control algorithms in many of the more complex powertrains can also lead to performance and efficiency degradation if the composition is incorrectly modelled. One such example is investigated in [13], where a turbocharged flex fuel engine is used in a hybrid power train with an optimal Equivalent fuel Consumption Minimisation Strategy (ECMS) controller based on Pontryagin's Minimum Principle.

Figure 2.2 shows the fuel consumption maps mapped using E5 and E85 fuels for the test engine. In this instance, it was found that incorrect assumptions of the fuel consumption could lead to fuel efficiency degradation of the hybrid vehicle of up to 30 % relative to the best possible performance obtained when the fuel map was known perfectly and available to the ECMS-based hybrid power train controller.

For fixed fuel operation, the traditional approach to engine control is to apply a look up table approach whereby the inputs are predetermined for each engine operating point using a lengthy calibration procedure. While many proposed engine control approaches use dynamic engine models in the controller [22, 27], they typically maintain look up tables to capture properties relating to in-cylinder dynamics such as indicated efficiency. These multidimensional surfaces are obtained during a separate calibration procedure. In both situations, if the fuel composition varies from that

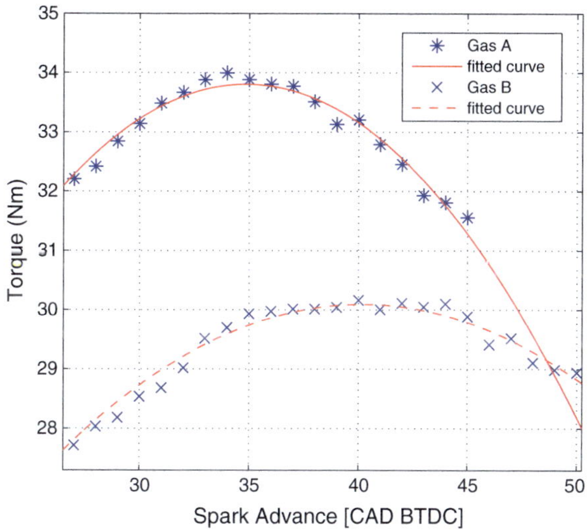

Fig. 2.1 Result of open loop spark sweeps with *Gas A* (pure methane) and *Gas B* (80 % methane, 9 % CO_2, 8.5 % N_2, 2 % ethane and 0.5 % propane) at 1,500 rpm and low load

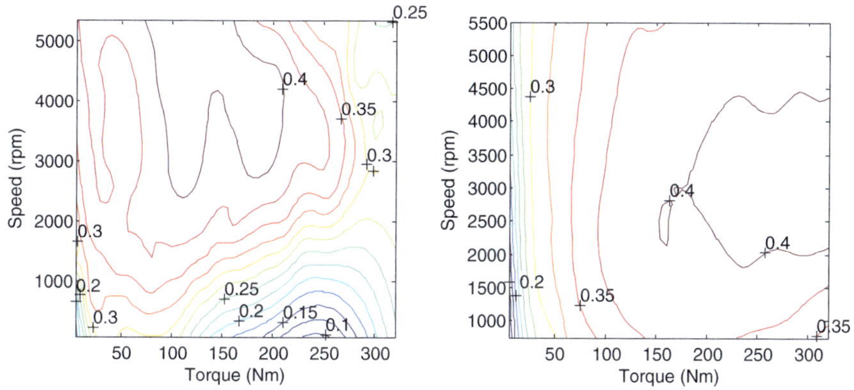

Fig. 2.2 Engine efficiency maps for a flex fuel engine running with (*left*) E05 and (*right*) E85 fuel

used during the calibration, the engine controller performs suboptimally and overall engine performance may degrade.

As a consequence, it would appear there is a need to implement some form of online optimisation to maximise the benefits promised by alternative fuels. However, the inclusion of any such adaptive capability must be in conjunction with rigorous guarantees on the performance of the closed loop system. With this in mind, recent developments in extremum seeking methods appear a good potential solution candidate, and three main categories of extremum seeking algorithms are reviewed in the subsequent section.

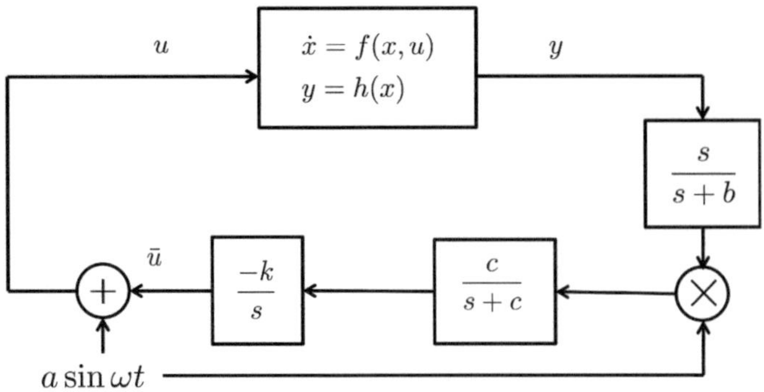

Fig. 2.3 Basic SISO extremum seeking scheme exhibiting sinusoidal perturbation to plant input, with high and low pass filters either side of demodulation step

This is then followed by a discussion of some of the automotive implementations of extremum seeking algorithms currently reported in the literature, as well as an outline of future possibilities for future research directions from both a theoretical and application based perspective.

2.2 Review of Extremum Seeking

The first known examples of extremum seeking techniques date back to 1922 [12], with several practical examples seen up until the 1950s, however the lack of formal proofs and performance guarantees led to the approach being largely set aside of several decades. This changed around the turn of the century with the development of local stability results in [11] for the basic extremum seeker shown in Fig. 2.3.

This approach essentially uses a sinusoidal perturbation to perturb the input to a dynamic plant with output y. The output is high pass filtered to remove any DC offset, before being multiplied by the dither to demodulate the gradient estimate. The gradient information can now be isolated by low pass filtering to remove all components of the plant output that are harmonics of the dither frequency. The resulting gradient estimate is then used in a gradient descent algorithm to push the plant towards its minimum. Provided the plant has a smooth output function $h(x)$ to which the output converges uniquely then semi-global practical stability is achieved with appropriate tuning of the parameters a, b, c, k and ω.

Following these initial results, many implementations of extremum seeking techniques have followed and further refinements of the theoretical foundations have been made. A comprehensive review of both the theoretical developments and applications of extremum seeking over the period 1922–2010 is provided in [30].

The extremum seeking literature can now be broadly classified into three groups: black-box approaches; grey-box approaches; and sampled data approaches. These

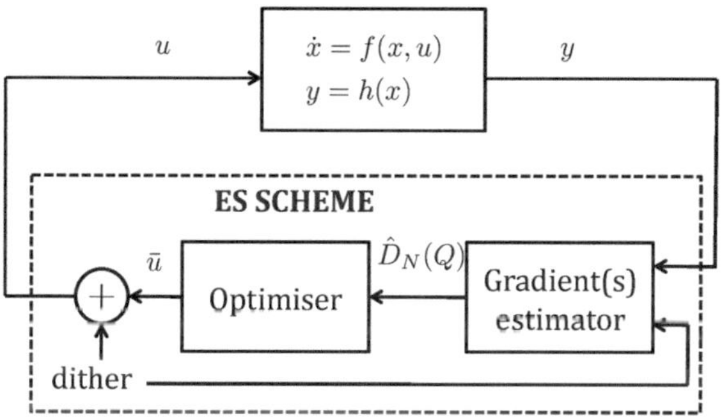

Fig. 2.4 Generalised *black* box extremum seeking framework

have different implementations and subsystem requirements which will be discussed in the following sections.

2.2.1 *Black-Box Extremum Seeking*

A guiding principle of the basic extremum seeking algorithm outlined above is that no model information about the plant is required, leading to it often being referred to as a black box scheme. Other approaches reported in the literature have considered alternative components in place of the filters and gradient descent in Fig. 2.3, with the essence of the majority of these black box approaches captured in Fig. 2.4, which shows the dynamic plant connected to a gradient estimator, and finally the input to the plant updated through an appropriate optimisation algorithm.

Although more formally stated in [18], the following requirements are placed on the components of Fig. 2.4 in a general setting.

Plant: The plant dynamics, $f(x, u)$ have an asymptotically stable equilibrium described for each plant input u by the surface $x = l(u)$. The input–output map of the plant at equilibrium $Q(u) := h(l(u))$ is continuous and has a unique global maximum, u^*.

Gradient estimator: Consider the first N-derivatives of $Q(u)$ as represented in the vector $D_N(Q)$, i.e.:

$$D_N(Q) := \left[\frac{dQ}{du}, \ldots, \frac{d^N Q}{du^N} \right]^T \tag{2.1}$$

The gradient estimator and dither signal must contains sufficient excitation of the plant to provide sufficiently accurate estimation of $D_N(Q)$ over a finite time interval. The dynamics of the gradient estimator may be represented as:

$$\dot{\hat{D}}_N(Q) = \varepsilon_1 F_1(\hat{D}_N(Q), P, y) \tag{2.2}$$
$$\dot{P} = \varepsilon_1 F_2(P, y) \tag{2.3}$$

Here P represents the presence of potential auxiliary states in the estimator.

Unlike in the extremum seeking scheme of Fig. 2.3, the dither need not be sinusoidal, and is only necessary to ensure the first N gradients of the input–output plant map can be estimated (assuming the chosen optimiser requires N derivatives). Consequently, various dither alternatives such as square wave, triangular wave and stochastic signals have been used. As with the dither signal, numerous gradient derivative estimators have been proposed and successfully deployed, ranging from combinations of first order filters [3, 19], through to Luenberger observers [15] and Kalman filters [32]. The dynamics of the gradient estimator are controlled through the tuning parameter ε_1, representing for example the cut-off frequency of the high pass filter or the gain used by a Luenberger observer. The suitable choice of ε_1 typically delivers time scale separation of the plant and estimator dynamics.

Optimiser: Consider the continuous optimisation algorithm operating on known derivatives $D_N(Q)$ of a static map $Q(z)$:

$$\dot{z} = \varepsilon_2 F_3(z, D_N(Q(z))) \tag{2.4}$$

The chosen optimisation algorithm ensures that the output of the static map $Q(z)$ converges to the optimum value, $Q(z^*)$, with some degree of robustness. Again there are many possible choices for the optimiser, ranging from the gradient descent approach shown in Fig. 2.3, to higher order optimisers such as Newton step [15] and other variants, although all typically have tunable dynamics represented by ε_2. As with the gradient estimator, the choice of ε_2 is used to ensure time scale separation, this time between the gradient estimator and the optimiser.

This decomposition of the closed loop extremum seeking scheme enables the practitioner to independently select the optimiser and gradient estimator from families of possible options satisfying the requirements stated above. The overall convergence of the system is then guaranteed by ensuring that the plant, gradient estimator and optimiser occupy different time scales (formally stated in Theorem 1 of [18]). This is achieved by selecting the gains such that the gradient estimator is sufficiently slower than the plant, and the optimiser is sufficiently slower than the gradient estimator to ensure time scale separation—therefore enabling the plant to be 'seen' as a static map by the gradient estimator, and both components to be 'seen' as static by the optimiser.

There is a tradeoff to be balanced in selecting the gains and dither, as smaller values yield slower convergence but guarantee the convergence of the output will be to a smaller vicinity of the optimum of the static map, $Q(u)$.

The generality of the framework presented above is useful in guaranteeing convergence and developing tuning rules despite very little knowledge of the plant. However, one potential drawback of this generality is the conservativeness of the result particularly in terms of convergence rate, to ensure time scale separation between

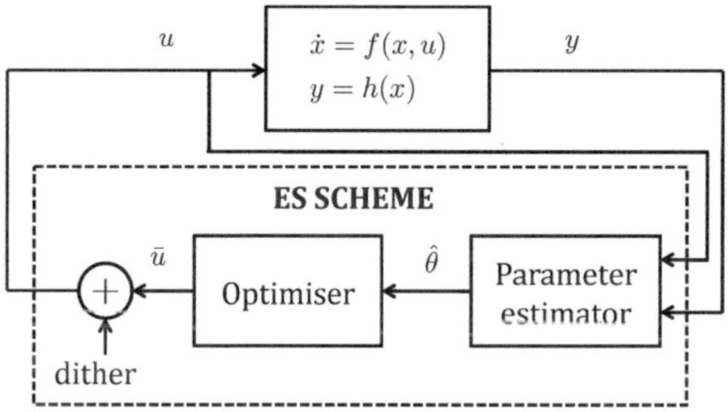

Fig. 2.5 Generalised it grey box extremum seeking framework

the major system elements. With greater specificity about the plant structure, and by utilising particular schemes for gradient estimation and optimisation, faster convergence properties can be obtained. Such approaches have been described for Wiener-Hammerstein plants [14], and principally rely on plant input and output filters to remove the need to wait for plant dynamics to settle before updating the extremum seeking output. This allows for high frequency dithers to be employed and arbitrarily fast (in the absence of noise) convergence to be achieved.

2.2.2 Grey-Box Extremum Seeking

The fast extremum seeking approaches of the previous section utilise knowledge about the plant dynamics in the design of the input and output filters. In many practical applications however, there is knowledge of the basic form of the optimisation surface, $Q(u)$, to which the dynamics converge. The utilisation of this knowledge in an extremum seeking context falls largely in the domain of the so-called grey box approaches. Here, the surface $Q(u)$, is parameterised in terms of a vector of unknown parameters, θ, i.e.:

$$Q(u, \theta) = \Psi(u)^T \theta \tag{2.5}$$

Thus, by estimating θ potentially non-local information about the surface can be obtained. In [1, 2], this approach was explored for specific instances of the parameter estimator and optimiser, and then further generalised in [16] leading to a description of the closed loop system in the form shown in Fig. 2.5.

The parallels with the generalised black box framework of Fig. 2.4 are clear, and there exist similar requirements on the plant and optimiser in the grey box scheme, although the following requirement for the parameter estimator replaces that for the gradient estimator given above.

Parameter estimator: The parameter estimator can be represented in the general form as states directly relevant to the parameter estimates and additional states within the parameter estimator, i.e.:

$$\dot{\hat{\theta}} = \varepsilon_1 G_1(\hat{\theta}, P, y, u) \tag{2.6}$$

$$\dot{P} = \varepsilon_1 G_2(P, y, u) \tag{2.7}$$

The parameter estimator and dither signal combination needs to contains sufficient excitation of the plant to provide estimation of the parameters within a finite time interval. A thorough analysis of a number of different parameter estimation schemes was conducted in [16], with possibilities for deployment including gradient algorithms with and without integral costs along with least squares estimation variants.

By incorporating the parameter estimator in place of the gradient estimator, the optimiser can potentially utilise differently structured optimisers that may take advantage of the estimated map, while still satisfying the time scale separation of the black box scheme. This reliance on time scale separation implies that substantial increases in convergence rates over the black box approach are not necessarily forthcoming.

With this in mind, preliminary research has been made into algorithm specific grey box approaches that lack the generality of the framework described above but may have potential along the lines of the fast black box approaches described in the previous section [26].

2.2.3 Sampled Data Approaches

The previous two categories of extremum seeking algorithms use continuous time optimisation algorithms, and consequently only draw upon the discrete time optimisation field indirectly through the time scale separation providing the optimiser can satisfy the requirements. An alternative viewpoint, drawing more directly on nonlinear programming techniques, was first presented in [31], and is illustrated in Fig. 2.6.

In this approach, the dynamic plant is treated as a system to be sampled, with a finite number of sampled plant outputs used to estimate derivatives of the static map before the control input is updated, and enabled the use of many discrete time optimisation algorithms such as Finite Differencing and Simultaneous Perturbation Stochastic Approximation (SPSA) [29]. The work was generalised further in [9], where a different style of proof was used and required the discrete time optimisation algorithm to be uniformly attractive with respect to small additive disturbances, rather than asymptotically stable as in [31]. The later approach also opens the possibility for non-gradient based global optimisation algorithms such as Piyavskii-Shubert and DIRECT to be rigorously deployed [8, 17] thereby reducing the requirement that the plant have only one global optimum—albeit at significant convergence time penalties relative to the local optimisation techniques.

As seen with the continuous extremum seeking approaches, there may be advantages in forsaking generality of the sampled data approach and focusing on a specific combination of discrete time optimiser and plant. In [28], a discrete time Hammerstien plant is subjected to an algorithm using a square wave dither with a two step

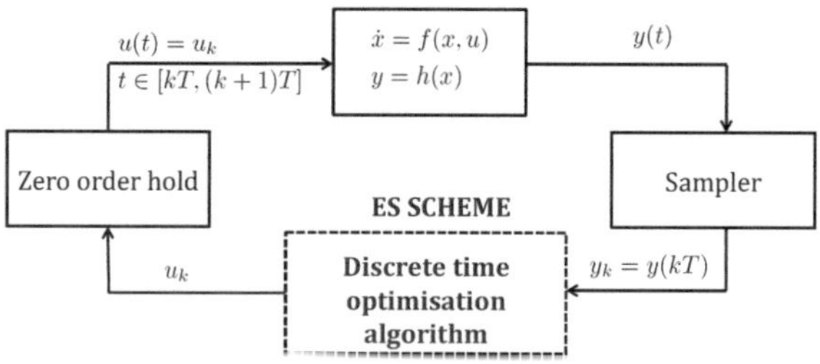

Fig. 2.6 Generalised sampled data extremum seeking framework

averaging filter with the specific nature of the system enabling LPV techniques to be used to generate an exponential stability result.

2.3 Application to Automotive Engine Calibration

One of the earliest applications of extremum seeking in the automotive calibration context was in [25], where a grey box extremum seeking architecture was designed and implemented for spark control on a gasoline spark ignition engine. The approach assumed a quadratic map between the spark angle and indicated torque, similar to Fig. 2.1. The parameters of the quadratic were then estimated using a recursive least squares estimator, although the optimiser used attempted to immediately drive the control input to the optimum level as calculated by the estimated parameters, and consequently the time scale separation required for convergence guarantees in the grey-box schemes of Sect. 2.2.2 was not present. Thus while successful results were reported, the initial conditions of the parameter estimator need to be sufficiently close to the true values for convergence to occur.

Addressing the multi-variable calibration problem from an extremum seeking perspective was explored in [21]. This approach considered a sampled data implementation of extremum seeking (as in Sect. 2.2.3), with the actuation variables of intake and exhaust valve timing along with the spark timing. From arbitrary initial conditions, the calibration process using extremum seeking was found to take around 15 min to locate the optimum, largely impacted by the multivariable nature of the problem and the noise associated with torque measurements requiring long averages of measured data.

In-service alternative fuelled engines may experience regular fuel composition changes, yet the opportunity to undertake manual recalibration is not present. The growing interest in these engines has renewed interest in online calibration. One recent implementation considered the situation for flex-fuelled engines [6] considers a ES implementation in the class of black box systems described in Sect. 2.2.1 to maximise fuel economy by adjusting the spark. This approach utilises a discrete time version of the black box approach of the form given in Fig. 2.3, with a square wave perturbation. Unlike in [21], only the spark is changed and this single variable nature of the optimisation coupled with the fact that the optimisation problem is effectively 'hot-started' as the optimal spark will not vary significantly between different ethanol-gasoline blends means the outcome is more positive in terms of real world deployability.

In a similar vein, but with compressed natural gas blends as the fuel source, different implementations of the grey box extremum seeking framework falling under the framework of [16] were investigated as possible spark optimisation strategies. Experiments are carried out for two blended pure methane and a blended gas consisting of 80 % methane, 9 % carbon dioxide, 8.5 % nitrogen, 2 % ethane and 0.5 % propane. As with [25] and shown in Fig. 2.1, open loop tests demonstrate that a quadratic polynomial approximation seems a good representation of the data, allowing the following model to be used relating torque, τ, and spark α:

$$\tau(\alpha) = \lambda_1 \alpha^2 + \lambda_2 \alpha + \lambda_3 \tag{2.8}$$

Defining the regressor vector $\phi = [\alpha^2 \;\; \alpha \;\; 1]^T$, the grey box approach then involves the selection and tuning of appropriate parameter estimator for $\theta := [\lambda_1 \;\; \lambda_2 \;\; \lambda_3]^T$, and an optimiser to drive τ towards τ^*.

To demonstrate the flexibility afforded by the framework approach, experiments were conducted using two different estimator-optimiser combinations. The first consisted of recursive least squares parameter estimator and gradient based optimiser, while the second consisted of a gradient based parameter estimator and a Jacobian-matrix transpose optimisation metric. All parameter estimators and optimisers had previously been shown to satisfy the theoretical requirements of the grey box framework in [16], and the tuned algorithms are repeated below in (2.9)–(2.13). The discrete nature of the presented algorithms reflects an emulation of the continuous time versions of (2.6)–(2.7) and (2.4).

Gradient-based parameter estimator after tuning:

$$\hat{\theta}_{k+1} = \hat{\theta}_k - [0.002 \;\; 0.05 \;\; 1]^T (\tau_k - \phi_k^T \hat{\theta}_k) \tag{2.9}$$

Recursive least squares parameter estimator after tuning:

$$\hat{\theta}_{k+1} = \hat{\theta}_k + P_k \phi_k (\tau_k - \phi_k^T \hat{\theta}_k) \tag{2.10}$$

$$P_{k+1} = P_k + (0.9 P_k - P_k \phi_k^T \phi_k P_k) \tag{2.11}$$

Gradient based optimiser after tuning:

$$\hat{\alpha}^*_{k+1} = \hat{\alpha}^*_k + 10(2\hat{a}_k \hat{\alpha}^*_k + \hat{b}_k) \tag{2.12}$$

Jacobian matrix transpose optimiser after tuning:

$$\hat{\alpha}^*_{k+1} = \hat{\alpha}^*_k - 75\hat{a}_k (2\hat{a}_k \hat{\alpha}^*_k + \hat{b}_k) \tag{2.13}$$

The applied spark advance is then the current estimate of the optimal spark advance, $\hat{\alpha}_k^*$, perturbed by a sinusoidal dither of amplitude one crank angle degree, i.e.:

$$\alpha_k = \hat{\alpha}_k^* + \sin(0.1kT) \tag{2.14}$$

All experiments to test these algorithms were conducted in the test cell of the ACART Laboratory at the University of Melbourne. The engine tested a six cylinder 4L Ford Falcon BF MY2006 gasoline engine which is converted to operate with natural gas.

In order to keep the speed and load of the engine constant, an eddy current dynamometer is used which can only work as a brake and is not capable of motoring the engine. The air-fuel ratio was maintained at an approximately stoichiometric condition by adjusting the injection duration with feedback from a wide-band exhaust oxygen gas sensor. The proposed algorithms were implemented in MATLAB. The output of the MATLAB program were sent in realtime directly to the engine control unit (ECU) via ATI Vision software, thereby adapting the stored calibration. The delays in communication between the different software programs were measured at approximately 6 ms, which was considered negligible in the context of this application. Feedback torque was obtained through measurements from a load cell on the dynamometer, although in the future could be replaced by in-cylinder pressure sensors and appropriate combustion analysis. The torque measurement was averaged over a period of three seconds to minimize the effects of combustion variability and measurement noise. The sample rate used by the controller was set nominally to 5 s so as to be longer than the torque measurement time.

The engine control unit was initially calibrated using the blended methane gas, leading to an initial estimate of MBT at approximately 40° BTDC. The actual fuel used in the engine was pure methane, and consequently the extremum seeking controllers were required to adapt the spark to find the new MBT spark, which lies at approximately 33° BTDC, although may vary slightly with engine temperature. The adapted spark and resulting engine torque for each of the two extremum seeking combinations are shown in Figs. 2.7 and 2.8.

In both instances, (although not shown) the parameter estimates also converge to a vicinity of the 'true' values, and the spark converges to a close vicinity of the optimum. To quantify the gains in efficiency the incorporation of these extremum seeking approaches strategies may provide, the fuel flow rate of the engine running with the initial spark advance at the specified operating point was compared to fuel flow rate after convergence of the spark to the optimal value. This latter value was corrected to allow for the torque difference although in practice this could be achieved through modification to the throttle angle. Consequently it was found that the fuel economy improved by approximately 3 % at this static operating condition.

To further demonstrate the flexibility of viewing extremum seeking as a framework approach, a sampled data approach is also presented for the same engine. In this approach, an ES scheme using a simple alternating dither signal $a(-1)^k$ is added to the current estimate for the optimising value of the system input, and used as the applied spark in place of (2.14). The resulting torque (after the two second averaging)

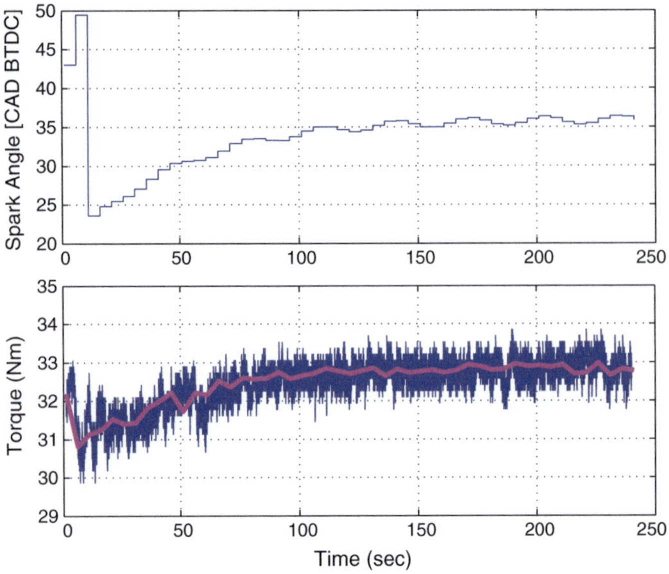

Fig. 2.7 *Grey* box convergence for a recursive least squares parameter estimator and gradient based optimiser (*top*) spark advance (*bottom*) measured and 2 s averaged engine torques

is multiplied by the signal $(-1)^k$ and passed through a two-step moving-average FIR filter and discrete-time integrator. This advances the estimate for the optimising value according to an approximate gradient ascent law, based on a two-point central difference approximation applied to the engine spark-to-torque mapping. The entire closed loop scheme is represented by the block diagram shown in Fig. 2.9.

Implementing this sampled data extremum seeking approach on the same engine set up as described previously, with an initial estimate of MBT spark of 22° leads to the results shown in Fig. 2.10. As previously observed for the grey box approaches, convergence to the vicinity of the optimum spark and torque occurs and there is a subsequent improvement in fuel economy of approximately 3 % at this operating point relative to the case of no adaptation of the spark.

The convergence speed of the algorithms tested in this section warrants discussion. As shown in Figs. 2.7–2.10, the schemes take of order 100 s to converge to the optimum at a fixed operating condition. On its own, this convergence time is not of significant concern as the rate of change of fuel composition is much slower thereby allowing even a 100 s transient to be deemed negligible if steady state operation is considered.

During transient engine operation, such as might be considered during urban driving, it is however unlikely that the engine will remain at a constant, non-idle operating condition for periods of this duration. The convergence rate is a consequence of the nature of the time scale separation requirement of the plant, estimator and optimiser for the extremum seeking framework-based theory, which is

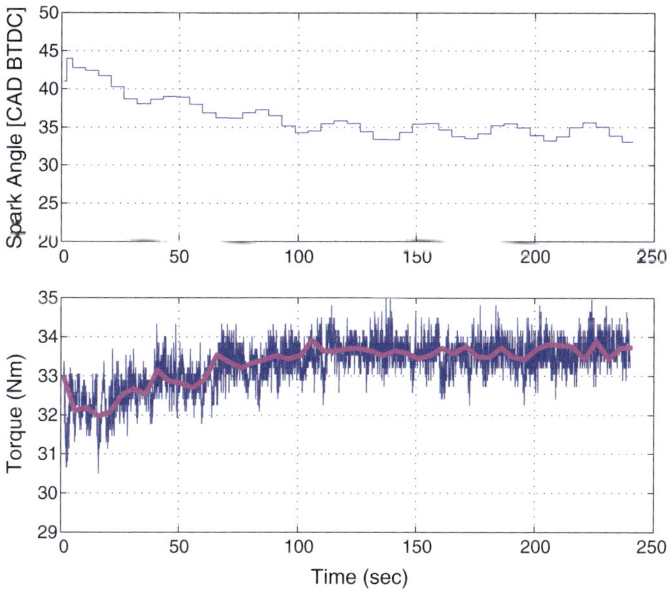

Fig. 2.8 *Grey* box convergence for a gradient based parameter estimator and Jacobian matrix transpose (*top*) spark advance (*bottom*) measured and 2 s averaged engine torques

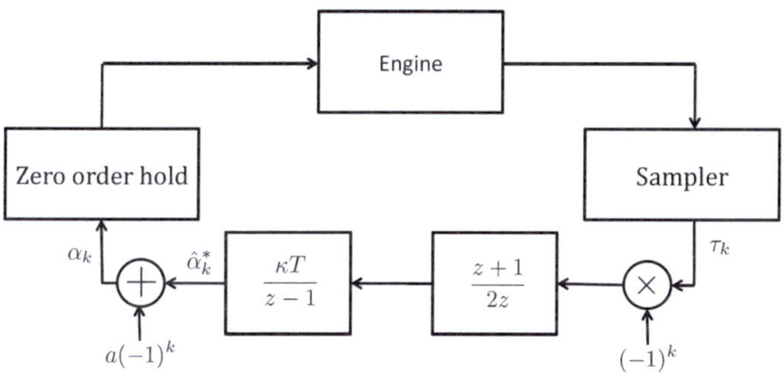

Fig. 2.9 Proposed sampled data extremum seeking approach for optimal spark estimation on CNG engine

principally centred on guaranteeing convergence using very relaxed requirements on the components of the closed loop scheme. By being more restrictive in the selection of these components, the tuning requirements can be modified and the convergence rate can potentially be sped up using the faster extremum seeking approaches in the vein of [14].

Finally, for the purposes of demonstration of the extremum seeking techniques, brake torque has been directly measured and used as the feedback to be optimised in these experiments. Such a measurement is clearly not directly available in an

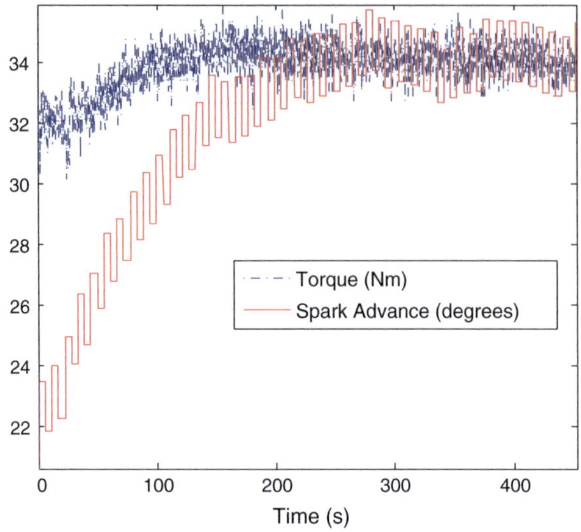

Fig. 2.10 Spark and torque progression under the proposed sampled data extremum seeking approach with $a = 1°$

on-road application, and so surrogate measurements must be employed depending on the available sensor set.

2.4 Incorporation of Constraints

Not discussed in the methodology so far is the issue of constraint management. For input constraints, if the plant inputs, $\bar{u}$, are required to lie within a set, U_c, the optimiser can explicitly take this into account by projecting the new optimiser output, $\bar{u}_{proj}$ onto the Pontryagin difference of the constraint set and the dither set, D i.e.:

$$U_c \backsim D \triangleq \{\bar{u} \in R^n \mid \bar{u} + d \in U_c, \forall d \in D\} \tag{2.15}$$

$$\bar{u}_{proj} = \text{proj}(\bar{u}, U_c \backsim D) \tag{2.16}$$

This ensures the dither is still able to persistently excite the system, even approaching the constraint boundary, and the gradient or parameter estimates in the black or grey box schemes are maintained.

In the context of the engine calibration problem, this might mean for example the physical actuator limits are captured so that, for example, the spark is constrained to occur within the compression stroke of that cylinder. These types of constraints are rigorously enforceable.

On the other hand, state constraints are not so easily dealt with as generally there is no concept of 'state' in the model used in an extremum seeking controller. In this case, an approximate solution is that the state constraints must be mapped to input constraints, so that U_c in (2.15)–(2.16) becomes time varying and most likely requires online estimation.

Again in the context of the online calibration problem, the spark timing in a natural gas engine is often constrained by knock limitations. The occurrence of knock is related to the fuel composition, which can be viewed as an internal state and is clearly unknown from the problem definition. Knock detection algorithms (see e.g. [7] and the references within) can be used to continually estimate the knock limit on spark advance, which may then be used to update the constraint set U_c.

2.5 Summary and Future Opportunities

The recent development of extremum seeking frameworks has delivered considerable flexibility into the deployment of different algorithms to achieve convergence to optimal performance in many applications. To the automotive community, this appears to be highly relevant as the industry continues a progression towards alternative fuels exhibiting variable composition, thereby necessitating some form of closed loop calibration being conducted during regular vehicle operation.

There remain a number of theoretical challenges and opportunities for research in extremum seeking algorithms. These include research into increasing the convergence rate without unduly compromising the region of attraction; dealing with map uncertainty in grey box frameworks; handling state constraints within the various frameworks; and identifying when certain frameworks might lead to better closed loop performance.

Similarly, there are also application-centric issues for automotive calibration including the deployment of the algorithms in transient driving conditions; implementing some of the novel theoretical developments promising faster convergence for online multivariable calibration; consideration of emissions in the cost function; and the integration with model based techniques for faster offline calibration particularly in highly actuated engines.

Acknowledgments The support of the Advanced Centre for Automotive Research and Testing (ACART, http://www.acart.com.au) and the Australian Research Council (ARC) is gratefully acknowledged.

References

1. Adetola V, DeHaan D, Guay M (2009) Adaptive model predictive control of constrained nonlinear systems. Syst Control Lett 58:320–326
2. Adetola V, Guay M (2006) Adaptive output feedback extremum seeking receding horizon control of linear systems. J Process Control 16:521–533
3. Ariyur KB, Krstic M (2003) Real time optimisation by extremum seeking control. Wiley, New York

4. Beer T, Grant T, Williams D, Watson H (2002) Fuel-cycle greenhouse gas emissions from alternative fuels in australian heavy vehicles. Atmos Environ 36:753–763
5. Californian Air Resources Board. Estimate of Premature Deaths Associated with Fine Particle Pollution (PM2.5) in California Using a U.S. Environmental Protection Agency, Methodology. August, 2010
6. Hellström E, Lee D, Jiang L, Stefanopoulou AG, Yilmaz H (2013) On-board calibration of spark timing by extremum seeking for flex-fuel engines. IEEE Trans Control Syst Technol 21(6):2273–2279
7. Jones JCP, Frey J, Muske KR, Scholl DJ (2010) A cumulative-summation-based stochastic knock controller. In: Proceedings of the institution of mechanical engineers part D: journal of automobile engineering 224:969–983
8. Khong SZ, Nesic D, Manzie C, Tan Y (2013) Multidimensional global extremum seeking via the direct optimisation algorithm. Automatica 49(7):1970–1978
9. Khong SZ, Nesic D, Tan Y, Manzie C (2013) Unified frameworks for sampled-data extremum seeking control: global optimisation and multi-unit systems. Automatica 49(9):2720–2733
10. Kim K, Kim H, Kim B, Lee K (2009) Effet de la composition du gaz naturel sur les performances d'un moteur GNC. Oil Gas Sci Technol Rev IFP 64(2):199–206
11. Krstic M, Wang H-H (2000) Stability of extremum seeking feedback for general nonlinear dynamic systems. Automatica 36:595–601
12. Leblanc M (1922) Sur lõelectrification des chemins de fer au moyen de courants alternatifs de frequence elevee. Revue Generale de lõElectricite
13. Manzie C, Grondin O, Sciarretta A, Zito G (2013) Robustness of ECMS-based optimal control in parallel hybrid vehicles. In: 7th IFAC symposium on advances in automotive, control
14. Moase W, Manzie C (2012) Fast extremum—seeking for Wiener-Hammerstein plants. Automatica 48:2433–2443
15. Moase W, Manzie C, Brear M (2010) Newton—like extremum—seeking for the control of thermoacoustic instability. IEEE Trans on Autom Control 55:2094–2105
16. Nesic D, Mohammadi A, Manzie C (2013) A framework for extremum seeking control of systems with parameter uncertainties. IEEE Trans Autom Control 58:435–448
17. Nesic D, Nguyen T, Tan Y, Manzie C (2013) A non-gradient approach to global extremum seeking: an adaptation of the Shubert algorithm. Automatica 49(3):809–815
18. Nesic D, Tan Y, Manzie C, Mohammadi A, Moase W (2012) A unifying framework for analysis and design of extremum seeking controllers. In: 24th Chinese decision and control conference (CCDC)
19. Nesic D, Tan Y, Moase W, Manzie C (2010) A unifying approach to extremum seeking: adaptive schemes based on derivative estimation. In: IEEE conference on decision and control
20. Australian Government Department of Climate Change (2009) National greenhouse gas inventory. http://www.climatechange.gov.au/inventory
21. Popovic D, Jankovic M, Magner S, Teel A (2006) Extremum seeking methods for optimization of variable cam timing engine operation. IEEE Trans Control Syst Technol 14:398–407
22. Del Re L, Allgöwer F, Glielmo L, Guardiola C, Kolmanovsky I (eds) (2010) Automotive model predictive control: models, methods and applications. Springer, Heidelberg
23. Ristovski ZD, Jayaratne ER, Morawska L, Ayoko GA, Liml M (2005) Particle and carbon dioxide emissions from passenger vehicles operating on unleaded petrol and LPG fuel. Sci Total Environ 345:93–98
24. Saleh HE (2008) Effect of variation in LPG composition on emissions and performance in a dual fuel diesel engine. Fuel 87(13–14):3031–3039
25. Scotson P, Wellstead PE (1990) Self-tuning optimization of spark ignition automotive engines. IEEE Control Syst Mag 3:94–101
26. Sharafi J, Moase W, Shekhar RC, Manzie C (2013) Fast model-based extremum seeking on hammerstein plants. In: IEEE conference on decision and control
27. Sharma R, Nesic D, Manzie C (2013) Sampled data model predictive idle speed control of ultra-lean burn hydrogen engines. IEEE Trans Control Syst Technol 21:538–545

28. Shekhar RC, Moase W, Manzie C (2013) Semi-global stability analysis of a discrete-time extremum-seeking scheme using LDI methods. In: IEEE conference on decision and control
29. Spall J (2000) Adaptive stochastic approximation by the simultaneous perturbation method. IEEE Trans Autom Control 45:1839–1844
30. Tan Y, Moase WH, Manzie C, Nesic D, Mareels IMY (2010) Extremum seeking from 1922 to 2010. In: 29th Chinese control conference, pp 14–26
31. Teel AR, Popovic D (2001) Solving smooth and nonsmoooth multivariable extremum seeking problems by the methods of nonlinear programming. In: Proceedings of the American control conference, pp 2394–2399
32. Wiederhold O, King R, Noack B, Neuhaus L, Neise W, Enghardt L, Swoboda M (2009) Extensions of extremum—seeking control to improve the aerodynamic performance of axial turbomachines. In: 39th AIAA fluid dynamics conference
33. WTRG-Economics Oil price history and analysis (2013). http://www.wtrg.com/prices.htm

Chapter 3
Model Predictive Control of Autonomous Vehicles

Mario Zanon, Janick V. Frasch, Milan Vukov, Sebastian Sager and Moritz Diehl

Abstract The control of autonomous vehicles is a challenging task that requires advanced control schemes. Nonlinear Model Predictive Control (NMPC) and Moving Horizon Estimation (MHE) are optimization-based control and estimation techniques that are able to deal with highly nonlinear, constrained, unstable and fast dynamic systems. In this chapter, these techniques are detailed, a descriptive nonlinear model is derived and the performance of the proposed control scheme is demonstrated in simulations of an obstacle avoidance scenario on a low-fricion icy road.

3.1 Introduction

Due to the well known vehicle-road dynamics, Model Predictive Control (MPC) is an excellent tool for precise trajectory planning in autonomous vehicle guidance, which can be of great importance in dangerous driving situations. High sampling rates (i.e., in the range of tenths of Hertz) and long prediction horizons however, which are required for a safe operation, pose a computational challenge, particularly in combination with the involved nonlinear vehicle dynamics. Many recent chapter chose a two-level MPC approach to overcome this computational challenge, being composed of a coarse path planning algorithm with a long prediction horizon, and a higher-fidelity path following algorithm on a shorter horizon, cf. [1–4]. Only very

M. Zanon (✉) · J. V. Frasch · M. Vukov · M. Diehl
Electrical Engineering Department (ESAT-SCD) and the Optimization in Engineering Center (OPTEC), KU, Leuven, Belgium
e-mail: mario.zanon@esat.kuleuven.be

M. Diehl
Institute of Microsystems Engineering (IMTEK), University of Freiburg, Georges-Köhler-Allee 102, 79110 Freiburg, Germany

J. V. Frasch and S. Sager
Institute for Mathematics, University of Magdeburg , Magdeburg, Germany

H. Waschl et al. (eds.), *Optimization and Optimal Control in Automotive Systems*, Lecture Notes in Control and Information Sciences 455,
DOI: 10.1007/978-3-319-05371-4_3,

recently the computational feasibility of a challenging, realistic scenario using a single MPC controller with a detailed nonlinear model was demonstrated in [5], using auto-generated tailored C code based on the real-time iteration scheme [6] for Bock's multiple shooting method [7].

In the following, the results from [5] are extended and combined with the corresponding Moving Horizon Estimation (MHE) scheme [8] for full state and paramter observation in an realistic context. In particular we yet extend the vehicle model used in [5, 8] by introducing a suspension model for a more realistic representation of the driving behavior even in extreme situations. The MHE scheme has also been modified to detect sudden changes in the road friction condition fast and reliably.

The description of the mathematical problem formulation is given in Sect. 3.2. Section 3.3 provides a presentation of the real-time feasible algorithmic framework. In Sect. 3.4, the vehicle model is derived, and simulation results are presented in Sect. 3.5. Conclusions are drawn in Sect. 3.6.

3.2 Control and Estimation Problems

In order to formulate the control and estimation schemes let the system dynamics be described by ordinary differential equations (ODE)

$$\dot{\mathbf{x}} = \mathbf{f}(\mathbf{x}, \mathbf{u}). \tag{3.1}$$

where $\mathbf{x}(t)$ denotes the differential states and $\mathbf{u}(t)$ denotes the controls. The formulations proposed in the following of the chapter can straightforwardly be extended to systems governed by differential-algebraic equations DAE [9]. Parameters $\mathbf{x_p}$ can be considered as states with zero time derivative, i.e. $\dot{\mathbf{x}}_\mathrm{p} = 0$.

3.2.1 Nonlinear Model Predictive Control

Nonlinear Model Predictive Control (NMPC) is an advanced control technique that relies on the system model to predict the system trajectory and minimize its deviation from a given reference. The full nonlinearity of the model can be taken into account by NMPC and constraints depending on both states and controls can be easily enforced in the problem formulation.

NMPC consists of solving the following dynamic optimization problem at every time instant

$$\begin{aligned}
\underset{\mathbf{x},\mathbf{u}}{\text{minimize}} \quad & \|\mathbf{x}(T_\mathrm{c}) - \mathbf{x}^\mathrm{r}(T_\mathrm{c})\|^2_{P_\mathrm{C}} \\
& + \int_0^{T_\mathrm{c}} \|\mathbf{x}(t) - \mathbf{x}^\mathrm{r}(t)\|^2_{Q_\mathrm{C}} + \|\mathbf{u}(t) - \mathbf{u}^\mathrm{r}(t)\|^2_{R_\mathrm{C}} \, \mathrm{d}t && (3.2a) \\
\text{subject to} \quad & \dot{\mathbf{x}}(t) = \mathbf{f}(\mathbf{x}(t), \mathbf{u}(t)), && (3.2b) \\
& \mathbf{x}(0) = \hat{\mathbf{x}}(0), && (3.2c) \\
& \mathbf{q}(\mathbf{x}(t), \mathbf{u}(t)) > 0, \qquad t \in [0, T_\mathrm{c}], && (3.2d) \\
& \mathbf{x}(T_\mathrm{c}) \in \mathbb{X}_{T_\mathrm{c}}, && (3.2e)
\end{aligned}$$

where T_c is the prediction horizon. The objective function is usually formulated as a least-squares (LSQ) objective (3.2a) penalizing the deviation from a given reference $\mathbf{x}^\mathrm{r}(\cdot)$, $\mathbf{u}^\mathrm{r}(\cdot)$, where Q_C, $P_\mathrm{C} \succeq 0$ and $R_\mathrm{C} \succ 0$ are weighting matrices, to be selected as tuning parameters. The constraint (3.2b) enforces the system dynamics. The initial condition (3.2c) imposes that the initial state coincides with the current estimate $\hat{\mathbf{x}}(0)$ and additional path constraints (3.2d) can be enforced. Finally, a terminal constraint (3.2e) can also be enforced.

The stability of MPC has been first proven for a steady-state reference under the condition that $\mathbb{X}_t = \{\mathbf{x}^r(T_\mathrm{c})\}$. In this context, the terminal cost $\|\mathbf{x}(T_\mathrm{c}) - \mathbf{x}^\mathrm{r}(T_\mathrm{c})\|^2_{P_\mathrm{C}}$ does not appear in the formulation. In many practical cases, this formulation can be too restrictive and lead to infeasibility of the optimization problem, especially when the control horizon T_c becomes short. To increase feasibility, the terminal constraint can be relaxed to an ellipsoidal constraint centered around the reference and a terminal cost needs to be added to the problem formulation. Stability can be proven under some conditions on the choice of the weighting matrix P_C and the ellipsoidal terminal constraint $\mathbb{X}_{T_\mathrm{c}}$. An excellent survey on stability of MPC is given in [10].

In practice, MPC is often implemented without terminal constraint (3.2e) and, in many cases, also without terminal cost. In this case, stability has been proved in [11], provided that the prediction horizon T_c is long enough.

3.2.2 Moving Horizon Estimation

The problem of estimating the current state given a set of measurements can be formulated as an optimization problem. This idea is at the basis of the Kalman filter. Moving Horizon Estimation (MHE) can be seen as an extension of the Kalman filter that can take into account the full model nonlinearities and gives the opportunity to enforce constraints. It is important to stress though, that MHE relies on a deterministic model and it does not need any specific assumption on the probability distribution of the noise.

MHE consists in minimizing the mismatch between the measurements $\tilde{\mathbf{y}}(t)$ coming from the sensors and the ones predicted by the model measurement function $\mathbf{y}(\mathbf{x}(t), \mathbf{u}(t))$. This corresponds to the dynamic optimization problem

$$\underset{\mathbf{x},\mathbf{u}}{\text{minimize}} \quad \left\|\mathbf{x}(-T_\mathrm{e}) - \hat{\mathbf{x}}(-T_\mathrm{e})\right\|^2_{P_\mathrm{E}} + \int_{-T_\mathrm{e}}^{0} \|\mathbf{y}(\mathbf{x}(t),\mathbf{u}(t)) - \tilde{\mathbf{y}}(t)\|^2_{Q_\mathrm{E}}\,\mathrm{d}t \qquad (3.3a)$$

$$\text{subject to} \quad \dot{\mathbf{x}}(t) = \mathbf{f}(\mathbf{x}(t),\mathbf{u}(t)), \qquad (3.3b)$$

$$\mathbf{q}(\mathbf{x}(t),\mathbf{u}(t)) \geq 0, \qquad t \in [-T_\mathrm{e}, 0], \qquad (3.3c)$$

where T_e is the estimation horizon. The cost function (3.3a) is usually formulated as a least-squares term with the weighting matrix Q_E. Though MHE is a deterministic observer, the probabilistic insight is very valuable for the choice of Q_E. In a similar manner as for NMPC, the constraints (3.3b) enforce the system dynamics and additional path constraints (3.3c) can be enforced.

Because of actuator noise and inaccuracy, the control inputs computed by the controller $\bar{\mathbf{u}}$ may not be perfectly implemented by the system. Thus, in the proposed formulation, the control inputs $\mathbf{u}$ are included as decision variables and their deviation from $\bar{\mathbf{u}}$ is penalized, i.e. $\|\mathbf{u}-\bar{\mathbf{u}}\|^2_{Q^\mathbf{u}_\mathrm{E}}$ is added to the cost function. This can be achieved by adding pseudo-measurements to the measurement function $\mathbf{y}(\mathbf{x}(t),\mathbf{u}(t))$.

The so-called arrival cost [(first term in Eq. (3.3a)] has the important role of summarizing past information in a quadratic term which depends on the initial state $\mathbf{x}(-T_\mathrm{e})$. For more details on the arrival cost see [12, 13].

3.3 Efficient Algorithms for fast NMPC and MHE

The nature of the NMPC and MHE problems (3.2a)–(3.2e) and (3.3a)–(3.3c) is a dynamic optimization problem. The complexity of such problems makes it hard for general-purpose solvers to compute solutions fast enough for a real-time implementation. Tailored numerical algorithms are thus needed to overcome these challenges in a real-time application.

3.3.1 Online Solution of the Dynamic Optimization Problem

A variety of methods has been proposed for the efficient online (i.e., time-critical) solution of dynamic optimization problems [6, 14–16]. While most methods are targeted at problems from areas like chemical engineering with rather slow sampling rates, one algorithm that has shown to be effective also for applications with fast system dynamics, is the so called *real-time iteration scheme* [6]. This algorithm is based on Bock's direct multiple shooting method which was originally introduced for the offline solution of optimal control problems [7]. Parametrizing the control input functions by suitably chosen basis functions on a finite grid, the infinite-dimensional optimization problem is discretized yielding a finite-dimensional nonlinear programming problem (NLP). Initial value problems are solved on each of these so called *shooting intervals* by appropriate numerical integration routines. The resulting NLPs are highly structured and are originally solved using a sequential quadratic programming (SQP) approach including a *condensing* step for projection of the high-dimensional quadratic program (QP) on a significantly lower dimensional problem. Alternatively, structure exploiting QP solvers have been proposed recently [17]. In contrast to Bock's original multiple shooting method, the real-time iteration scheme performs only one linearization and one QP solution per sampling time. Thus, significantly higher sampling rates can be achieved, yet contractivity can still be guaranteed [18]. Furthermore, by performing the linearization based on the previous iterate even before observing the new system state/measurement (initial value embedding), feedback delays can be drastically reduced to simply the solution time of a parametric QP [17, 19]. Still, the feedback law is a guaranteed first order approximation of its converged optimal solution, even in the presence of an active set change. More details on the real-time iteration scheme are provided in [6, 18].

3.3.2 Fast Solvers Based on Automatic Code Generation

Automatic code generation of tailored solvers has recently shown to significantly reduce the computational times [20]. The `ACADO Code Generation` tool [21] is part of the open-source software package ACADO Toolkit [22]—a toolkit for automatic control and dynamic optimization. It implements the real-time iteration scheme. The user interface allows one to specify nonlinear dynamic model equations as well as general nonlinear objective and constraint functions. The code-generator exports a generalized Gauss-Newton method for nonlinear MPC [20], and nonlinear MHE [23].

The tool exploits problem structure and dimensions together with sparsity patterns to remove all unnecessary computations and remove the need for dynamic memory allocation. The tool generates self-contained ANSI-C compliant code, which can be deployed on any platform supporting the standard C library. Branching in the exported code is minimized leading to improved code locality, thus faster execution times.

Recent extensions of the `ACADO Code Generation` tool include support for implicit integrators for ODEs and differential algebraic equations (DAEs) [24, 25]. One of the inherent properties of the multiple shooting algorithm is that model simulation and sensitivity generation can be performed on each shooting interval independently. In other words, integration can be easily parallelized, by applying the so called shared memory model. The toolkit can export the code which uses OpenMP framework for parallelization.

3.4 Vehicle Model

Both NMPC and MHE strongly rely on a mathematical model of the vehicle. Having a descriptive model is hence fundamental to ensure good control and estimation performance. In this chapter, a multibody model is proposed in which the chassis is modeled as a rigid mass connected to the four wheels by suspensions. This model extends the model previously proposed in [8].

The chassis position and orientation are defined in the X–Y plane of an absolute reference frame E, while the velocities are given in the local $x - y - z$ frame e. The heave motion of the chassis is neglected and the four wheels are modeled as independent bodies with only spinning inertia. Throughout the chapter, when referring to quantities related to the wheels, subscripts fl, fr, rl, rr denote quantities corresponding respectively to the front left, front right, rear left and rear right wheel. For ease of notation, let $\mathscr{F} := \{\mathrm{f}, \mathrm{r}\}$, $\mathscr{S} := \{\mathrm{l}, \mathrm{r}\}$ and $\mathscr{W} := \mathscr{F} \times \mathscr{S} = \{\mathrm{fl}, \mathrm{fr}, \mathrm{rl}, \mathrm{rr}\}$.

The control inputs are the steering rate $\dot{\delta}$, the accelerating torque T^{a} and the four braking torques of each wheel $T^{\mathrm{b}}_{\diamond\star}$, $\forall \diamond\star \in \mathscr{W}$.

3.4.1 Chassis Dynamics

The equations of motion are written with respect to the vehicle's center of gravity (CoG). Reference frames E and e are chosen orthonormal, right-handed with the z-axis pointing up, and the y-axis pointing left. The chassis equations of motion thus are

$$
\begin{aligned}
m\dot{v}^x &= mv^y\dot{\psi} + F^x_{\mathrm{fr}} + F^x_{\mathrm{fl}} + F^x_{\mathrm{rr}} + F^x_{\mathrm{rl}} + F_D, && (3.4a)\\
m\dot{v}^y &= -mv^x\dot{\psi} + F^y_{\mathrm{fr}} + F^y_{\mathrm{fl}} + F^y_{\mathrm{rr}} + F^y_{\mathrm{rl}}, && (3.4b)\\
I^z\ddot{\psi} &= a(F^y_{\mathrm{fl}} + F^y_{\mathrm{fr}}) - b(F^y_{\mathrm{rl}} + F^y_{\mathrm{rr}}) + c(F^x_{\mathrm{fr}} - F^x_{\mathrm{fl}} + F^x_{\mathrm{rr}} - F^x_{\mathrm{rl}}), && (3.4c)\\
I^y\ddot{p} &= T^y_{\mathrm{s}}, && (3.4d)\\
I^x\ddot{r} &= T^x_{\mathrm{s}}, && (3.4e)\\
\dot{X} &= v^x\cos\psi - v^y\sin\psi, && (3.4f)\\
\dot{Y} &= v^x\sin\psi + v^y\cos\psi, && (3.4g)
\end{aligned}
$$

where m denotes the mass and I^x, I^y, I^z the moments of inertia of the chassis. The distances of the tires from the vehicle's CoG are characterized by a, b and c, cf. Fig. 3.1. The CoG is assumed to be located halfway between the left and right side of the car. The vehicle's yaw angle ψ is obtained by direct integration of $\dot{\psi}$ as is the steering angle δ from input $\dot{\delta}$. The drag force due to air resistance is denoted by F_D, while $F^x_{\cdot\cdot}$, $F^y_{\cdot\cdot}$ denote the components of the tire contact forces along the vehicle's local x and y axis. The suspension torques are defined as T^x_{s} and T^y_{s}.

The considered vehicle has front steering, thus

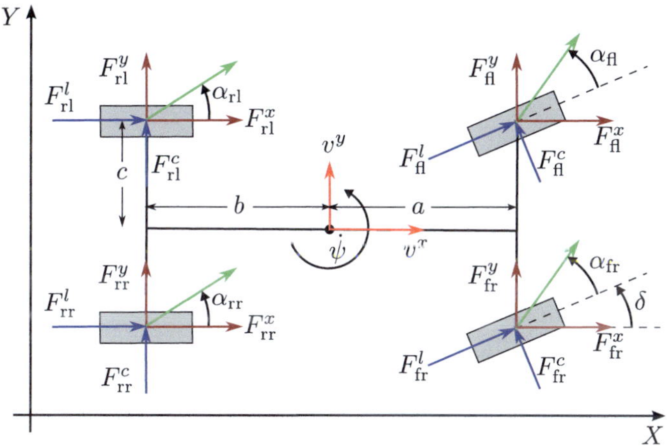

Fig. 3.1 Tire forces and slip angles of the 4-wheel vehicle model in inertial coordinates. The tires' directions of movement are indicated by *green vectors*

$$
\begin{aligned}
F^x_{\mathrm{f}\star} &= F^l_{\mathrm{f}\star}\cos\delta - F^c_{\mathrm{f}\star}\sin\delta, \qquad & F^y_{\mathrm{f}\star} &= F^l_{\mathrm{f}\star}\sin\delta - F^c_{\mathrm{f}\star}\cos\delta, \qquad & \forall \star \in \mathscr{S},\\
F^x_{\mathrm{r}\star} &= F^l_{\mathrm{r}\star}, & F^y_{\mathrm{r}\star} &= F^c_{\mathrm{r}\star}, & \forall \star \in \mathscr{S},
\end{aligned}
$$

where $F^l_{..}$, $F^c_{..}$ denote the longitudinal and cornering tire forces respectively.

3.4.2 Tire Contact Forces: Pacejka's Magic Formula

In this chapter it is proposed to compute the tire forces using Pacejka's Magic Formula: a very accurate semi-empirical nonlinear tire model that is commonly used for automotive applications. The Magic Formula allows to compute the longitudinal and cornering forces as a function of the longitudinal slip and slip angle, while taking into account the effect of combined slip. The self-aligning torque $M^z_{..}$ has a significant contribution only at low speeds [26] and is assumed to be negligible in this chapter. Longitudinal and cornering forces are thus computed as

$$
\left[F^l_{\diamond\star}, F^c_{\diamond\star}\right] = f_{\mathrm{P}}(\alpha_{\diamond\star}, \kappa_{\diamond\star}, \mu, F^z_{\diamond\star}), \quad \forall \diamond\star \in \mathscr{W},
$$

where $f_{\mathrm{P}}(\cdot)$ denotes Pacejka's tire model. The inputs to the Magic formula are: (a) the side slip angle $\alpha_{..}$, defined, as displayed in Fig. 3.1, as the angle between the wheel's orientation and it's velocity, (b) the longitudinal slip $\kappa_{..}$, (c) the tire-road friction coefficient μ and (d) the vertical load on the wheel $F^z_{..}$. The longitudinal slip is defined as

$$\kappa_{\diamond\star} = \frac{\omega_{\diamond\star} R_{\mathrm{e}} - v_{\diamond\star}}{v_{\diamond\star}},$$

where $v_{\diamond\star}$ is defined as the wheel velocity, $\omega_{\diamond\star}$ as the wheel rotational speed and R_{e} as the effective tire radius.

More details on the computation of slip angles and Pacejka forces can be found in [26–28]; the precise model implementation used for this chapter, including all parameters, can be found in [29].

3.4.3 Wheel Dynamics

The wheels being modeled as separate bodies with only one rotational degree of freedom, the dynamic equations only depend on the accelerating and breaking torques $T^{\mathrm{a}}_{\cdot\cdot}$ and $T^{\mathrm{b}}_{\cdot\cdot}$ and on the longitudinal force $F^{l}_{\cdot\cdot}$. The rotational accelerations are given by

$$\dot{\omega}_{\diamond\star} = \tfrac{1}{I^{\mathrm{w}}}(T^{\mathrm{a}}_{\diamond\star} + T^{\mathrm{b}}_{\diamond\star} - R_{\mathrm{e}} F^{l}_{\diamond\star}), \quad \forall \diamond\star \in \mathscr{W},$$

where individual wheel braking is considered. For the acceleration torque a model of the differential is considered, which, assuming rear-wheel drive, yields

$$\begin{aligned} T^{\mathrm{a}}_{\mathrm{f}\star} &= 0, \quad \forall \star \in \mathscr{S}, \\ T^{\mathrm{a}}_{\mathrm{r}\star} &= T^{\mathrm{a}}\left(1 - \frac{\omega_{\mathrm{r}\star}}{\omega_{\mathrm{rl}} + \omega_{\mathrm{rr}}}\right), \quad \forall \star \in \mathscr{S}. \end{aligned}$$

3.4.4 Vertical Forces and Suspension Model

In this chapter, the suspension of the vehicle is assumed to only act on the roll and pitch motions of the chassis. The rotation of the chassis is defined as

$$R = R_y(p) R_x(r) = \begin{bmatrix} \cos p & 0 & \sin p \\ 0 & 1 & 0 \\ -\sin p & 0 & \cos p \end{bmatrix} \begin{bmatrix} 1 & 0 & 0 \\ 0 & \cos r & \cos r \\ 0 & \sin r & -\sin r \end{bmatrix},$$

where r and p denote the roll and pitch angles respectively.

The forces of the suspension due to the spring elasticity and damping are defined respectively as

$$F^{\mathrm{el}}_{\diamond\star} = -k_{\diamond} \Delta_{\diamond\star}, \qquad F^{\mathrm{d}}_{\diamond\star} = -D_{\diamond} \dot{\Delta}_{\diamond\star}, \quad \forall \diamond\star \in \mathscr{W},$$

where $k_{\cdot}$ and $D_{\cdot}$ denote the elastic and damping constants of the suspension spring. The suspension displacement is defined as $\Delta_{\diamond\star} = R\zeta_{\diamond\star} - \zeta_{\diamond\star}, \forall \diamond\star \in \mathscr{W}$, where $\zeta_{\cdot\cdot}$ denotes the wheel position in the body reference frame e.

Denoting the rest vertical forces by $\bar{F}^z_{\cdot\cdot}$, the vertical forces $F^z_{\cdot\cdot}$ are given by

$$F^z_{\diamond\star} = \bar{F}^z_{\diamond\star} + F^{\mathrm{el}}_{\diamond\star} + F^{\mathrm{d}}_{\diamond\star}, \quad \forall \diamond\star \in \mathscr{W}.$$

The torques acting on the chassis are given by

$$\begin{aligned} T^y_{\mathrm{s}} &= -2\left((k_{\mathrm{f}} + k_{\mathrm{r}})c^2 \sin r + c^2(D_{\mathrm{f}} + D_r)w_x\right), \\ T^x_{\mathrm{s}} &= -2\left((k_{\mathrm{f}}a^2 + k_{\mathrm{r}}b^2)\sin p + (D_{\mathrm{f}}a^2 + D_{\mathrm{r}}b^2)w_y\right). \end{aligned}$$

3.4.5 Spatial Reformulation of the Dynamics

For a natural formulation of obstacles and general road bounds under varying vehicle speed one can reformulate the model dynamics in the curvilinear coordinate defined by the track centerline $\sigma(s) = [X^\sigma(s), Y^\sigma(s)]^T$, where $s \in [s_0, s_f]$ is a curve parameterization of constant speed $\|\frac{d\sigma}{ds}\| := 1$. In particular, the global vehicle coordinates X, Y, and Ψ are replaced by offset coordinates $e^y := \left\|[X, Y]^T - [X^\sigma, Y^\sigma]^T\right\|_2$ and $e^\psi := \psi - \psi^\sigma$ in distance and orientation from σ, with

$$\psi^\sigma(s) = \operatorname{atan2}\left(\frac{dY^{\sigma(s)}}{ds}, \frac{dX^{\sigma(s)}}{ds}\right).$$

Instead of time t, the independent variable of the dynamic system is taken to be the parametrization $s \in [s_0, s_f]$ of the reference curve σ.

The coordinate transformation mapping s: $[t_0, t_f] \to [s_0, s_f]$ is implicitly defined by the vehicle's velocity along σ. From geometric considerations, for the vehicle velocity in σ-direction $v^\sigma = v^x \cos(e^\psi) - v^y \sin(e^\psi)$ and its projection onto the reference curve $\dot{\sigma}$, it holds that $\frac{\dot{\sigma}}{v^\sigma} = \frac{\rho^\sigma}{\rho^\sigma - e^y}$, where ρ^σ is the radius of local curvature of σ at s. From $\frac{d\sigma}{ds} = 1$, it follows $\dot{\sigma} = \frac{d\sigma}{ds} \cdot \frac{ds}{dt} = \dot{s}$, and therefore the coordinate transformation is defined by

$$\dot{s} = \frac{\rho^\sigma}{\rho^\sigma - e^y}\left(v^x \cos(e^\psi) - v^y \sin(e^\psi)\right).$$

For sufficiently small deviations e^y from the centerline (more precisely, $e^y(s) < \rho^{\sigma(s)}$) the coordinate mapping is monotonous if $v^\sigma > 0$, i.e. if the vehicle is driving forward, and the vehicle state ξ is uniquely determined in the spatial coordinate system for each $s \in [s_0, s_f]$. The spatial dynamics of the state vector ξ can be expressed in relation to the time dependent dynamics through

$$\xi' := \frac{d\xi}{ds} = \frac{d\xi}{dt} \cdot \frac{dt}{ds} = \dot{\xi} \cdot \frac{1}{\dot{s}},$$

The last equality holds by the inverse function theorem, which can be applied due to monotonicity of the coordinate mapping $s(t)$.

More details of the spatial coordinate transformation are provided in [5]. Note that time information may be recovered by integrating $\frac{dt}{ds} = \frac{1}{\dot{s}}$ along σ, and inertial coordinates are given by:

$$\begin{aligned} X &= X^{\sigma} - e^{y} \sin(\psi^{\sigma}) \\ Y &= Y^{\sigma} + e^{y} \cos(\psi^{\sigma}) \\ \psi &= \psi^{\sigma} + e^{\psi}. \end{aligned}$$

3.5 Control of Autonomous Vehicles

In this section, the MHE and NPMPC schemes used for the simulations are presented. Both schemes are based on a piecewise constant control parametrization and the system dynamics $\mathbf{f}(\mathbf{x}, \mathbf{u})$ are discretized over the shooting intervals using an implicit Runge-Kutta method of order 2 [25].

3.5.1 MHE Formulation

The estimation horizon for MHE has been selected as $S_{\mathrm{E}} = 10$ m, divided into $N = 10$ control intervals of uniform duration $S_{\mathrm{c}} = S_{\mathrm{E}}/N$.

The available measurements come from an inertial measurement unit (IMU), a GPS, force sensors on the suspensions, and encoders on the wheels and the steering wheel. They are summarized in Table 3.1, together with their standard deviation σ. The weighting matrix Q_{E} was chosen diagonal, with all diagonal elements matching the square of the inverse of the standard deviation σ_i they correspond to, i.e. $Q_{\mathrm{E}i,i} = (\sigma_i)^{-2}$.

The arrival cost has been computed in a similar way as in [13], where the Kalman update is computed in an efficient way and it is ensured that the norm of the arrival cost weighting matrix P_{E} is bounded from above.

Table 3.1 Available measurements

Sensor	Measurements	Standard deviation σ
IMU	Linear acceleration	10^{-2} m/s^2
IMU	Angular velocity	0.1 rad/s
GPS	Position	10^{-2} m
Force sensor	Vertical forces	5×10^2 N
Encoder	Wheel rotational velocity	10^{-3} rad/s
Encoder	Steering angle	10^{-3} rad

3.5.1.1 Friction Coefficient Estimation

In order to accurately estimate the friction coefficient μ, the model needs to account for sudden changes in the road friction. This can be achieved by making the friction coefficient time varying, using a first order model $\dot{\mu} = u_\mu$.

Penalization of the variable u_μ is needed in order to ensure that the estimate of μ is not strongly affected by sensor noise. Penalizing u_μ in a quadratic (L^2) norm has the effect of filtering out noise, but does not allow for fast detection of large jumps in the friction coefficient. Jumps are better detected when penalizing the absolute value of u_μ (L^1 norm), as large changes are penalized less than with a L^2 penalty. In this case, though, small variations of μ are filtered out together with the noise. The Huber penalty combines the benefits of the L^1 and the L^2 penalty. It is defined by

$$H(x) = \begin{cases} \frac{1}{2}x^2 & |x| \leq \rho \\ \rho(|x| - \frac{1}{2}\rho) & |x| \geq \rho \end{cases}. \tag{3.5}$$

We refer to [30] for an algorithmically differentiable implementation using slack variables.

3.5.2 MPC Formulation

The control horizon for NMPC has been selected as $S_C = 20$ m, divided into $N = 20$ control intervals of uniform duration $S_c = S_C/N$. This longer horizon has been chosen, to guarantee that the obstacles are seen sufficiently in advance to allow for avoidance maneuvers, including stopping the vehicle in extreme conditions.

The weights Q and R have been chosen as diagonal matrices, with each element selected in accordance with Table 3.2. The units of measure of the weights are selected so as to yield a dimensionless cost. The terminal cost matrix P_C has been taken as the solution to the discrete algebraic Riccati equation using the proposed weighting matrices Q and R.

Table 3.2 Weights for NMPC

State or control	e_y, e_ψ	$v_x, v_y, \dot{\psi}$	r, p	ω_x, ω_y	$\omega_{\cdot\cdot}$	$T^a, T^b_{\cdot\cdot}$	$\dot{\delta}$
Associated weight	1	10	1	1	1	10^{-4}	$1e2$

3.5.2.1 Constraints with Feasibility Guarantee

For the inputs, the following constraints have been selected

$$0 \le T^a \le \overline{T}^a, \qquad \underline{T}^b \le T^b_{\diamond\star} \le 0 \qquad \forall \diamond\star \in \mathscr{W}.$$

By using the spatial reformulation of the model, the obstacle avoidance constraints become simple bounds, defined as

$$\underline{e}_y \le e_y \le \overline{e}_y. \tag{3.6}$$

As this obstacle avoidance constraint drives the trajectory away from the reference, the NMPC scheme avoids the obstacle by steering the vehicle as close to it as possible. Bounds (3.6) thus become active. For a real system, the state estimate will always be noisy due to measurement noise or model inaccuracy. Even the smallest violation of constraint (3.6) might yield an infeasible NLP and make the controller unreliable. Feasibility of the NLP can be guaranteed by reformulating the obstacle avoidance constraints (3.6) using non-negative slack variables as

$$\begin{aligned} e_y &\le \overline{e}_y + u^U_{e_y}, \quad u^U_{e_y} \ge 0, \\ e_y &\ge \underline{e}_y - u^L_{e_y}, \quad u^L_{e_y} \ge 0\,. \end{aligned}$$

In the proposed formulation, the slack variables $\mathbf{u}_{\mathbf{e_y}} = \begin{bmatrix} u^L_{e_y} \\ u^U_{e_y} \end{bmatrix}$ are introduced on each interval. They can be seen as a measure of the constraint violation corresponding to each interval.

To penalize the constraint violation, two terms are added to the cost function J_{NMPC}, which becomes

$$\begin{aligned} J_{\text{NMPC}} = \|\mathbf{x}(T_c) - \mathbf{x}^r(T_c)\|^2_{P_C} + \int_0^{T_c} \|\mathbf{x}(t) - \mathbf{x}^r(t)\|^2_{Q_C} + \|\mathbf{u}(t) - \mathbf{u}^r(t)\|^2_{R_C} \\ + \mathbf{w}_1{}^T \mathbf{u}_{\mathbf{e_y}}(t) + \|\mathbf{u}_{\mathbf{e_y}}(t)\|_{W_2} \mathrm{d}t\,. \end{aligned}$$

The proposed penalty on the slack variables implements the sum of an L^1 and an L^2 norm, using positive (definite) weights $\mathbf{w}_1$ and W_2, respectively. This choice allows to add a stronger penalty for large constraint violations (effect of the L^2 norm)

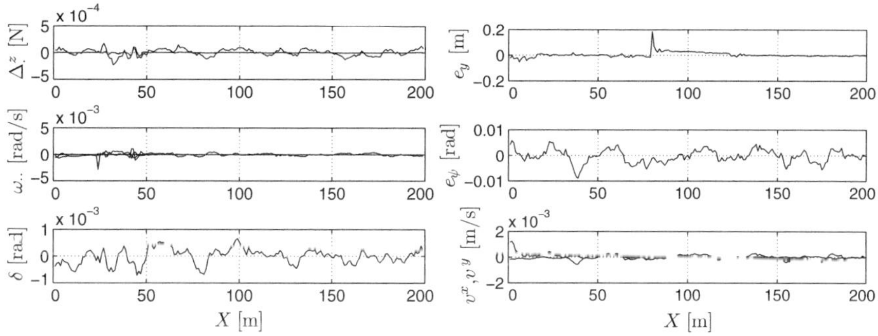

Fig. 3.2 Vehicle trajectory for a straight reference (*thin line*) and obstacles (*thick line*). The MHE estimates are shown in *circles*. The estimated friction coefficient is displayed in the bottom figure (*thick line*), together with the actual friction coefficient (*thin line*)

while always having a nonzero gradient even when the constraints are not violated (effect of the L^1 norm).

3.5.3 Simulation Results

An obstacle avoidance simulation has been run to demonstrate the performance of the proposed control scheme. The NMPC and MHE schemes have been implemented using the code generation tool of ACADO [20].

As displayed in Fig. 3.2, the vehicle is required to travel at a reference speed $v_x = 10\,\text{m/s}$ while avoiding two 6 m long obstacles positioned at $s = 43$ m and $s = 123$ m on a 200 m long straight road. The first obstacle has to be avoided on the left and is 2 m wide, while the second one needs to be avoided on the right and is 0.8 m wide. The road surface has a very low friction coefficient $\mu = 0.3$, corresponding to a snow-covered or icy road. After 80 m, the friction coefficient increases to $\mu = 0.5$.

The trajectory obtained by applying the proposed control scheme to this scenario is displayed in Fig. 3.2, top graph, where it can be seen that the controller is able to avoid the obstacles. In Fig. 3.2, bottom graph, the estimated friction coefficient is displayed. It can be noted that the Huber penalty (3.5) successfully rejects the measurement noise, but still allows to detect the jump in the friction coefficient. The jump is detected in an approximate way immediately after it occurs. After this detection, MHE slowly corrects the inaccuracy. The MHE estimation error for the state vector is displayed in Fig. 3.3, where it can be noted that all quantities are well estimated. The greatest error is relative to e_y, as the selected GPS noise is relatively high.

All simulations have been run on an Intel Xeon CPU E5520 at 2.27 GHz. The computational times are reported in Table 3.3.

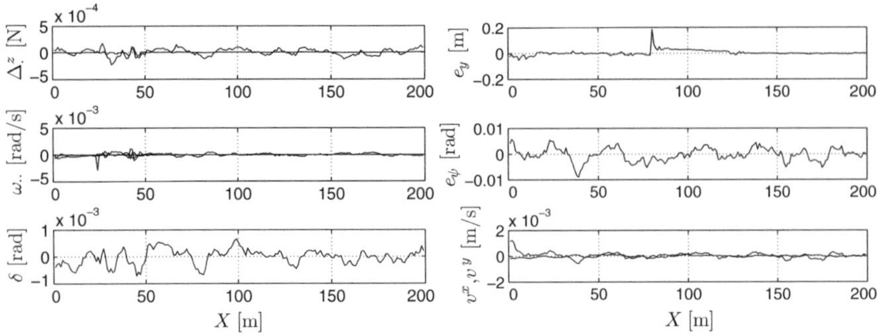

Fig. 3.3 *Left figure* estimation error for the load transfer $\Delta^z_{\cdot}$, the rotational velocities $\omega_{\cdot\cdot}$ and the steering angle δ. *Right figure* estimation error for the distance from the reference e_y and e_ψ, and the longitudinal and lateral velocities v^x and v^y

Table 3.3 Computational times of preparation and feedback phase

	Preparation (ms)	Feedback (ms)
MPC	15.2	5.8
MHE	16.6	5.0

3.5.4 Treating Gear Shifts

An important control in autonomous driving is the gear choice. It implies the introduction of an integer valued control function $\eta(s) \in \{1, 2, \ldots, n_{\text{gears}}\}$ and correspodingly a switched dynamic system $f(x(s), u(s), \eta(s))$, usually involving gear specific transmission ratios or degrees of efficiency. From the variety of approaches that has been proposed to solve offline control problems with gear choices [31–33] the one that is best suited in this online setting is described in [34]. It builds on a *partial outer convexification* of the dynamics [35], i.e., the introduction of convex multiplier controls $\omega_i(s) \in \{0, 1\}$, $\sum_{i=1}^{n_{\text{gears}}} \omega_i(s) = 1$ and new dynamics

$$\dot{x}(s) = \sum_{i=1}^{n_{\text{gears}}} \omega_i(s)\; f(x(s), u(s), i).$$

This reformulation allows to relax integral gear choices to continuous inputs $\omega_i(s) \in [0, 1]$ which can be treated by the presented software framework. Although $\omega_i(s)$ can take any value in the interval $[0, 1]$, the optimal relaxed solution is often of bang-bang type and hence a feasible input. The reason is the outer convexification which favors the most effiicent gear with respect to acceleration both in an energy-optimal as in a time-optimal setting, as explained in [32]. A possible exception are intervals with a mismatch between the actuator time grid and optimal switching points, where a rounding approach can be applied. For sufficiently fast actuators both ε-optimal [36] and stable behavior [34] have been shown.

If engine speed constraints need to be incorporated, a more elaborated approach is necessary, see [37] for a survey of possible problem formulations and methods.

3.6 Conclusions and Outlook

This chapter proposed a framework for combined state and parameter estimation and control of autonomous vehicles based on NMPC and MHE. The proposed schemes have been tested in simulations and have shown to effectively control a stiff, highly nonlinear model with 15 states and 6 controls on a low-friction road in an obstacle-avoidance scenario.

Recent algorithmic developments made it possible to run those advanced control and estimation techniques in real-time, with computational times in the order of 20 ms on a standard CPU.

The road-tire friction coefficient has been estimated by penalizing its variation with a Huber norm and feasibility of path constraints subject to noise has been guaranteed through a slack reformulation.

Future research will aim at extending the proposed framework to more complex vehicle models that include gear shifts.

Acknowledgments This research was supported by Research Council KUL: PFV/10/002 Optimization in Engineering Center OPTEC, GOA/10/09 MaNet and GOA/10/11 Global real-time optimal control of autonomous robots and mechatronic systems. Flemish Government: IOF / KP / SCORES4CHEM, FWO: PhD/postdoc grants and projects: G.0320.08 (convex MPC), G.0377.09 (Mechatronics MPC); IWT: PhD Grants, projects: SBO LeCoPro; Belgian Federal Science Policy Office: IUAP P7 (DYSCO, Dynamical systems, control and optimization, 2012-2017); EU: FP7-EMBOCON (ICT-248940), FP7-SADCO (MC ITN-264735), ERC ST HIGHWIND (259 166), Eurostars SMART, ACCM.

References

1. Falcone P, Borrelli F, Asgari J, Tseng H, Hrovat D (2008) Low complexity MPC schemes for integrated vehicle dynamics control problems. 9th international symposium on advanced vehicle, control
2. Gao Y, Gray A, Frasch JV, Lin T, Tseng E, Hedrick J, Borrelli F (2012) Spatial predictive control for agile semi-autonomous ground vehicles. In: Proceedings of the 11th international symposium on advanced vehicle, control
3. Gao Y, Lin T, Borrelli F, Tseng E, Hrovat D (2010) Predictive control of autonomous ground vehicles with obstacle avoidance on slippery roads. In: Dynamic systems and control conference
4. Gray A, Gao Y, Lin T, Hedrick J, Tseng E, Borrelli F (2012) Predictive control for agile semi-autonomous ground vehicles using motion primitves. In: American control conference, ACC
5. Frasch JV, Gray AJ, Zanon M, Ferreau HJ, Sager S, Borrelli F, Diehl M (2013) An auto-generated nonlinear MPC algorithm for real-time obstacle avoidance of ground vehicles. In: Proceedings of the European control conference (2013)

6. Diehl M, Bock H, Schlöder J, Findeisen R, Nagy Z, Allgöwer F (2002) Real-time optimization and nonlinear model predictive control of processes governed by differential-algebraic equations. J Process Control 12(4):577–585
7. Bock H, Plitt K (1984) A multiple shooting algorithm for direct solution of optimal control problems. In: Proceedings 9th IFAC World Congress Budapest. Pergamon Press, NY, pp 242–247
8. Zanon M, Frasch J, Diehl M (2013) Nonlinear moving horizon estimation for combined state and friction coefficient estimation in autonomous driving. In: Proceedings of the European control conference
9. Zanon M, Gros S, Diehl M (2013) Airborne wind energy. Control of rigid-airfoil airborne wind energy systems. Springer, Berlin
10. Mayne D, Rawlings J, Rao C, Scokaert P (2000) Constrained model predictive control: stability and optimality. Automatica 26(6):789–814
11. Grüne L (2012) NMPC without terminal constraints. In: Proceedings of the IFAC conference on nonlinear model predictive, control 2012
12. Rao C (2000) Moving horizon estimation of constrained and nonlinear systems. Ph.D. Thesis, University of Wisconsin-Madison
13. Kühl P, Diehl M, Kraus T, Schlöder JP, Bock HG (2011) A real-time algorithm for moving horizon state and parameter estimation. Comput Chem Eng 35(1):71–83. doi:10.1016/j.compchemeng.2010.07.012
14. Biegler L, Rawlings J (1991) Optimization approaches to nonlinear model predictive control. In: Ray W, Arkun Y (eds) Proceedings of 4th international conference on chemical process control-CPC IV. AIChE, CACHE, pp 543–571
15. de Oliveira N, Biegler L (1995) An extension of Newton-type algorithms for nonlinear process control. Automatica 31(2):281–286
16. Mayne DQ, Michalska H (1990) Receding horizon control of nonlinear systems. IEEE Trans Autom Control 35(7):814–824
17. Frasch JV, Sager S, Diehl M (2013) A parallel quadratic programming method for dynamic optimization problems. Submitted
18. Diehl M, Findeisen R, Allgöwer F (2007) A stabilizing real-time implementation of nonlinear model predictive control. In: Biegler L, Ghattas O, Heinkenschloss M, Keyes D, van Bloemen Waanders B (eds) Real-time and online PDE-constrained optimization, SIAM, pp 23–52
19. Ferreau HJ, Bock HG, Diehl M (2008) An online active set strategy to overcome the limitations of explicit MPC. Int J Robust Nonlinear Control 18(8):816–830. doi:10.1002/rnc.1251
20. Houska B, Ferreau H, Diehl M (2011) An auto-generated real-time iteration algorithm for nonlinear MPC in the microsecond range. Automatica 47(10):2279–2285. doi:10.1016/j.automatica.2011.08.020
21. Houska B, Ferreau H, Diehl M (2011) ACADO toolkit—An open source framework for automatic control and dynamic optimization. Optimal Control Appl Methods 32(3):298–312. doi:10.1002/oca.939
22. ACADO Toolkit Homepage. http://www.acadotoolkit.org (2009–2013)
23. Ferreau H, Kraus T, Vukov M, Saeys W, Diehl M (2012) High-speed moving horizon estimation based on automatic code generation. In: Proceedings of the 51th IEEE conference on decision and control (CDC 2012)
24. Quirynen R, Gros S, Diehl M (2013) Fast auto generated ACADO integrators and application to MHE with multi-rate measurements. In: Proceedings of the European control conference
25. Quirynen R, Vukov M, Diehl M (2012) Auto generation of implicit integrators for embedded NMPC with microsecond sampling times. In: Lazar M, Allgöwer F (eds) Proceedings of the 4th IFAC nonlinear model predictive control conference
26. Pacejka HB (2006) Tyre and vehicle dynamics. Elsevier, Amsterdam
27. Kehrle F, Frasch JV, Kirches C, Sager S (2011) Optimal control of formula 1 race cars in a VDrift based virtual environment. In: Bittanti S, Cenedese A, Zampieri S (eds) Proceedings of the 18th IFAC World Congress, pp 11,907–11,912
28. Kiencke U, Nielsen L (2005) Automotive control systems. Springer, Berlin

29. Gray A, Zanon M, Frasch J (2012) Parameters for a Jaguar X-Type. http://www.mathopt.de/RESEARCH/obstacleAvoidance.php
30. Boyd S, Vandenberghe L (2004) Convex optimization. Cambridge University Press, Cambridge
31. Gerdts M (2006) A variable time transformation method for mixed-integer optimal control problems. Optim Control Appl Methods 27(3):169–182
32. Kirches C, Sager S, Bock H, Schlöder J (2010) Time-optimal control of automobile test drives with gear shifts. Optimal Control Appl Methods 31(2):137–153
33. Terwen S, Back M, Krebs V (2004) Predictive powertrain control for heavy duty trucks. In: Proceedings of IFAC symposium in advances in automotive control, Salerno, Italy, pp 451–457
34. Kirches C (2011) Fast numerical methods for mixed-integer nonlinear model-predictive control. Advances in numerical mathematics. Springer, Wiesbaden
35. Sager S, Reinelt G, Bock H (2009) Direct methods with maximal lower bound for mixed-integer optimal control problems. Math Prog 118(1):109–149
36. Sager S, Bock H, Diehl M (2012) The integer approximation error in mixed-integer optimal control. Math Prog A 133(1–2):1–23
37. Jung M, Kirches C, Sager S (2013) On perspective functions and canishing constraints in mixed-integer nonlinear optimal control. In: Facets of combinatorial optimization—Festschrift for Martin Grötschel (2013, to appear)

Chapter 4
Approximate Solution of HJBE and Optimal Control in Internal Combustion Engines

Mario Sassano and Alessandro Astolfi

Abstract Optimal control problems naturally arise in several kinds of applications, including automotive systems. Unfortunately, the solution of such problems—which hinges upon a partial differential differential equation, the so-called Hamilton-Jacobi-Bellman (HJB) pde—might be hard or even impossible to determine in practice. Herein, introducing the notion of Dynamic Value function, we propose a novel technique that consists in the *immersion* of the given model into an extended state-space in which the solution may be defined in a constructive manner. This leads to a dynamic control law that approximates the optimal policy. The proposed approach is validated by means of a case study arising from the field of combustion engines, namely optimal control of the torque and the speed of a test bench.

4.1 Introduction

Optimal control consists in steering the state of a dynamical system towards a desired configuration while simultaneously minimizing a given criterion of optimality [3, 8]. Since the resulting control law allows to stabilize a desired equilibrium of the closed-loop system—a task of paramount importance *per se* in the nonlinear context—as well as to consider optimization objectives, such problems are rather pervasive in control design applications, obviously including mechanical and automotive systems among the others [5]. It is well-known that the optimal control action is

M. Sassano (✉)
Dipartimento di Ingegneria Civile e Ingegneria Informatica, University of "Tor Vergata", Via del Politecnico 1, 00133 Rome, Italy
e-mail: mario.sassano@uniroma2.it

A. Astolfi
Department of Electrical and Electronic Engineering, Imperial College London , Exhibition road SW7 2AZ, London, UK
e-mail: a.astolfi@ic.ac.uk

H. Waschl et al. (eds.), *Optimization and Optimal Control in Automotive Systems*, 59
Lecture Notes in Control and Information Sciences 455,
DOI: 10.1007/978-3-319-05371-4_4, © Springer International Publishing Switzerland 2014

typically provided in terms of the solution of a partial differential equation, to so-called Hamilton-Jacobi-Bellman (HJB) pde [2]. The analytic solution of such a (nonlinear) pde might be difficult or even impossible to determine outside classes of rather simple academic models. Therefore, several methods have been proposed in the literature to *approximate*, at least in a neighborhood of the desired equilibrium, the optimal policy [1, 6, 9].

The main contribution of the chapter consists in the definition of the notion of Dynamic Value function. These functions, similarly to the classical value functions, are shown to be strongly related to the solution of an *auxiliary* optimal control problem, which approximates that of the original problem. Interestingly, the construction makes use of a dynamic extension to the state of the system, hence yielding a dynamic control law in place of the classical static optimal feedback.

The rest of the chapter is organized as follows. In Sect. 4.2 some basic definitions and results concerning optimal control and its solution are reviewed. The notion of Dynamic Value function is introduced in Sect. 4.3 together with some motivations for its definition. The concept of (matrix) *algebraic $\bar{P}$ solution*—which is instrumental to the construction of a class of canonical Dynamic Value functions without involving the solution of any pde—is discussed in the same section. Finally, the performances of the proposed control law are assessed by means of a practically motivated application within the automotive framework. In particular, we consider the optimal control problem for the torque and the speed of an internal combustion engine test bench driven by an electric dynamometer.

4.2 Hamilton-Jacobi-Bellman Equation and Optimal Control

To provide the reader—who might not be familiar with the standard definitions and notation of the classical optimal control problem—with a comprehensive framework, we briefly review the basic ideas and results concerning the aforementioned optimal control problem. Towards this end, consider a nonlinear dynamical system described by equations of the form

$$\dot{x} = f(x) + g(x)u, \tag{4.1}$$

with $f : \mathbb{R}^n \to \mathbb{R}^n$ and $g: \mathbb{R}^n \to \mathbb{R}^{n \times m}$ smooth mappings, where $x(t) \in \mathbb{R}^n$ denotes the state of the system and $u(t) \in \mathbb{R}^m$ the input. The infinite horizon optimal control problem with stability consists in finding a control action u that minimizes the cost functional

$$J(u) = \frac{1}{2}\int_0^\infty (q(x(t)) + u(t)^\top u(t))dt, \tag{4.2}$$

where $q : \mathbb{R}^n \to \mathbb{R}_+$ is a positive semi-definite function, subject to the dynamical constraint (4.1), the initial condition $x(0) = x_0$ and the requirement that the zero equilibrium of the closed-loop system be locally asymptotically stable. Assuming that q is at least twice continuously differentiable guarantees the existence of a, possibly not unique, matrix-valued function $Q : \mathbb{R}^n \to \mathbb{R}^{n\times n}$, $Q(x) = Q(x)^\top \geq 0$ for all $x \in \mathbb{R}^n$, such that $q(x) = x^\top Q(x)x$ for all $x \in \mathbb{R}^n$.

Assumption 1 The vector field f is such that $f(0) = 0$, *i.e.* $x = 0$ is an equilibrium point for the system (4.1) when $u(t) = 0$ for all $t \geq 0$.

Assumption 2 The nonlinear system (4.1) with output $y = q(x)$ is zero-state detectable, *i.e.* $y(t) = 0$ and $u(t) = 0$ for all $t \geq 0$ imply that the state $x(t)$ converges to zero as time tends to infinity.

As a consequence of Assumption 1, there exists a, possibly not unique, continuous matrix-valued function $F : \mathbb{R}^n \to \mathbb{R}^{n\times n}$ such that $f(x) = F(x)x$, for all $x \in \mathbb{R}^n$. The classical optimal control design methodology for system (4.1) relies on the solution of the well-known Hamilton-Jacobi-Bellman (HJB) partial differential equation [2, 3, 8]

$$\frac{\partial V}{\partial x}(x)f(x) - \frac{1}{2}\frac{\partial V}{\partial x}(x)g(x)g(x)^\top\frac{\partial V}{\partial x}(x)^\top + \frac{1}{2}q(x) = 0, \tag{4.3}$$

for all $x \in \mathbb{R}^n$, together with the boundary condition $V(0) = 0$. The solution of the HJB Eq. (4.3), if it exists, is the *value function* of the optimal control problem, *i.e.* it is a function which associates to every point in the state space, x_0, the optimal cost of the trajectory of system (4.1) with $x(0) = x_0$. The knowledge of the value function on the entire state space permits the construction of the minimizing input, which is defined in terms of the static state feedback $u_o = -g(x)^\top(\partial V(x)/\partial x)^\top$.

Finally, it appears useful at this preliminary stage to recall that in the linearized setting the solution of the optimal control problem is provided by a linear static state feedback control law of the form $u_o = -B^\top\bar{P}x$, where $\bar{P} \in \mathbb{R}^{n\times n}$ denotes the symmetric positive definite solution of the *algebraic Riccati equation* $\bar{P}A + A^\top\bar{P} - \bar{P}BB^\top\bar{P} + \bar{Q} = 0$, where the matrices $A \in \mathbb{R}^{n\times n}$, $B \in \mathbb{R}^{n\times m}$ and $Q \in \mathbb{R}^{n\times n}$ are defined as

$$A \triangleq \left.\frac{\partial f}{\partial x}\right|_{x=0} = F(0),\ B \triangleq g(0),\ \bar{Q} \triangleq \left.\frac{1}{2}\frac{\partial^2 q}{\partial x^2}\right|_{x=0} = Q(0). \tag{4.4}$$

4.3 Dynamic Value Function and Algebraic $\bar{P}$ Solution

In this chapter we consider a *modified* definition of optimal control problem and therefore an alternative notion of its solution, as detailed in the following statements.

Problem 1 Consider system (4.1), with Assumptions 1 and 2, together with the cost functional (4.2). The *regional dynamic optimal control* problem with stability consists in determining an integer $\tilde{n} \geq 0$, a dynamic control law of the form

$$\begin{aligned}\dot{\xi} &= \alpha(x, \xi),\\ u &= \beta(x, \xi),\end{aligned} \tag{4.5}$$

with $\xi(t) \in \mathbb{R}^{\tilde{n}}$, $\alpha : \mathbb{R}^n \times \mathbb{R}^{\tilde{n}} \to \mathbb{R}^{\tilde{n}}$, $\beta : \mathbb{R}^n \times \mathbb{R}^{\tilde{n}} \to \mathbb{R}^m$, $\alpha(0,0) = 0$, $\beta(0,0) = 0$, smooth mappings and a set $\bar{\Omega} \subset \mathbb{R}^n \times \mathbb{R}^{\tilde{n}}$ containing the origin of $\mathbb{R}^n \times \mathbb{R}^{\tilde{n}}$ such that the closed-loop system

$$\begin{aligned} \dot{x} &= f(x) + g(x)\beta(x, \xi), \\ \dot{\xi} &= \alpha(x, \xi), \end{aligned} \tag{4.6}$$

has the following properties.

(i) The zero equilibrium of the system (4.6) is asymptotically stable with region of attraction containing $\bar{\Omega}$.
(ii) For any $\bar{u}$ and any initial condition (x_0, ξ_0) such that the trajectory of the system (4.6) remain in $\bar{\Omega}$ the inequality $J(\beta) \leq J(\bar{u})$ holds, where β is defined in (4.5).

◇

Note that Problem 1 may be interpreted as an *auxiliary* optimal control problem associated to the extended system (4.1)–(4.5), with state (x, ξ), with respect to the cost functional (4.2). Herein, we consider an approximate solution to the above auxiliary optimal control problem.

Problem 2 Consider system (4.1), with Assumptions 1 and 2. The *approximate regional dynamic optimal control* problem with stability consists in determining an integer $\tilde{n} \geq 0$, a dynamic control law described by (4.5), a set $\bar{\Omega} \subset \mathbb{R}^n \times \mathbb{R}^{\tilde{n}}$ containing the origin of $\mathbb{R}^n \times \mathbb{R}^{\tilde{n}}$ and a function $c : \mathbb{R}^n \times \mathbb{R}^{\tilde{n}} \to \mathbb{R}_+$ such that the regional dynamic optimal control problem is solved with respect to the modified cost functional

$$J(u) = \frac{1}{2} \int_0^\infty (q(x(t)) + u(t)^\top u(t) + c(x(t), \xi(t)))dt. \tag{4.7}$$

◇

4.3.1 Definition of Dynamic Value Function

From the above statements it appears evident that the approximation is twofold. On one hand, we introduce a *dynamic extension* ξ, which leads to a dynamic state feedback in place of the classical static state feedback, while on the other hand we allow for an additional cost c in (4.7), which imposes a *penalty* on the state of the dynamic extension as well as the state of system (4.1). A straightforward consequence of the latter relaxation is that partial differential inequalities may be solved in place of equations. Therefore, an alternative notion of value function is defined accordingly.

Definition 1 Consider system (4.1), with Assumptions 1 and 2, together with the cost functional (4.2). A *Dynamic Value Function* $\mathcal{V}$ is a pair (D_α, V) defined as follows.

i) D_α is the dynamical system

$$\dot{\xi} = \alpha(x, \xi)\,,$$
$$u = -g(x)^\top \frac{\partial V}{\partial x}(x, \xi)^\top, \tag{4.8}$$

with $\xi(t) \in \mathbb{R}^n$, $\alpha : \mathbb{R}^n \times \mathbb{R}^n \to \mathbb{R}^n$, $\alpha(0, 0) = 0$, sufficiently smooth mapping and such that the zero equilibrium of $\alpha(0, \xi)$ is locally asymptotically stable.

ii) $V : \Omega \subseteq \mathbb{R}^n \times \mathbb{R}^n$ is positive definite around the origin of $\mathbb{R}^n \times \mathbb{R}^n$ and such that

$$\mathscr{H}(x, \xi) \triangleq \frac{\partial V}{\partial x} f(x) + \frac{\partial V}{\partial \xi}\alpha(x, \xi) + \frac{1}{2}q(x) - \frac{1}{2}\frac{\partial V}{\partial x} g(x)g(x)^\top \frac{\partial V}{\partial x}^\top \leq 0, \tag{4.9}$$

for all $(x, \xi) \in \Omega$.

To put the above definitions into perspective, in the following lemma we investigate and characterize the relation between a Dynamic Value function and the solution of the approximate regional dynamic optimal control problem.

Lemma 1 *Consider system (4.1), with Assumptions 1 and 2, together with the cost functional (4.2). Let* $\mathscr{V} = (D_\alpha, V)$ *be a Dynamic Value function for the system (4.1) with respect to some non-empty open set* Ω *containing the origin of* $\mathbb{R}^n \times \mathbb{R}^n$*. Then the dynamical system (4.8) solves the approximate regional dynamic optimal control problem with stability, namely Problem 2, for all* $(x, \xi) \in \tilde{\Omega}$*, where* $\tilde{\Omega}$ *is the largest level set of the function V contained in* Ω. △

Proof The claim follows straightforwardly by noting that, by (4.9), V is a continuously differentiable *value function* for the closed-loop system (4.1)–(4.8), with respect to the cost functional (4.7). The additional cost $c : \mathbb{R}^n \times \mathbb{R}^n \to \mathbb{R}_+$ is defined, by (4.9), as $c(x, \xi) = -2\mathscr{H}(x, \xi) \geq 0$ for all $(x, \xi) \in \Omega$. Moreover, asymptotic stability of the zero equilibrium of the closed-loop system (4.1)–(4.8) is proved by letting V be a *candidate* Lyapunov function. In fact, the time derivative of the function V along the trajectories of the system (4.1)–(4.8) yields

$$\frac{dV}{dt} \leq -\frac{1}{2}q(x) - \frac{1}{2}\frac{\partial V}{\partial x}(x, \xi)g(x)g(x)^\top \frac{\partial V}{\partial x}(x, \xi)^\top \leq 0\,, \tag{4.10}$$

where the first inequality is obtained by (4.9), which shows that all the trajectories of the closed-loop system (4.1)–(4.8) remain bounded for all $t \geq 0$. Moreover, by LaSalle's invariance principle the latter trajectories converge to a set, containing the origin, such that

$$\{(x, \xi) \in \mathbb{R}^n \times \mathbb{R}^n : g(x)^\top \frac{\partial V}{\partial x}(x, \xi)^\top = 0\}\,, \tag{4.11}$$

in which the system (4.1) reduces to $\dot{x} = f(x)$. It can be shown now that $x(t)$ asymptotically converges to zero by combining (4.11) with Assumption 2. The proof is concluded by showing that $\xi(t)$ tends to zero by the stability properties of the vector field $\alpha(0, \xi)$ and by standard arguments on interconnected systems. □

Remark 1 Lemma 1 entails that the knowledge of a dynamic value function provides a solution, i.e. the control law (4.8), to Problem 2, which is strongly related to the original optimal control problem. Interestingly, the structure of the control law (4.8) sheds a different light on the notion of solution of an optimal control problem by allowing for dynamic state feedbacks which approximate or, in some cases, provide the optimal solutions.

4.3.2 A Class of Canonical Dynamic Value Functions

While the *interpretation* of the concept of Dynamic Value function has already been provided to the reader in the previous section, the *motivation* of the above discussions will become clear in the rest of the chapter. This is, in fact, achieved by showing that a class of *canonical* Dynamic Value functions may be constructively defined without relying on the analytic solution of any partial differential equation or inequality. Towards this end, we now formally introduce the notion of (matrix) *algebraic* $\bar{P}$ solution, which is instrumental for the above construction.

Definition 2 Consider system (4.1) together with the cost functional (4.2). Let $\Sigma : \mathbb{R}^n \to \mathbb{R}^{n\times n}$, $\Sigma(x) = \Sigma(x)^\top \geq 0$ for all $x \in \mathbb{R}^n$. A continuous matrix-valued function $P : \mathbb{R}^n \to \mathbb{R}^{n\times n}$, $P(x) = P(x)^\top$ for all $x \in \mathbb{R}^n$, is said to be a *matrix* $\mathscr{X}$*-algebraic* $\bar{P}$ solution of Eq. (4.3) if

$$P(x)F(x) + F(x)^\top P(x) + Q(x) - P(x)g(x)g(x)^\top P(x) + \Sigma(x) = 0\,, \quad (4.12)$$

for all $x \in \mathscr{X} \subseteq \mathbb{R}^n$, and $P(0) = \bar{P}$. If Eq. (4.12) holds for all $x \in \mathbb{R}^n$, *i.e.* $\mathscr{X} = \mathbb{R}^n$, then P is a *matrix algebraic* $\bar{P}$ solution.

Despite the fact that (4.12) follows the spirit of the so-called *State Dependent Riccati Equation*, see [4] for a complete survey, the *use* that it is made of the resulting solution P is significantly different from that of the SDRE approach. Nonetheless, the Eq. (4.12) and its solution enjoy the desirable properties usually associated to a state dependent solution of the Riccati equation. For instance, the solution $P(x)$ can be shown to be a positive definite matrix for all $x \in \mathbb{R}^n$ provided that the pairs $(F(x), Q(x)^{1/2})$ and $(F(x), g(x))$ are *point-wise* observable and controllable, respectively, namely for each fixed value of the state $x \in \mathbb{R}^n$. Note that the matrix $A_{cl}(x) = F(x) - g(x)g(x)^\top P(x)$ denotes the state dependent representation of the closed-loop system when only the *algebraic input* $u = -g(x)^\top P(x)x$ is implemented.

Consider a *matrix algebraic* $\bar{P}$ solution P introduced in Definition 2. It appears evident that the mapping $x \mapsto P(x)x$ may not be integrable in the standard sense. In fact, its Jacobian, which is different in general from the matrix P(x), may not be a symmetric matrix. Therefore, exploiting the degree-of-freedom provided by the dynamic extension in Definition 1, we consider herein an alternative notion of *integration* of the above mapping, yielding

$$V(x,\xi) \triangleq \frac{1}{2}x^\top P(\xi)x + \frac{1}{2}\|x-\xi\|_R^2 , \tag{4.13}$$

with $R = R^\top > 0$ and $\xi \in \mathbb{R}^n$. Note that a simple Schur complement argument shows that the function V in (4.13) is, at least locally, positive definite around the origin, for any matrix $R > 0$. Moreover, the function is indeed globally positive definite provided that the matrix $P(x)$ is positive definite for all $x \in \mathbb{R}^n$, as discussed in the above comment.

The following result characterizes a class of canonical Dynamic Value functions, whose construction relies on the notion of (matrix) algebraic $\bar{P}$ solution. To provide a concise statement, let $\Psi : \mathbb{R}^n \times \mathbb{R}^n \to \mathbb{R}^{n\times n}$ denote the jacobian matrix of the mapping $(x,\xi) \mapsto 1/2P(\xi)x$ with respect to ξ whereas the matrix-valued function $\Phi : \mathbb{R}^n \times \mathbb{R}^n \to \mathbb{R}^{n\times n}$ is such that $(x-\xi)^\top \Phi(x,\xi)^\top = x^\top (P(x) - P(\xi))$, for all $(x,\xi) \in \mathbb{R}^n \times \mathbb{R}^n$. Finally, consider the matrix-valued function $\Delta : \mathbb{R}^n \times \mathbb{R}^n \to \mathbb{R}^{n\times n}$ defined as $\Delta(x,\xi) = (R - \Phi(x,\xi))R^{-1}\Psi(x,\xi)$.

Theorem 1 *Consider system (4.1), with Assumptions 1 and 2, together with the cost functional (4.2). Let*[1] *P be a matrix* algebraic *$\bar{P}$ solution of (4.1). Let the matrix $R = R^\top > 0$ be such that*

$$\frac{1}{2}A_{cl}(x)^\top \Delta(x,\xi) + \frac{1}{2}\Delta(x,\xi)^\top A_{cl}(x) < \Sigma(x) + \frac{1}{2}\Delta(x,\xi)^\top g(x)g(x)^\top \Delta(x,\xi) , \tag{4.14}$$

for all $(x,\xi) \in \Omega \setminus \{0\}$, where $\Omega \subseteq \mathbb{R}^n \times \mathbb{R}^n$ is a non-empty open set containing the origin. Then, there exists $\bar{k} \geq 0$ such that V in (4.13) satisfies the Hamilton-Jacobi-Bellman inequality (4.9) for all $(x,\xi) \in \Omega$ and for all $k > \bar{k}$, with $\dot{\xi} = -k(\partial V(x,\xi)/\partial\xi)^\top$. Hence $\mathscr{V} = (D_\alpha, V)$, with D_α defined as

$$\begin{aligned} \dot{\xi} &= -k\left(\Psi(x,\xi)x - R(x-\xi)\right) , \\ u &= -g(x)^\top \left(P(x)x + (R - \Phi(x,\xi))(x-\xi)\right) , \end{aligned} \tag{4.15}$$

and V as in (4.13), is a Dynamic Value function for the system (4.1) and, by Lemma 1, (4.15) solves the approximate regional dynamic optimal control problem for all $(x,\xi) \in \bar{\Omega}$, where $\bar{\Omega}$ is the largest level set of V contained in Ω. △

Proof A detailed proof of Theorem 1 may be obtained by directly adapting the arguments of the proof of Theorem 3 in [7]. □

---

[1] Without loss of generality we suppose that P is a matrix algebraic $\bar{P}$ solution. In fact, if the Eq. (4.12) holds for some $\mathscr{X} \subset \mathbb{R}^n$ the statement may be straightforwardly modified accordingly.

Remark 2 The control law u in (4.15) is given by the sum of the *algebraic input* discussed above and a dynamic compensation term, which deals with the mismatch between the *desired* partial derivative $P(x)$ and the actual partial derivative with respect to x of the function V in (4.13). ▲

Remark 3 Let $R = \Phi(0,0) = \bar{P}$, then clearly $\Delta(0,0) = 0$. Note that the above choice of the matrix R is such that V is, at least locally, positive definite. Provided that $\Sigma(0) > 0$, since $\Delta(0,0) = 0$ and by continuity of the functions in the inequality (4.14), there exists a non-empty open neighborhood of the origin $\hat{\Omega}$ such that (4.14) holds for all $(x,\xi) \in \hat{\Omega}$. ▲

Remark 4 An alternative definition of the algebraic Eq. (4.12) may be given, which allows to consider a single scalar algebraic inequality, in n unknowns, in place of the matrix algebraic Eq. (4.12), in $n(n+1)/2$ independent unknowns. More precisely, we may consider the nonlinear system (4.1) and let $\sigma : \mathbb{R}^n \to \mathbb{R}^{n\times n}$, $\sigma(0) > 0$, $x^\top \sigma(x) x \geq 0$ for all $x \in \mathbb{R}^n$. Then, a continuously differentiable mapping $p : \mathbb{R}^n \to \mathbb{R}^{n\times n}$ is said to be an *algebraic* $\bar{P}$ solution of Eq. (4.3) if

$$p(x)f(x) + \frac{1}{2}q(x) - \frac{1}{2}p(x)g(x)g(x)^\top p(x)^\top + \frac{1}{2}x^\top \sigma(x) x \leq 0\,, \tag{4.16}$$

together with the *tangency* condition $\partial p(x)/\partial x|_{x=0} = \bar{P}$. The corresponding Dynamic Value function must be adapted to the alternative definition, yielding the function $V_s(x,\xi) \triangleq p(\xi)x + (1/2)\|x-\xi\|_R^2$, and the dynamical system

$$\begin{aligned} \dot{\xi} &= -k\left(\frac{\partial p}{\partial \xi}(\xi)x - R(x-\xi)\right), \\ u &= -g(x)^\top \left(p(\xi) + R(x-\xi)\right). \end{aligned} \tag{4.17}$$

Interestingly, the results of Theorem 1 may be equivalently proved by replacing (4.13) with V_s and (4.15) with (4.17), respectively. ▲

4.3.3 Minimization of the Extended Cost

Two different—and in some respect alternative—ways to reduce the *approximation error* of the dynamic solution (4.15) with respect to the optimal control law are explored in this section. In particular, these two directions may be explored by considering the actual *value* of the cost paid by the optimal solution, on one hand, and the *structure* of the cost which is minimized along the trajectories of the system, on the other hand.

To begin with, note that, by definition of value function, $V(x_0,\xi_0)$, where the function V is defined in (4.13), provides the minimum value of the cost functional (4.7) evaluated along the trajectories of the closed-loop system (4.1)–(4.15), initialized at $(x(0),\xi(0)) = (x_0,\xi_0)$. If, on one hand, the initial condition x_0 of the original plant (4.1) is typically given *a priori* in practical applications, the initial condition ξ_0 of the dynamic extension, on the other hand, may be arbitrarily selected. Therefore, in order to minimize the actual cost *paid* by the dynamic solution (4.15), for a fixed initial condition x_0 of system (4.1), the initial condition ξ_0 should be selected as $\xi_0^* = \arg\min_\xi V(x_0, \xi)$.

The implications of the above considerations deserve particular attention, since they provide the motivations for the second approach to reduce the approximation error. As a matter of fact, the definition of value function entails that all the possible trajectories for $\xi(t)$, obtained for instance by modifying the gain k, yield the same minimum value of the cost functional (4.7) provided that they share the same initial condition ξ_0^*. Nevertheless, it might be desirable to *select* one of these trajectories according to some specific criterion. In particular, the following result explains how the additional cost c can be minimized by letting the gain k in (4.15) be a function of the states, *i.e.* $k(x, \xi)$. Towards this end, define, for $\varepsilon > 0$, the open set $\mathcal{M}_\varepsilon \triangleq \{(x, \xi) \in \mathbb{R}^n \times \mathbb{R}^n : \|\Psi(x, \xi) - R(x - \xi)\| < \varepsilon\}$ and consider the continuous function

$$\mathrm{sat}_\delta(x) = \begin{cases} x, & 0 \le |x| \le \delta \\ \mathrm{sign}(x)\delta, & |x| > \delta \end{cases} \tag{4.18}$$

with $x \in \mathbb{R}$ and $\delta > 0$.

Theorem 2 *Consider system (4.1), with Assumptions 1 and 2, together with the cost functional (4.2). Suppose that P is a (matrix) algebraic $\bar{P}$ solution of the Eq. (4.3) and let R be such that (4.14) holds in some non-empty open set $\Omega \subseteq \mathbb{R}^n \times \mathbb{R}^n$. Consider the function V as in (4.13) with ξ obtained as the solution of (4.15) for some constant $\hat{k} > \bar{k}$. Let $\varepsilon > 0$ and $k(x, \xi)$ in (4.15) be defined as*

$$k(x, \xi) = \mathrm{sat}_{\hat{k}}\left(m(x, \xi)\right), \tag{4.19}$$

with

$$m(x, \xi) \triangleq \left(\frac{\partial V}{\partial \xi}\frac{\partial V}{\partial \xi}^\top\right)^{-1}\left(\frac{\partial V}{\partial x} f(x) + \frac{1}{2} q(x) - \frac{1}{2}\frac{\partial V}{\partial x} g(x) g(x)^\top \frac{\partial V}{\partial x}^\top\right). \tag{4.20}$$

Then the Dynamic Value function[2] $\mathcal{V}^\varepsilon = (D_\alpha^\varepsilon, V^\varepsilon)$ *is such that*

$$\begin{cases} c^\varepsilon(x, \xi) = 0 & (x, \xi) \in \Omega \setminus \mathcal{M}_\varepsilon, \\ c^\varepsilon(x, \xi) \le c(x, \xi) & (x, \xi) \in \mathcal{M}_\varepsilon. \end{cases} \tag{4.21}$$

Proof. The claim is proved in two steps. First it is shown that the function m is not saturated for all $(x, \xi) \in \Omega \setminus \mathcal{M}_\varepsilon$, then that (4.21) holds. To begin with, rewrite the function (4.19) as

---

[2] The notation $(D_\alpha^\varepsilon, V^\varepsilon)$ and $c^\varepsilon(x, \xi)$ describes the differential Eq. (4.15) and the function $V(x, \xi)$ in (4.13) with ξ obtained as the solution of (4.15)–(4.19) and the corresponding approximation error defined as in Lemma 1, respectively.

$$k(x,\,\xi) = \begin{cases} m(x,\,\xi), & (x,\,\xi) \in \Omega \setminus \mathscr{M}_\varepsilon \\ \text{sat}_{\hat{k}}\,(m(x,\,\xi))\,, & (x,\,\xi) \in \mathscr{M}_\varepsilon \end{cases} \tag{4.22}$$

and note that the function m is smaller than or equal to $\hat{k}$ for all $(x,\,\xi) \in \Omega \setminus \mathscr{M}_\varepsilon$. In fact in this set V is such that $\|\partial V(x,\xi)/\partial\xi\|^2 \geq \varepsilon^2$ and

$$\frac{\partial V}{\partial x} f(x) + \frac{1}{2} q(x) - \frac{\partial V}{\partial x} g(x) g(x)^\top \frac{\partial V}{\partial x}^\top - m(x,\xi) \frac{\partial V}{\partial \xi} \frac{\partial V}{\partial \xi}^\top = 0\,, \tag{4.23}$$

where the equality is obtained by substitution of the function m as in (4.19). Moreover, by Theorem 1,

$$\frac{\partial V}{\partial x} f(x) + \frac{1}{2} q(x) - \frac{\partial V}{\partial x} g(x) g(x)^\top \frac{\partial V}{\partial x}^\top - \hat{k} \frac{\partial V}{\partial \xi} \frac{\partial V}{\partial \xi}^\top \leq 0\,, \tag{4.24}$$

for all $(x,\,\xi) \in \Omega$. Therefore, subtracting (4.23) from (4.24) yields, in $\Omega \setminus \mathscr{M}_\varepsilon$, $-(\hat{k} - m(x,\,\xi))\|\partial V(x,\xi)/\partial\xi\|^2 \leq -(\hat{k} - m(x,\,\xi))\varepsilon^2 \leq 0$, hence $m(x,\,\xi) \leq \hat{k}$ and $k(x,\,\xi)$ is not saturated outside the set $\mathscr{M}_\varepsilon$. To show that (4.21) holds note that, by (4.23), the function V satisfies the partial differential inequality (4.9) with the equality sign for all $(x,\,\xi) \in \Omega \setminus \mathscr{M}_\varepsilon$, hence $c^\varepsilon = 0$. Moreover, by the definition of $k(x,\,\xi)$ as in (4.19), $k(x,\,\xi) \leq \hat{k}$ for all $(x,\,\xi) \in \mathscr{M}_\varepsilon$. Finally, the proof is concluded noting that since $0 \leq c^\varepsilon(x,\,\xi) \leq c(x,\,\xi)$ for all $(x,\,\xi) \in \Omega$, then $\mathscr{V}^\varepsilon = (D_\alpha^\varepsilon,\, V^\varepsilon)$ is indeed a Dynamic Value function. □

As a final remark, it is interesting to point out that the solution resulting from the choice of k as in (4.19) minimizes the desired cost functional (4.2) for all the values of $(x,\xi) \in \mathbb{R}^n \times \mathbb{R}^n$ such that $c(x,\xi) = 0$, hence recovering the optimal solution to the *original* problem.

4.4 Optimal Control in Internal Combustion Engine Test Benches

Combustion engines are operated at test benches in the same way as in a passenger car or a heavy-duty truck, since the former allow to generate the same load a combustion engine would undergo in normal operations. The crucial advantage in the use of a test bench resides in the possibility of reproducing desired conditions in terms of temperature and pressure, and, consequently, of drastically reducing the cost and time required for development and configuration. In a vehicle the velocity and, by means of the transmission, the rotational speed ω_E result from the engine and the load torques. For this reason the engine torque T_E as well as the engine speed ω_E need to be controlled to operate a combustion engine on a test bench. In industrial practice a test bench is usually controlled by means of two separate control loops:

the torque is often influenced by the accelerator pedal position α of the engine under test, while the speed is controlled by the loading machine.

In the standard setting of a test bench, the combustion engine is connected via a single shaft to a different main power unit. The latter may either be a purely passive brake or an electric machine, which offers the possibility of an active operation. In this situation, the accelerator pedal position α of the combustion engine and the set value $T_{D,\,set}$ of the dynamometer torque provide the inputs to the test bench. A *simplified* model of the entire mechanical system can be described by Gruenbacher [5]

$$\dot{T}_E = -\left(c_0 + c_1\omega_E + c_2{\omega_E}^2\right) T_E + \tau\left(\omega_E, T_E, \alpha\right), \tag{4.25}$$

$$\Delta\dot{\varphi} = \omega_E - \omega_D, \tag{4.26}$$

$$\theta_E\dot{\omega}_E = {}_E - c\Delta\varphi - d\left(\omega_E - \omega_D\right), \tag{4.27}$$

$$\theta_D\dot{\omega}_D = c\Delta\varphi + d\left(\omega_E - \omega_D\right) - T_D, \tag{4.28}$$

where $\Delta\varphi$ is the torsion of the connection shaft while θ_E and θ_D denotes the inertias of the combustion engine and the dynamometer, respectively. The contributions due to the inertias of the adapter flanges, the damping element, the shaft torque measurement device and the flywheel are already included in these values. The parameter c characterizes the stiffness of the connection shaft, whereas d describes its damping. Moreover, $c_i > 0$, $i = 1, \ldots, 3$ are constant parameters and $\tau : \mathbb{R} \times \mathbb{R} \times \mathbb{R} \to \mathbb{R}$ is a nonlinear static function.

The employed electric dynamometer is modeled as a second order low-pass filter, with dynamics significantly faster than those of the other components of the test bench, hence they can be neglected in the design. Within the range of maximum torque and maximum rate of change, the torque of the dynamometer can be described by $T_D = T_{D,\,set}$. Letting $v = \tau\left(\omega_E, T_E, \alpha\right)$, the system (4.25)–(4.28) can be rewritten as

$$\dot{x} = Ax + f(x) + Bu \tag{4.29}$$

with

$$A = \begin{pmatrix} -c_0 & 0 & 0 & 0 \\ 0 & 0 & 1 & -1 \\ \frac{1}{\theta_E} & -\frac{c}{\theta_E} & -\frac{d}{\theta_E} & \frac{d}{\theta_E} \\ 0 & \frac{c}{\theta_D} & \frac{d}{\theta_D} & -\frac{d}{\theta_D} \end{pmatrix}, \quad B = \begin{pmatrix} 1 & 0 \\ 0 & 0 \\ 0 & 0 \\ 0 & \frac{1}{\theta_D} \end{pmatrix}$$

$$f(x) = \left(-\left(c_1\, x_3 + c_2\, {x_3}^2\right) x_1 0\,0\,0\right)^{\top},$$

where $x(t) \in \mathbb{R}^4$, $x = \left(x_1\ x_2\ x_3\ x_4\right)^{\top} = \left(T_E\ \Delta\varphi\ \omega_E\ \omega_D\right)^{\top}$, denotes the state of the system and $u(t) \in \mathbb{R}^2$, $u = \left(v\ T_{D,\,set}\right)^{\top}$, the input. The actual control input to apply in order to generate the desired signal v, namely α, is then obtained in practice by an approximate inversion of the nonlinear function τ.

Towards the construction of the dynamic control law (4.17), we preliminary let e_i denote the regulation error of the component x_i with respect to the corresponding reference value, namely $e_i = x_i - x_i^*$, $i = 1, \ldots, 4$, where $x(t) \in \mathbb{R}^4$, $x = \left(x_1\ x_2\ x_3\ x_4\right)^\top = \left(T_E\ \Delta\varphi\ \omega_E\ \omega_D\right)^\top$, describes the state of the system. To motivate the *infinite-horizon* scenario we suppose that the transient response of system (4.25)–(4.28) in closed-loop with the dynamic control law to be designed is significantly faster than the occurrence of step-changes in the desired reference values.

Then, in the *error coordinates*, let $q(e) = e^\top Qe$ in the cost functional (4.2), where the positive definite matrix $Q \in \mathbb{R}^{4\times 4}$ weights the relative regulation errors. Without loss of generality, let the *algebraic* $\bar{P}$ *solution* of (4.16) be of the form $p(e) = e^\top \bar{P} + \mathscr{P}(e)$, where $\mathscr{P} : \mathbb{R}^4 \to \mathbb{R}^{1\times 4}$ contains higher-order polynomials of the error variable e. The matrix $\bar{P}$ is the symmetric and positive definite solution of the algebraic Riccati equation associated to the error system linearized around the desired working point and the quadratic cost $q(e)$. The proposed structure of the *algebraic* $\bar{P}$ *solution* P intuitively suggests that the linear solution $e^\top \bar{P}$ is modified, by the addition of the term $\mathscr{P}$, in order to compensate for the nonlinear terms in (4.25)–(4.28).

Exploiting the specific structure of the vector field f in system (4.29), we let $\mathscr{P}$ be defined as $\mathscr{P} = (\mathscr{P}_1,\ 0,\ 0,\ 0)$. It is worth noting that this specific choice is arbitrary and alternative choices may be explored. Then, the inequality (4.16) is solved with respect to the unknown $\mathscr{P}_1$ obtaining a solution of the form $\mathscr{P}_1(e) = \mathscr{N}(e)\mathscr{D}(e)^{-1}$, with the function $\mathscr{D} : \mathbb{R}^4 \to \mathbb{R}$ strictly positive for all the values of *interest*, namely around the desired operating range, of the state variable x. Finally the control law (4.17), with the matrix R selected according to the arguments of Remark 3, is validated on a high quality test bench simulator developed by the Institute for Design and Control of Mechatronical Systems, Kepler University, Linz. In addition to the dynamics of the entire mechanical description the simulator also includes a more accurate, genuinely nonlinear, data-based model of the combustion engine, limitations of the dynamometer as well as disturbance effects. The combustion engine model takes for instance the dynamics of the accelerator pedal and combustion oscillations into account. Measurement noise similar to the one observed on an actual combustion engine test bench is superimposed to all relevant values.

Since (4.17) has been designed for a rather *simplified* model, the control action needs to be modified in order to cope with some of the nonlinearities of the combustion engine. Thus, letting ν_i, $i = 1, 2$, denote the i-th component of (4.17) we define the actual control inputs implemented on the combustion engine test bench as $u_i = \nu_i(e, \xi) + k_i \int_0^t \nu_i(e(\tau), \xi(\tau))d\tau$, for $i = 1, 2$. Additionally we let the gain k_2 be a function of the derivative—which is implemented as $\frac{s}{\delta s+1}$, with $\delta \gg 0$—of the regulation error for the speed of the engine, namely $k_2(\dot{e}_3)$. In particular, the gain is defined such that, when the regulation error of the engine speed changes *too fast*, the integral action is negligible compared to the other components of the control signal. This choice is reasonable, as shown in the simulations, and needed in order to avoid an excessively aggressive reference profile for the torque of the dynamometer. Large

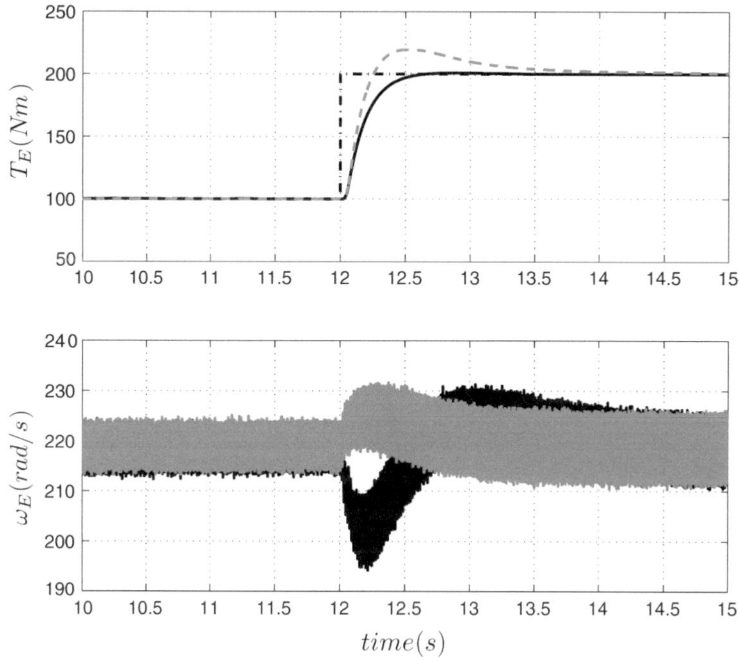

Fig. 4.1 *Top graph* time histories of the engine torque T_E determined by the control law (4.17) (*solid line*) and with the control law in [5] (*dashed line*), together with the desired reference value (*dash-dotted line*). *Bottom graph* time histories of the engine speed ω_E determined by the control law (4.17) (*dark line*) and with the control law in [5] (*grey line*)

errors between the desired and the measured engine speed and torque, due to changes in the references, are mainly, and relatively rapidly, compensated by the action provided by the dynamic control law (4.17). Note that, locally around the origin, the linear part of the dynamic control law is dominant with respect to higher-order terms, hence the integrals provide a *standard* integral action.

In the simulations the control law (4.17) is compared with the control law developed in [5], which already shows significantly improved performances with respect to existing standard implementations based on two separate control loops. The top graph of Fig. 4.1 shows the time histories of the engine torque T_E determined by (4.17) (solid line) and by the control law in [5] (dashed line). The bottom graph of Fig. 4.1 displays the time histories of the engine speed ω_E determined by (4.17) (solid line) and by the control law in [5] (dashed line). Employing the controller developed herein leads to a significant reduced overshooting of the engine torque T_E when a change of the operation point is required. Although the engine torque T_E shows a slightly increased rise time, the final value is reached approximately four times faster. The coupling effect is also improved by the dynamic control law proposed, as it can be appreciated from the bottom graph of Fig. 4.1. Note that the speed is corrupted by additive disturbance caused by the resolution of the shaft encoders and by combustion

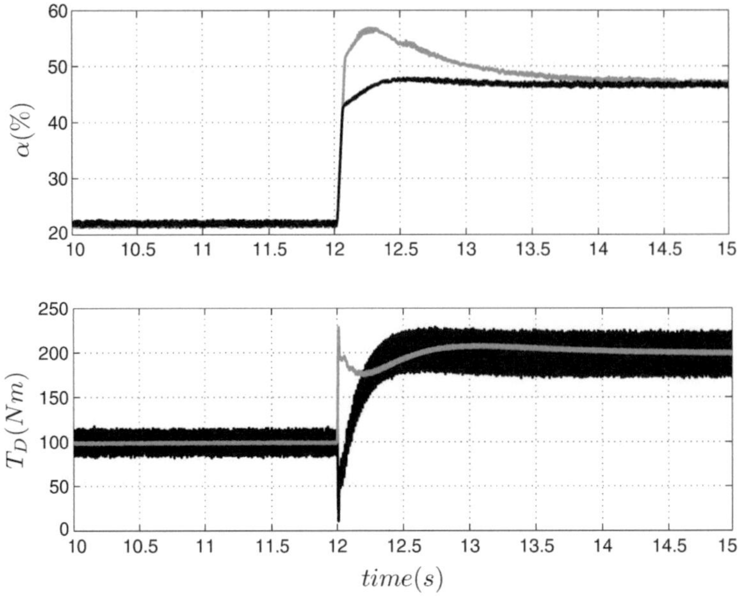

Fig. 4.2 *Top graph* time histories of the accelerator pedal position α determined by the control law (4.17) (*solid line*) and by the control law in [5] (*dashed line*). *Bottom graph* time histories of the torque of the dynamometer T_D determined by the control (4.17) (*dark line*) and by the control law in [5] (*grey line*)

oscillations. Differences are also visible in the input signals generated by the control laws, see Fig. 4.2. In fact at time $t = 12s$, the control (4.17) avoids an overshoot in the accelerator pedal position α, which is an advantage from the polluting emissions point of view, while requiring, on the other side, a more demanding behavior to the electric dynamometer.

4.5 Conclusions

In this chapter we have approached the optimal control problem for nonlinear systems, the classical solution of which is provided in terms of the well-known Hamilton-Jacobi-Bellman partial differential equation. We have introduced and discussed the notion of Dynamic Value function which permits the construction of a dynamic control law that approximates the optimal policy. The approximation error may be minimized by initializing opportunely the internal state of the dynamic control law, on one hand, and by steering the evolution of the dynamic extension in such a way that the actual minimized cost resembles the desired cost. Interestingly, a class of such functions may be constructed without involving the solution of any pde. Finally, the applicability of the proposed dynamic control law has been tested and validated on an interesting control problem within the automotive framework.

References

1. Bardi M, Capuzzo Dolcetta I (1997) Optimal control and viscosity solutions to Hamilton-Jacobi-Bellman equations. Birkhauser, Boston
2. Bertsekas DP (1995) Dynamic programming. Athena Scientific, Belmont
3. Bryson AE, Ho YC (1975) Applied optimal control: optimization estimation and control. Taylor and Francis, UK
4. Çymen T (2008) State-dependent Riccati equation (SDRE) control: a survey. In: Proceedings of the 17th IFAC world congress, Seoul, Korea
5. Gruenbacher E (2005) Robust inverse control of a class of nonlinear systems. PhD thesis, Kepler University Linz, Austria
6. Lukes DL (1969) Optimal regulation of nonlinear dynamical systems. SIAM J Control 7:75–100
7. Sassano M, Astolfi A (2012) Dynamic approximate solutions of the HJ inequality and the HJB equation for input affine nonlinear systems. IEEE Trans Autom Control 57(10):2490–2503
8. Vinter RB (2000) Optimal control. Birkhauser, Boston
9. Wernli A, Cook G (1975) Successive control for the nonlinear quadratic regulator problem. Automatica 11:75–84

Part II
Inter and Intra Vehicle System Optimization

Chapter 5
Intelligent Speed Advising Based on Cooperative Traffic Scenario Determination

Rodrigo H. Ordóñez-Hurtado, Wynita M. Griggs, Kay Massow and Robert N. Shorten

Abstract A novel system for safe speed recommendation, based on a cooperative method for vehicular density estimation and on the intelligent determination of the traffic scenario, is presented.

5.1 Introduction

At present, Intelligent Speed Adaptation (ISA) systems, as a part of Advanced Driver Assistance Systems (ADASs), have become a fundamental part in the designing of safe vehicle operation systems, with the aim of improving driver/pedestrian safety using environmentally friendly applications [1]. Statistically, ISA systems always represent an improvement in the reduction of CO_2 emissions and fuel consumption, and on saving/prediction of accidents (fatal, serious and slight) [1]. Advisory systems rely on the calculation of safe recommended parameters to be presented to the driver using an appropriate display system [2]. Thus, in general, advisory ISA methodology involves less algorithmic and analytical complexity, and constitutes the first step towards more comprehensive (mandatory) systems, and the enhancement of Adaptive Cruise Control (ACC) algorithms [3]. ISA systems can be greatly

R. H. Ordóñez-Hurtado (✉) · W. M. Griggs · R. N. Shorten
National University of Ireland Maynooth, Maynooth, Co. Kildare, Ireland
e-mail: Rodrigo.Ordonez@nuim.ie

W. M. Griggs
e-mail: Wynita.Griggs@nuim.ie

R. N. Shorten
IBM Research Ireland, Dublin, Ireland
e-mail: Robert.Shorten@nuim.ie

K. Massow
Fraunhofer Institute for Open Communication Systems FOKUS, Berlin, Germany
e-mail: kay.massow@fokus.fraunhofer.de

H. Waschl et al. (eds.), *Optimization and Optimal Control in Automotive Systems*, Lecture Notes in Control and Information Sciences 455,
DOI: 10.1007/978-3-319-05371-4_5, © Springer International Publishing Switzerland 2014

improved by including relevant information from different sources such as environmental (weather, visibility, etc) and road (vehicular density, speed limits, etc) information, thus resulting in more reliable systems [4]. A recent application based on weather information can be found in [5]. Regarding road information, vehicular density represents a very important factor in designing systems for safe speed advising, because from it, it is possible to obtain a more realistic awareness of the general traffic situation [6]. In this sense, most ADASs involving vehicular density estimation techniques are based on the use of loop detectors [7, 8]. However, there are many drawbacks in using this kind of dedicated infrastructure device: (1) vehicular density is computed only for fixed road sections (between two consecutive loop stations), i.e. the available information is space-discontinuous; and (2) density variations cannot be properly detected at each location with a low density of loop detector stations (but a high density of them is not desirable from the monetary point of view) [7]. Moreover, if we think in terms of a decentralisd scheme, then vehicular density should be estimated for each node belonging to the vehicular ad-hoc network (VANET), making even less feasible the use of loop detectors. Thus, more practical ways to estimate vehicular (traffic) density are required, such as the one presented in [6]. With the above in mind, we propose a two-stage methodology for intelligent speed advising: the first stage concerns traffic scenario determination based on a cooperative methodology using vehicle-to-vehicle (V2V) communication for vehicular density estimation, and a rule-based system (Sect. 5.4), and the second stage concerns the calculation of safe parameters based on the proposed traffic scenario determination (Sect. 5.5). Experimental validation is presented in Sect. 5.6, and we conclude the chapter with Sect. 5.7.

5.2 Intelligent Speed Adaptation System

ISA systems can be classified as either static or dynamic. A static ISA system is a system where the recommender is supported only on fixed/localised speed limits, whereas a dynamic ISA system also uses environmental information to update the recommended speed. ISA systems can also work in advisory, voluntary or mandatory modes. In an advisory mode, the function of the ISA is to recommend a speed to the driver, and in mandatory mode a control action is used to enforce the advised speed. For dynamic-mandatory cases it has been shown that ISA systems are able to provide safety benefits in terms of a reduction of up to 44 % in fatality [5]. Recent developments in ITS infrastructure have made possible the development of more advanced ISA systems. In this chapter we describe one such a system. Our system allows the inclusion of relevant available information using current traffic information and road speed limits for calculating the recommended speed. We propose to use V2V communication as a main tool for vehicular density calculation rather than loop detectors, with the aim of obtaining space-continuous information through a cheaper approach (as opposed to using dedicated infrastructure devices). Consequently vehicular density is used as one of the inputs of a rule-base reasoning engine designed to determine the current traffic scenario; the scenario that will be used to dynamically calculate the final recommended speed. The main advantage of using such an inference engine is the possibility of including expert knowledge in an easy and intuitive way via IF-THEN rules. Finally, we assess our speed adaptation scheme using a traditional safe policy applied to the resulting inter-vehicle distances.

5.3 Procedure

We are proposing a two-stage methodology for intelligent speed advising: (a) the first stage focuses on traffic scenario determination, and (b) the second stage regards safe parameters calculation. In the first stage we consider the problem from a spatial-temporal perspective. We begin this process by defining a point of reference that represents a point along the future trajectory of the vehicle for which the recommender is being constructed. Thus, the following concepts arise:

- the *Host Vehicle* (HV) is the vehicle for which the recommendation is being constructed.
- the *Next Point of Interest* (NPI) is a coordinate in the near future of the Host Vehicle's evolution, i.e. a point located in the future trajectory of the Host Vehicle.
- the *Next Vehicle* (NV) is the (potentially virtual) vehicle that is currently closest to the NPI.

Once the Next Vehicle has been selected, the calculation of the vehicular density is obtained as the vehicle density in some prespecified area around the NPI. Due to the spatial-temporal nature of the problem, vehicular density is calculated for both the Host Vehicle and the NPI.

The current traffic scenario is determined by using the calculated vehicular density and speed for both the Host and Next Vehicles, and the variation of the Host Vehicle speed as the inputs of an inference engine. Our inference engine is made up by a set of 28 IF-THEN rules.

For the second stage of the procedure, we propose to calculate the recommended speed using a weighted formula that combines both the Host and Next Vehicle speeds, as well as density information.

Finally, once the recommended speed is obtained, we use a widely known policy for safe recommended distance.

For ease of exposition, we make the following assumptions:

- our road setup is as depicted in Fig. 5.1:
  - a five-section (S1–S5), two-lane (L1–L2), one traveling direction straight road;
  - a 2D Cartesian system for spatial representation of the road (top view) where the x axis is the direction of travel;
  - a stationary bottleneck represented by the narrowing of the road (S3), emulating a closed lane e.g. due to an on-road accident;

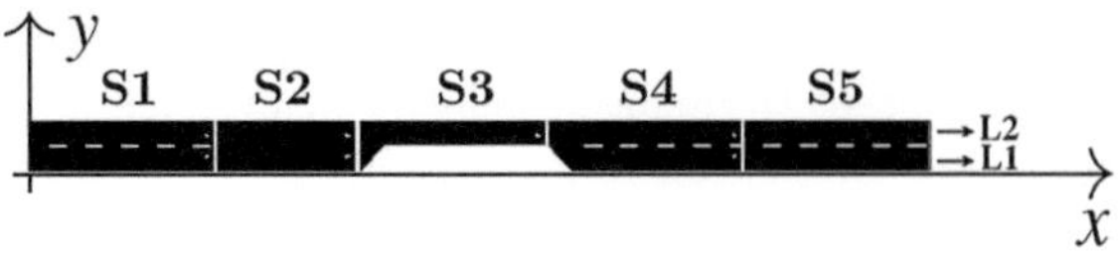

Fig. 5.1 Road setup used

- and

A1 all vehicles belonging to the VANET have a compatible V2V system;
A2 all required information can be acquired using suitable devices/techniques, and then transmitted using such a V2V system;
A3 processing times for V2V communication and outputs calculation are greatly shorter than intervals between instants for speed recommendation.

5.4 Methodology: First Stage

In the first stage of the proposed ISA methodology, the basic idea is to use V2V communication to obtain an estimation of the vehicular density. With such information, in addition to speed values and other relevant data, we determine the current traffic scenario using a rule-based system.

However, as the traffic scenario determination is a spacial-temporal problem, other sources of information must be considered in addition to the Host Vehicle information. This demands the selection of a vehicle placed at a point in the ahead road-section in which the Host Vehicle is traveling on, in order to represent a point along the future trajectory of the Host Vehicle. Hereafter, such a point is the NPI, and the vehicle representing the NPI will be referred to as the Next Vehicle (which is not necessarily the vehicle immediately preceding the Host Vehicle).

5.4.1 Selection of the Next Point of Interest and the Next Vehicle

The NPI is a reference placed at a distance x_{ahead} in front of the Host Vehicle. As we are considering a straight road collinear to the x axis, we define the NPI at $(x_H + x_{ahead}, y_H)$, where (x_H, y_H) is the position of the Host Vehicle.

In order to select the Next Vehicle to represent the NPI, we look inside a circle with radius r_N centered at $(x_H + x_{ahead}, y_H)$ as shown in Fig. 5.2. If no vehicles are inside the circle (see Fig. 5.2b), then we let the Next Vehicle be a virtual vehicle located at $(x_N, y_N) = (x_H + x_{ahead}, y_H)$; otherwise the closest vehicle to the NPI is selected (see Fig. 5.2a).

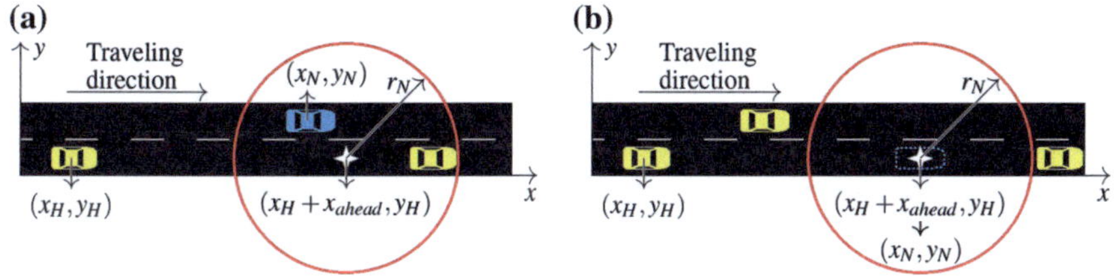

Fig. 5.2 NPI location: **a** the next vehicle is the nearest vehicle to $(x_H + x_{ahead}, y_H)$, i.e. the *blue* one, and **b** the next vehicle is a virtual vehicle located at $(x_H + x_{ahead}, y_H)$

5.4.2 Vehicular Density Estimation

The vehicular density estimation for any sampling node in the VANET can be carried out based on [7], as follows: (1) the sampling node broadcasts a poll message, (2) all nodes receiving the poll message respond to the sampling node with a reply message, and (3) vehicular density δ for the sampling node is given by

$$\delta(t) = \frac{n_r + 1}{A}, \qquad A = \begin{cases} \pi r_D^2, & \text{if } 2r_D \leq W_R = W_L N_L. \\ 2r_D W_L N_L, & \text{otherwise.} \end{cases},$$

where n_r is the number of returned replies inside the polling area A, W_R is the road's width, W_L is the lane's width, and N_L the total amount of lanes. Note that the factor $+1$ is added to the factor n_r to include the sampling node into the density equation. However, if the sampling node is a virtual vehicle, then the vehicular density is not $\frac{1}{A}$ but rather zero (see Fig. 5.2b).

5.4.3 Traffic Scenario Determination

Once the vehicular density for both Host/Next Vehicles is calculated, we can use that information in addition to the Host/Next Vehicle speeds in order to determine the traffic scenario. In this chapter we proposed to use an inference engine for that purpose, as explained in the following subsections.

5.4.3.1 Inference Engine Design

The inference engine consists of a (user-defined) knowledge base for assigning values to the outputs according to the values of the inputs. Let us define the inputs/outputs variables, to latter define the base of rules that relates them.

Table 5.1 Membership functions

Type	Variable	Set	Membership function
Input	$\bar{V}_{H,N}, \bar{\delta}_{H,V}$	L	[0 0 0.1 0.8]
		H	[0.1 0.8 1 1]
	$\Delta V_H(t)$	N	[−100 − 100 − 7.5 − 2.5]
		Z	[−7.507.5]
		P	[2.5 7.5 100 100]
Output	FT, AC, CT, PB, LC	N	[0 0 1]
		Y	[0 0.8 1 1]

Inputs/Outputs Definition

Five variables are chosen as inputs: the normalised Host Vehicle's velocity ($\bar{V}_H$); the normalised Next Vehicle's velocity ($\bar{V}_N$); the normalised vehicular density for the Host Vehicle ($\bar{\delta}_H$); the normalised vehicular density for the Next Vehicle ($\bar{\delta}_N$); and the variation on the Host Vehicle's velocity $\Delta V_H(t) = V_H(t) - V_H(t-1)$. In addition, five variables are chosen as outputs: Free Traffic (FT); Approaching Congestion (AC); Congested Traffic (CT); Passing Bottleneck (PB); and Leaving Congestion (LC).

The sets Low (L) and High (H) are for the inputs $\bar{V}_{H,N}$ and $\bar{\delta}_{H,V}$, the sets Negative (N), Zero (Z) and Positive (P) are for the input $\Delta V_H(t)$, and the sets Not (N) and Yes (Y) are for all the outputs. Membership functions are shown in Table 5.1.

Finally, the traffic scenario is classified according to the following equation:

$$T(t) = \text{argmax}(\text{FT}(t), \text{AC}(t), \text{CT}(t), \text{PB}(t), \text{LC}(t)).$$

Base of Rules

The rules R_k for relating the five inputs to the five outputs are of the form

$$\underline{\text{R}_k}: \text{ IF } input_1 = \bullet \text{ AND} \dots input_i = \bullet, \text{ THEN } \left(output_1 = \bullet \text{ AND} \dots output_j = \bullet\right) * w_k,$$

according to values in Table 5.2, which are supported by applying both traffic flow theory [9] and common sense to each particular case, and considering the values taken for each input.

5.4.3.2 Normalisation of Variables

In order to provide a general interpretation of the rules, we use normalised values instead of raw ones. Such a normalisation process depends on each kind of input, as presented in the following subsections.

Table 5.2 Base of rules for traffic scenario determination

Rule	Inputs $\left[\bar{V}_H, \bar{\delta}_H, \bar{V}_N, \bar{\delta}_N\right]$	ΔV_H	Outputs [FT, AC, CT, PB, LC]	Weight
1	[L, L, L, L]	N	[Y, N, N, N, N]	0.6
2		Z	[Y, N, N, N, N]	0.6
3		P	[N, N, N, N, Y]	1.0
4	[L,L,L,H]	–	[N, Y, N, N, N]	1.0
5	[L, L, H, L]	N	[Y, N, N, N, N]	1.0
6		Z	[N, N, N, N, Y]	1.0
7		P	[N, N, N, N, Y]	1.0
8	[L, L, H, H]	–	[N, Y, N, N, N]	1.0
9	[L, H, L, L]	N	[N, N, Y, N, N]	1.0
10		Z	[N, N, N, Y, N]	1.0
11		P	[N, N, N, Y, N]	1.0
12	[L, H, L, H]	–	[N, N, Y, N, N]	1.0
13	[L, H, H, L]	–	[N, N, N, Y, N]	1.0
14	[L, H, H, H]	–	[N, N, Y, N, N]	1.0
15	[H, L, L, L]	N	[N, Y, N, N, N]	1.0
16		Z	[Y, N, N, N, N]	1.0
17		P	[Y, N, N, N, N]	1.0
18	[H, L, L, H]	–	[N, Y, N, N, N]	1.0
19	[H, L, H, L]	N	[Y, N, N, N, N]	1.0
20		Z	[Y, N, N, N, N]	1.0
21		P	[N, Y, N, N, N]	1.0
22	[H, L, H, H]	–	[N, Y, N, N, N]	1.0
23	[H, H, L, L]	N	[N, N, Y, N, N]	0.8
24		Z	[N, N, Y, N, N]	0.8
25		P	[N, N, N, Y, N]	1.0
26	[H, H, L, H]	–	[N, N, Y, N, N]	1.0
27	[H, H, H, L]	–	[N, N, N, Y, N]	1.0
28	[H, H, H, H]	–	[N, N, Y, N, N]	1.0

Velocity Normalisation

This normalisation depends on the raw value of the velocity, the Maximum Individual Speed (MIS) of the vehicle, and the Road Speed Limit (RSL) of the road section in which the vehicle is traveling on. The normalised velocity is given by

$$\bar{V}_{H,N}(t) = \min\left(\alpha_{speed} * \tilde{V}_{H,N}(t), 1\right),$$

$$\begin{aligned}\alpha_{speed} &= \frac{1}{\max\left(MIS_{H,N}, RLS_{H,N}\right)}, \tilde{V}_{H,N}(t) \\ &= \begin{cases} V_{H,N}(t), & \text{if } MIS_{H,N} > RLS_{H,V} \\ f_1\left(RLS_{H,V}, MIS_{H,N}, V_{H,N}(t)\right) & \text{otherwise.}\end{cases},\end{aligned}$$

Table 5.3 Maximum allowed density (f_2) given r_D

r_D	7.5	8.5	9.5	11	13	15	19.5	21
$f_2(r_D)$	7	6	5.5	5	4.5	4.2	3.9	3.8

where f_1 provides a value corresponding to the linear interpolation of the $V_{H,N}(t)$ using a curve given by $\{(0, RLS_{H,V}), (0, MIS_{H,N})\}$. Note that min $(\bullet, 1)$ ensures that the maximum value of $\bar{V}_{H,N}(t)$ is 1 even when the velocity overcomes the corresponding RLS.

Vehicular Density Normalisation

This normalisation depends on the raw value of the vehicular density, the polling radio r_D, and the Maximum Allowed Density (MAD) curve (constructed from Table 5.3). The normalised vehicular density can be calculated as

$$\bar{\delta}_{H,N}(t) = \min\left(\alpha_{density} * \delta_{H,N}(t), 1\right), \qquad \alpha_{density} = \frac{1}{f_2(r_D)},$$

where the value of $f_2(r_D)$ (representing the MAD given r_D) is calculated according to a linear interpolation using data in Table 5.3 (obtained from simulation tests). Again, min $(\bullet, 1)$ assures that the maximum value of $\bar{\delta}_{H,N}(t)$ is 1 even when the vehicular density overcomes the corresponding estimated MAD.

5.5 Methodology: Second Stage

Once the traffic scenario is determined, we can use such information to design our Advisory ISA methodology. However, since the definition of the recommended speed V_R (and as a consequence, the recommended distance D_R) should be based upon both the determined traffic scenario and the Next Vehicle's velocity (V_N), then we first introduce a model for updating V_N in cases of a virtual Next Vehicle.

5.5.1 Updating Speed in Virtual Next Vehicles

If a virtual Next Vehicle is chosen, then both the location and velocity of the Next Vehicle have to be calculated from other sources rather than a real vehicle on the road. Recall that we already assigned the location of such a virtual Next Vehicle as $(x_H + x_{ahead}, y_H)$ (see Sect. 5.4.1), but a model for its velocity updating is still missing. Thus, we propose a way to update V_N similar to

Table 5.4 Decision matrix for $\underline{V}_N$, α_{NV} and α_R

Traffic scenario	FT	AC	CT	PB	LC
Value of $\underline{V}_N$	0.3	0.2	0.1	0.1	0.3
Value of α_{NV}	1.4	0.7	0.9	0.9	1.4
Value of α_R	0.7	0.7	0.7	0.45	0.7

$$V_N(t) = \alpha_{NV}(t) * V_N(t-1),$$

where $\alpha_{NV}(\bullet)$ is the evolution parameter, but including some particular considerations. Our way for updating the normalised virtual Next Vehicle speed $\bar{V}_N$ is then given by

$$\bar{V}_N(t) = \min\left(\alpha_{NV}(T(t-1)) * f\left(\bar{V}_N(t-1)\right), 1\right), \tag{5.1}$$

$$f\left(\bar{V}_N(t-1)\right) = \max\left(\bar{V}_N(t-1), \underline{V}_N(T(t-1))\right), \tag{5.2}$$

where $T(\bullet)$ is the determined traffic scenario, and $\underline{V}_N(\bullet)$ is the minimum allowed normalised speed for a virtual Next Vehicle. The inclusion of this minimum limit speed, which is greater than zero, is to avoid $\bar{V}_N = 0$ which represents that the Host Vehicle is approaching a stopped vehicle, which is not true because there are no real vehicles around the NPI (see Fig. 5.2). Then, min $(\bullet, 1)$ in (5.1) is to guarantee that $\bar{V}_N$ never exceeds the maximum normalisation value 1, and max $\left(\bullet, \underline{V}_N(\bullet)\right)$ in (5.2) is to guarantee that the virtual Next Vehicle is always moving at least at $\underline{V}_N$. Both α_{NV} and $\underline{V}_N$ are design parameters, and reference values (obtained from simulation tests) are given in Table 5.4.

Justification for values in Table 5.4 are as follows:

- Given previous *Free-Traffic/Leaving-Congestion* scenarios, it is assumed that the virtual Next Vehicle can accelerate without problems. Thus, $\alpha_{NV} = 1.4$ represents an increase of 40 % in V_N, and $\underline{V}_N = 0.3$ sets a minimum value for $\bar{V}_N$ at 0.3.
- Given previous *Congested-Traffic/Passing-Bottleneck* scenarios, it is assumed that the virtual Next Vehicle could not have accelerated, and probably could have had a small deceleration. Thus, $\alpha_{NV} = 0.9$ represents a decrease of 10 % in V_N, and $\underline{V}_N = 0.1$ sets a minimum value for $\bar{V}_N$ at 0.1.
- Given a current *Approaching-Congestion* scenario, it is assumed that the virtual Next Vehicle is indeed decelerating. Thus, $\alpha_{NV} = 0.7$ represents a decrease of 30 % in that V_N, and $\underline{V}_N = 0.2$ sets a minimum value for $\bar{V}_N$ at 0.2.

5.5.2 *Proposed Recommended Speed Scheme*

We propose a calculation method for the recommended cruise speed similar to the one presented in [5], i.e. a convex linear combination with time-variant coefficients.

However, here we calculate the recommended speed by combining two terms, the Host Vehicle speed V_H and the Next Vehicle speed V_N, as follows

$$V_R(t) = (\alpha_R(T(t))) * V_N(t) + (1 - \alpha_R(T(t))) * V_H(t), \tag{5.3}$$

where $\alpha_R(\bullet)$ is a time-variant weighting factor calculated from the decision matrix presented in Table 5.4. Note that α_R is a design parameter, so values in Table 5.4 were tuned from simulation tests. The justification for values of α_R is as follows:

- *Approaching-Congestion/Congested-Traffic* scenarios should force the Host Vehicle to slow down in order to travel at most around V_N, which in general is a real vehicle inside a dense platoon immediately ahead of the Host Vehicle. Then, $\alpha_R = 0.7$ means that the recommended speed depends more upon V_N than V_H, causing V_N to act like an upper bound.
- *Free-Traffic/Leaving-Congestion* scenarios should force the Host Vehicle to speed up in order to reach V_N, which is expected to be traveling in a *Free-Traffic* scenario. Then, $\alpha_R = 0.7$ means that the recommended speed depends more upon V_N rather than V_H, causing V_N to act like a goal speed.
- The *Passing-Bottleneck* scenario is determined based on the existence of a Next Vehicle who is leaving the congestion with positive high ΔV_N. However, here the Host Vehicle is about to leave the traffic jam but is still inside it, so the recommended speed should depend more upon V_H than V_N. Then, $\alpha_R = 0.45$ causes V_H to act like an upper bound.

According to (5.3), it is observed that V_R always depends directly on both V_H and V_N, so any noisy behaviour in either of them will be directly reflected on V_R. Thus, two additional processes have to be added: (1) a quantisation process, to avoid a noisy recommended speed, and (2) a saturation process, to avoid recommending a speed greater than the speed limit of the road on which the Host Vehicle is traveling. With this we finally obtain $V_R \in \min(\{5 * n\}, RLS_H)$, with $n = 1, 2, \ldots$.

5.5.3 Proposed Recommended Distance Scheme

We can assess the performance of our recommended speed scheme by evaluating the usually adopted safe inter-distance policy [10]

$$D_R(t) = h_0 + h_1 V_f(t) + h_2 \left(V_f^2(t) - V_l^2(t) \right),$$

where D_R is the recommended (safe) distance, h_0 is the minimum safe distance to the preceding vehicle, h_1 is the minimal required headway time (usually set in $h_s = 0.6\,[s]$), h_2 is a problem-dependent weighting factor, V_f corresponds to the speed of the Host Vehicle, and V_l to the speed of the preceding vehicle. However, here the reference is the Next Vehicle, which does not necessarily coincide with the preceding vehicle. Thus, we have to take

$$V_l(t) = V_N(t),$$

and h_0 as the safe distance to the Next Vehicle, redefined as follows

$$h_0 = \left\lfloor \frac{X_N - (X_H + G_{min})}{L_V + G_{min}} \right\rfloor * (L_V + G_{min}) + G_{min},$$

where G_{min} is the minimum allowed gap (safe distance) between two consecutive vehicles, and L_V is the mean longitude of a vehicle in the network. Note that the *Approaching-Congestion*, *Congested-Traffic* and *Passing-Bottleneck* scenarios are of special interest, because only in these cases it is expected that there exists a high density of vehicles between the Host Vehicle and the Next Vehicle (i.e., a higher probability of collision).

Now, the recommended distance must be compared to the relative distance X_{rel}, measured as the difference between the Host Vehicle's position and the Next Vehicle's position

$$X_{rel} = X_N - X_H.$$

Then by defining $e = X_{rel} - D_R$, the case $e \geq 0$ means that there is a safe situation (the relative distance is greater than or equal to the recommended distance), and thus the case $e < 0$ means that there is a non-safe situation.

5.6 Validation

To validate the proposed methodology, we use SUMO to simulate thirty-one vehicles with properties as in Table 5.5, which travel according to a modified Krauss car-following model [11] on the road defined by Fig. 5.1 and Table 5.6. Vehicle 08 exhibits a special behaviour: it stops at $Distance = 296$ m (road section S4) for 100 s, after which it restarts its travel. The data obtained from SUMO was exported to the Matlab environment Version 7.12.0.635 (R2011a).

The idea behind using vehicles with very high deceleration abilities is to obtain data in extreme situations (i.e. the vehicles are very prone to having a collision), in order to evaluate the performance of the proposed methodology in recommending a safe speed early. In addition, speed restrictions on S2/S3 and S5 emulate realistic behaviours around a traffic bottleneck and the variety of piecewise constant speed limits along a same road, respectively.

Table 5.5 Properties of simulated vehicles used in tests

Attribute	Vehicle type		
	A	B	C
Vehicle's ID	03, 09, 11, 13, 15, 17, 19, 23, 25, 27, 29, 31	04, 05, 07, 10, 12, 14, 18, 20, 24, 26,28.	01, 02, 06, 08, 16, 21, 22, 30
Length [m]	4.4	4.0	4.2
Max Speed [m/s]	40	30	16.677
Acceleration [m/s^2]	3	2	1
Deceleration [m/s^2]	10	10	10
Minimum Gap [m]	2.5	2.5	2.5
Sigma	0.5	0.5	0.5

Table 5.6 Properties of road sections in Fig. 5.1

Section	Length (m)	Max speed (m/s)
S1	175	27.778
S2	5	0.7
S3	30	2.5
S4	235	27.778
S5	55	2.5

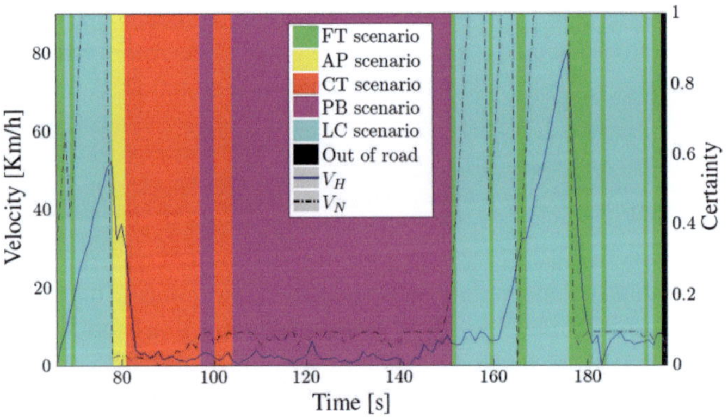

Fig. 5.3 Entire traffic scenario determination for vehicle 20 using LOM

5.6.1 *Traffic Scenario Determination*

The inference engine was implemented using the FL Toolbox for use with Matlab [12], and tested using $r_N = 4$ m, $r_D = 14$ m, $x_{ahead} = 32$ m, $W_L = 3.5$ m and $N_L = 2$. The obtained results for vehicle number 20 are presented in Fig. 5.3 using the LOM (Last Of Maximum) method to calculate the outputs. With this method, all the estimations have a certainty value of 1.0 (i.e. complete certainty).

The traffic scenario determination for the entire set of vehicles is shown in Fig. 5.4a, b. From Fig. 5.4a, it can be concluded that almost all traffic scenarios

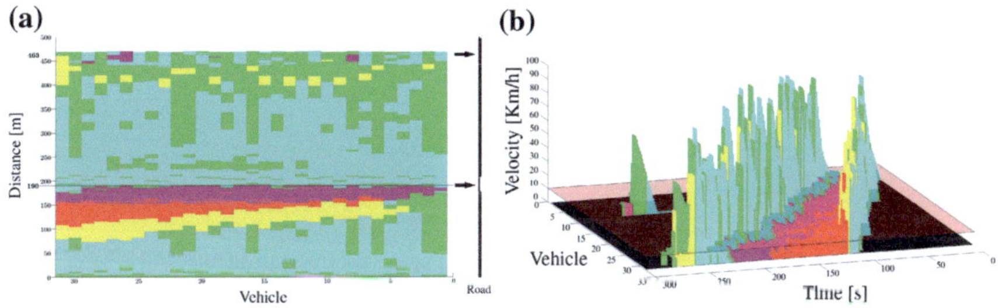

Fig. 5.4 Entire traffic scenario estimation for all vehicles using LOM. **a** Vehicle versus distance. **b** Vehicle versus velocity versus time

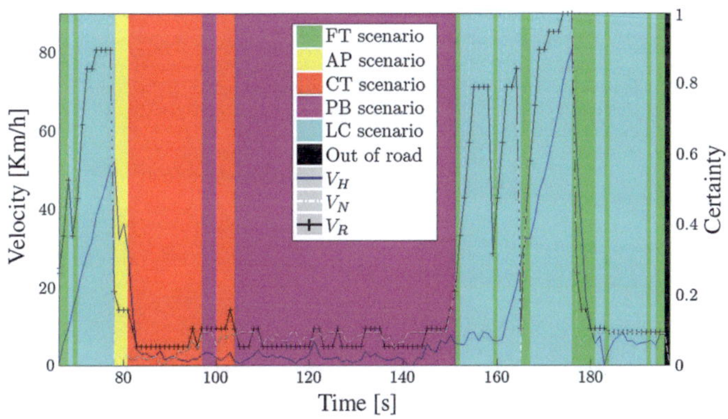

Fig. 5.5 Entire V_R profile for vehicle 20 using the proposed methodology

immediately below a cut at $Distance = 190$ m (the medium point of the bottleneck) are determined as a *Passing-Bottleneck* scenario (magenta), and that some vehicles detect the new speed limit around $Distance = 463$ m as a *Congested-Traffic* scenario (red) for a few seconds, just after the new speed limit's commencement at $Distance = 445$ m.

For its part, Fig. 5.4b shows that most of the velocities beneath a cut at $Velocity = 10\,[\text{km/h}]$ are successfully classified as either *Passing-Bottleneck* or *Congested-Traffic* scenarios. Moreover, increasing velocities are suitably classified as either *Free-Traffic* scenario (green) or *Leaving-Congestion* scenario (cyan), generally after *Passing-Bottleneck* or *Congested-Traffic* scenarios, as confirmed in Fig. 5.4a. Finally, decreasing velocities in Fig. 5.4b are successfully classified as an *Approaching-Congestion* scenario (yellow) when a *Passing-Bottleneck* or *Congested-Traffic* scenario is about to occur (also confirmed in Fig. 5.4a).

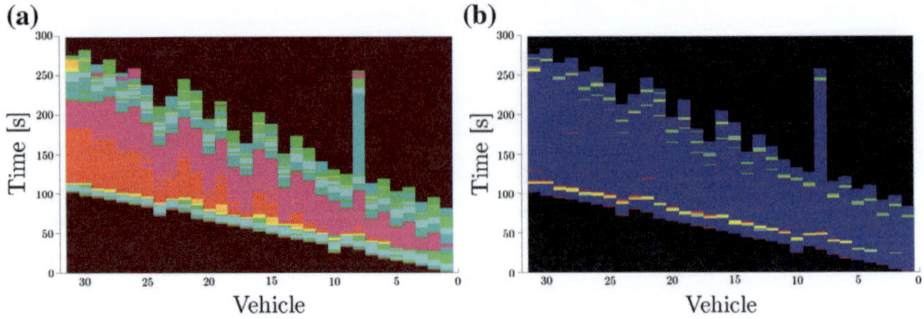

Fig. 5.6 Analysis of V_R for all vehicles in function of time. [*Black* sections: beyond road's length]. **a** HV-speed/trafic-scenario porfiles. **b** Prevalence of the V_R over V_H porfile: *Bule* means $V_R > V_H$

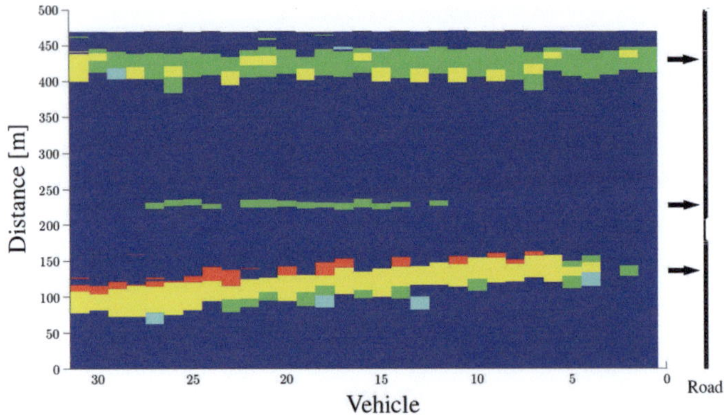

Fig. 5.7 Analysis of V_R for all vehicles in function of distance. Prevalence of the V_R profile over the V_H profile: *blue* means $V_R > V_H$

5.6.2 Recommended Speed

Results for the particular case of Vehicle 20 (Fig. 5.5) show that the critical *Approaching-Congestion* scenario is tackled properly by detecting the sudden (and maintained) decreasing of V_N at times 78 s and 177 s and then imposing an anticipated low V_R. With this, the Host Vehicle can be warned of the oncoming traffic jam early, and thus gains several seconds to perform a smoother braking action.

Another sudden decrease in V_N occurs at 165 s, and a low speed is recommended for a single instant. Such a decrease is not caused by any congestion, but by a stopped vehicle (Vehicle 08) in the middle of a *Free-Traffic* scenario. Then, the V_R profile is momentarily affected, indicating the existence of an isolated stopped vehicle (the estimated scenario remains as a *Free-Traffic* scenario during, and for some instants after, such a detection).

The performance of all vehicles can easily be analysed from Figs. 5.6 and 5.7 with a quick visual inspection, due to the colour-convention used: blue sections indicate

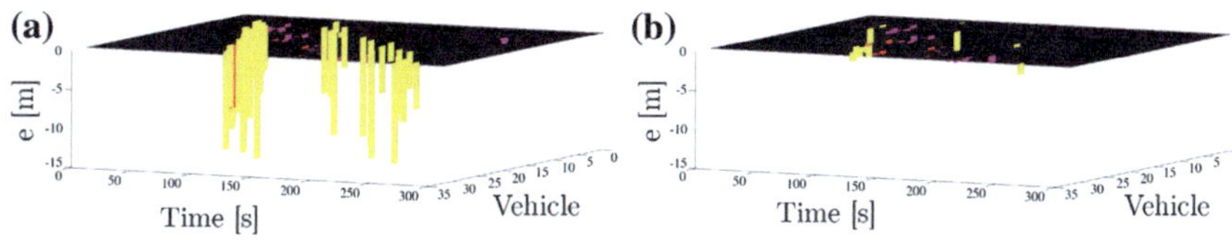

Fig. 5.8 3D analysis of $e = X_{rel} - D_r$ for all vehicles. Colored sections correspond to $e < 0$ (according the previous color convention). **a** X_{ret}, D_r obtained from $V_f = V_H$. **b** X_{ret}, D_r obtained from V_f in function of V_R

that $V_R \geq V_H$. Note that the section for $V_R < V_H$ corresponding to the lower arrow in Fig. 5.7 exhibits a pattern that in general coincides with *Approaching-Congestion* scenarios, which can be explained as the approaching of the oncoming *Congested-Traffic/Passing-Bottleneck* scenarios with a suitable safe (low) speed. Two other cases in which $V_R < V_H$ also happen can be better understood from Fig. 5.7: the middle arrow indicates the detection of an isolated vehicle which is stopped in the middle of the road (Vehicle 08); and the upper arrow represents the case in which a segment of the road with a lower speed limit will be promptly reached.

5.6.3 Recommended Distance

Our D_R scheme was tested with data from both:(1) the original set-up (without taking into account V_R), and (2) the improved set-up (manual and isolated adjustment of the speed according to our V_R scheme) with $h_2 = 0.01$, $L_V = 4.2$, and G_{min} [m] $=$ 2.5 [m].

In Fig. 5.8a (original setup) we can see that many of the $e < 0$ cases are produced in *Approaching-Congestion* scenarios. This is particularly interesting because, there, V_H is much faster than V_N, producing large negative values for e (Fig. 5.8a), thus resulting in a high probability of collision. In Fig. 5.8b (improved setup) we can see that most of those dangerous situations are suitably tackled, and just a few of minor $e < 0$ still remain.

Recall that in Cooperative ACC schemes the control law depends on the value of e [10]: the smaller e value, the weaker action control (breaking effort) in tracking the safety parameters. Thus, according to Fig. 5.8b, our recommended speed/distance schemes provide high performance in terms of travelling in safe conditions.

5.7 Conclusions and Future Work

A new scheme for safe speed advising based on a cooperative and decentralised methodology for traffic scenario determination was proposed. Its performance was assessed using safe policies and supported by experimental tests via SUMO package.

Currently, efforts are focused on evaluating the proposed ISA system beyond the used setup, i.e. using other realistic situations such as roads with curves and mobile bottlenecks.

An immediate future task is to use the proposed methodology to design a Cooperative ACC system by closing the speed/distance loops using the here proposed V_R and D_R schemes and a suitable controller. For such a mandatory ISA we have to be able to develop the corresponding analysis to guarantee string stability. In addition, other kinds of information can be used to improve the performance of the advisory system, such as meteorological (weather) and environmental (pollution) information.

Acknowledgments This work was in part supported by Science Foundation Ireland grant 11/PI/1177. This work was also supported by the European Commission under TEAM, a large scale integrated project part of the FP7-ICT for Cooperative Systems for energy-efficient and sustainable mobility. The authors would like to thank all partners within TEAM for their cooperation and valuable contribution.

References

1. Carsten O, Lai F, Chorlton K, Goodman P, Carslaw D, Hess S (2008) Speed limit adherence and its effect on road safety and climate change final report. University of Leeds, 2008
2. Adell E, Varhelyi A, Alonso M, Plaza J (2008) Developing human-machine interaction components for a driver assistance system for safe speed and safe distance. Intell Transport Syst 2(1):1–14
3. Hounsell N, Shrestha B, Piao J, McDonald M (2009) Review of urban traffic management and the impacts of new vehicle technologies. Intell Transport Syst 3(4):419–428
4. Tradisauskas N, Juhl J, Lahrmann H, Jensen C (2009) Map matching for intelligent speed adaptation. Intell Transport Syst 3(1):57–66
5. Gallen R, Hautiere N, Cord A, Glaser S (2013) Supporting drivers in keeping safe speed in adverse weather conditions by mitigating the risk level. IEEE Trans Intell Transportation Syst 99:1–14
6. Tyagi V, Kalyanaraman S, Krishnapuram R (2012) Vehicular traffic density state estimation based on cumulative road acoustics. IEEE Trans Intell Transportation Syst 13(3):1156–1166
7. Garelli L, Casetti C, Chiasserini C, Fiore M, (2011) Mobsampling: V2V communications for traffic density estimation. In 2011 IEEE 73rd vehicular technology conference (VTC Spring), Budapest, Hungary, May 15–18 2011, pp 1–5
8. Schakel W, Klunder G, van Arem B, Harmsen E, Hagenzieker M (2012) Reducing travel delay by in-car advice on speed, headway and lane use based on downstream traffic flow conditions-a simulation study. (2012)
9. Kerner BS (2009) Introduction to modern traffic flow theory and control: the long road to three-phase traffic theory. Springer London Limited, London
10. Martinez J-J, Canudas-de Wit C (2007) A safe longitudinal control for adaptive cruise control and stop-and-go scenarios. IEEE Trans Control Syst Technol 15(2):246–258
11. Behrisch M, Bieker L, Erdmann J, Krajzewicz D (2011) Sumo - simulation of urban mobility: An overview. In SIMUL 2011, the third international conference on advances in system simulation, Barcelona, Spain, October 2011, pp 63–68
12. MathWorks I, Wang W-C (1998) Fuzzy Logic Toolbox: for Use with MATLAB: User's Guide. MathWorks, Incorporated

Chapter 6
Driver Control and Trajectory Optimization Applied to Lane Change Maneuver

Patrick J. McNally

Abstract The problem of driver control and trajectory optimization for the lane change maneuver is approached through the application of the same design and simulation tools used in the development of modern automobiles. The vehicle model and driver control algorithms are combined with a genetic algorithm for trajectory optimization to determine an optimal path for achieving an objective measure of maximum speed. Results are compared with subjective results from a professional driver using a new driver-in-the-loop system. The conclusion is integrated objective-subjective simulation methods can be used earlier in the design to improve vehicle handling performance.

6.1 Background

The process of engineering complex systems including automobiles has undergone a dramatic change in the last two decades. A major reason for this fundamental change is the development of computer based tools for design drafting, digital mock-up, and simulation. Model-based engineering (MBE) elevates computer based models in the engineering process to a central and governing role in the specification, design, integration, validation, and operation of a system [4]. All major automotive companies have adopted MBE techniques to various degrees.

A Virtual Prototype is a computer simulation of a physical product that can be presented, analyzed, and tested from concerned product life-cycle aspects such as design/engineering [14].

Complete virtual prototyping environments which compare the simulated performance of an automobile against a set of objective metrics are now available and have become standard practice for handling performance, ride performance, and

P. J. McNally (✉)
VI-grade LLC, Dexter, MI, USA
e-mail: patrick.mcnally@vi-grade.com

H. Waschl et al. (eds.), *Optimization and Optimal Control in Automotive Systems*, Lecture Notes in Control and Information Sciences 455,
DOI: 10.1007/978-3-319-05371-4_6,

even durability [11, 12]. However, the subjective experience the individual feels while driving the vehicle is as important as the performance metrics printed in a magazine, and so a broader approach than virtual prototyping with objective measures is required. So far these subjective experiences have not been encapsulated in objective metrics.

6.1.1 Experiential Engineering

The field of Experiential Engineering attempts to re-create the human environment for driving the vehicle so that decisions requiring human perception (from emotions or other difficult to measure inputs) can be made about the design of the vehicle. To be successful, this environment must be utilized during the product development process, in parallel with the vehicle's technical development [8].

Our approach is to combine objective and subjective measurement into an integrated environment of engineering tools and processes, thereby accelerating the vehicle development process. We refer to the objective tests as conducted via "offline simulation" and subjective tests as conducted via "online simulation", which puts the vehicle operator "in-the-loop" of control.

6.1.2 Lane Change Problem

Determination of an optimal control strategy to conduct the ISO lane change maneuver or obstacle avoidance maneuver has been studied for some time [3, 7, 9]. An optimal preview control method is applied to the automobile path following problem in [9]. The method was applied to the closed loop simulation of an automobile driver/vehicle system during a lane-change maneuver. The computer simulation results are compared with equivalent vehicle test measurements. In [7], the authors consider open-loop and closed loop lane-change maneuvers and design time optimal steering controllers with nonlinear constraints. First, they generate a special open loop lane-change steering signal which minimizes the period of lane-change subject to constraints on the lateral acceleration and jerk magnitude. Then, they discuss how to implement those steering commands in the closed-loop system using a lane-following controller.

Most recently, the problem was approached with a reduced vehicle dynamics model and driver control model that is solved repeatedly on a moving prediction horizon to yield near optimal setpoint trajectories for the full model [3]. A NLP approach is used to solve the problem of finding the path and driver control for the autonomous guidance problem. We take a similar approach here with a reduced vehicle dynamics model that is solved repeatedly, however, the trajectory control approach we take is more general and the application is toward the automobile handling design problem. Also, we are looking for a solution to the design problem that leads to a vehicle that gives a good "subjective" feeling of control.

6.2 Model Based Engineering Environment for Objective Evaluation

ADAMS/Car is an industry standard virtual prototyping solution for the design of handling, ride, and durability characteristics of an automobile. The software efficiently organizes and solves a complete set of differential and algebraic equations in the time domain [1].

VI-CarRealTime is an ADAMS/Car derivative model, specifically developed to allow the application of reduced order ADAMS/Car models to realtime problems such as racing trackside simulation, fast optimization and design of experiments studies, and hardware and driver in the loop simulation. The VI-CarRealTime model is a 14 DOF model representing 6 DOF's for the vehicle and 2 DOF's for each wheel. A VI-CarRealTime model can be automatically extracted from an ADAMS/Car model and automatically compared with the ADAMS/Car results for a series of standard events [2].

6.2.1 Determination of Driver Controls

VI-Driver is a virtual prototype of a human driver that is designed to drive a simulated vehicle model in a very simple and efficient way [5, 10]. The driver model has been developed with the idea that the "driver" must be:

- robust enough to adapt to a wide range of vehicle characteristics
- simple to tune, and self-adapting whenever possible
- capable of driving on both limit and sub-limit maneuvers.

VI-Driver separates the lateral and longitudinal control problem into uncoupled controllers, as shown in Fig. 6.1. Each control loop has a predictive (feed-forward) and compensatory (feedback) control system. In the most general case, the lateral controller takes a trajectory to follow as input, and a speed profile to match with the longitudinal controller.

The target trajectory curve for lateral tracking is determined through the following process. Refer to the trajectory model shown in Fig. 6.2.

Given a target curve γ which represents the trajectory that must be followed, we define a connecting contour γ_c (in red) with the following constraints:

1. initial position (compatible with the vehicle position)
2. initial orientation (compatible with vehicle speed, vehicle side slip angle β)
3. final position (the reference trajectory evaluated at the preview distance D)
4. final orientation (smoothly joining the reference trajectory evaluated at the preview distance D).

To fit these constraints, a cubic polynomial is used, and the coefficients are calculated with the following relations:

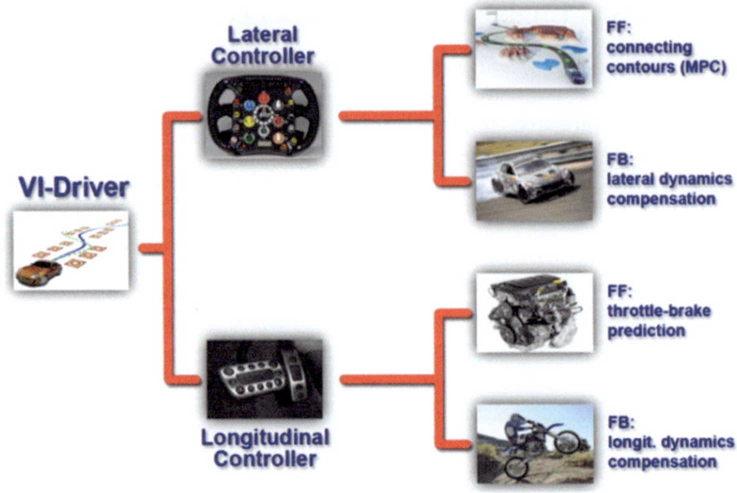

Fig. 6.1 VI-Driver lateral and longitudinal controllers

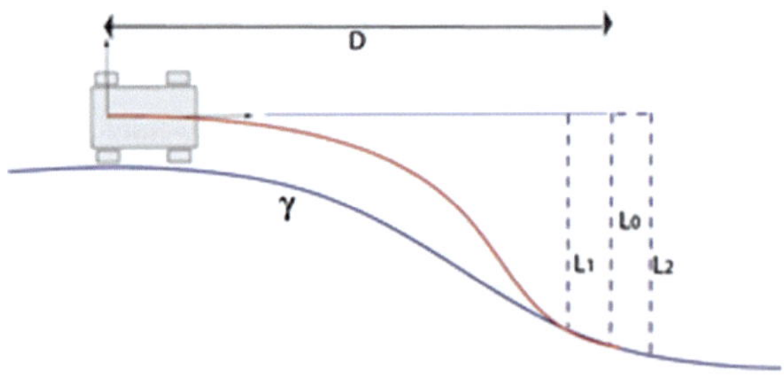

Fig. 6.2 Connecting contour for determining reference direction

$$\begin{bmatrix} \gamma_c(0,t) \\ \frac{\partial \gamma_c}{\partial x}(0,t) \end{bmatrix} = \begin{bmatrix} 0 \\ \beta \end{bmatrix}$$
$$\begin{bmatrix} \gamma(D,t) \\ \frac{\partial \gamma}{\partial x}(D,t) \end{bmatrix} = \begin{bmatrix} L_0 \\ L_2 - L_1 \end{bmatrix} \tag{6.1}$$

Using the differential flatness property, the connecting contour is used as the control input trajectory. The same property allows inverting that trajectory and calculating the appropriate steering control action. The preview distance parameter, D, can be adjusted to shape the input reference.

For directional control, the simplest vehicle model that captures the dynamical effects of interest is the classical bicycle treated as a nonholonomic model, as shown in Fig. 6.3.

VI-Driver's lateral controller uses the bicycle model as the basis for implementing a model predictive control technique [10]. Given the vehicle speed V, the side slip

Fig. 6.3 Classical bicycle model used as reference for lateral controller

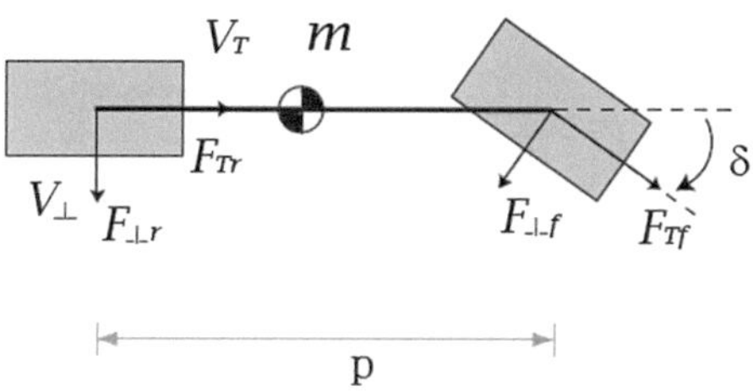

angle β, the preview time (t_p), and the preview distance D (computed as $V \cdot t_p$), the principle provides a very good approximation of the steering angle necessary to bring back the vehicle on the target path.

A final stage has been implemented, which compensates for the unmodeled lateral dynamics. This yaw rate controller uses the reference path curvature and the actual vehicle yaw rate to correct the steering action to bring the vehicle instantaneous curvature as close as possible to the reference path curvature.

Because of unmodelled dynamics, disturbances and numerical integration noise, the simulated vehicle almost always diverges from the global optimal trajectory target. Nevertheless, the simplicity and the natural robustness of the lateral controller provides correct steering action, and it has been observed to behave very realistically, even in cases where the target trajectory cannot be followed due to excess'ive side slip. More advanced MPC methods which include slip prediction have also been implemented for other work but at the sacrifice of computation speed [5].

6.2.2 Optimization Problem

The automotive manufacturer wishes to find the vehicle suspension and chassis design which executes the lane change maneuver in minimum time.

Formally, the design optimization problem is stated as:

$$Find\, \mathbf{u}^*, \mathbf{p}\, which\, causes$$
$$\dot{\mathbf{x}}(t) = \mathbf{a}(\mathbf{x}(t), \mathbf{u}(t), \mathbf{p})$$
$$To\, follow\, an\, admissible\, trajectory\, \mathbf{x}^*\, that\, minimizes\, the\, performance\, measure$$
$$\min(J) = t_f + w \cdot v$$

Where $\mathbf{u}(t)$ is our (admissible) driver control, $\mathbf{p}$ is a vector of suspension and chassis design parameters and the cost function is a weighted combination of final maneuver time t_f and vehicle speed v, a constant for our maneuver.

Towards this goal, we have constructed computational methods to solve the following two problems: (1) given a vehicle speed, find a feasible trajectory through the set of cones so that no cones are hit, and (2) find the minimum time to execute the lane change maneuver over all feasible trajectories through the set of cones.

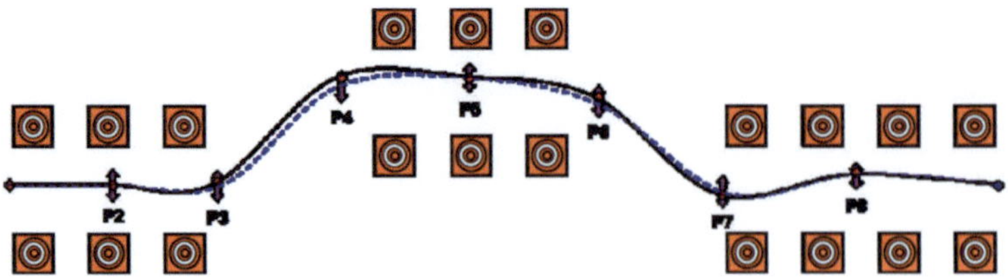

Fig. 6.4 Spline control points applied to ISO lane change trajectory

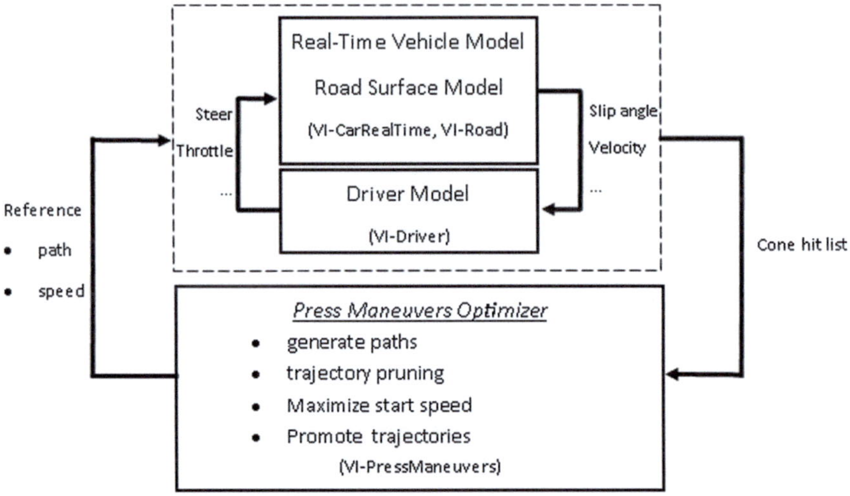

Fig. 6.5 Process flow for path and speed optimization

Problem 1 is setup with a population of trajectories, parameterized by a set of trajectory spline points, as shown in Fig. 6.4. A simulation is run for each candidate trajectory. Problem 2 is initiated by increasing the starting speed for the maneuver.

The two problems are solved automatically as one iteration. The number of total iterations is set by the user. Ultimate convergence to the final combination of maximum speed, minimum time, and correct trajectory is determined by reviewing successive run results.

The overall process of simulated vehicle control, trajectory optimization, and vehicle speed maximization is shown in Fig. 6.5. VI-Driver is used to control the path of the vehicle through the cones at a constant speed.

The Press Maneuvers Optimizer block uses a genetic algorithm to generate a set of target trajectories, parameterized by the spline points. For each iteration, a cone interference check is made to verify if the target trajectory is feasible. If a simulation results in a cone hit, all trajectories with the same spline segment are pruned out and the dynamic simulation is not run. All trajectories that result in a successful simulation without a cone hit are used as a starting set for the next generation.

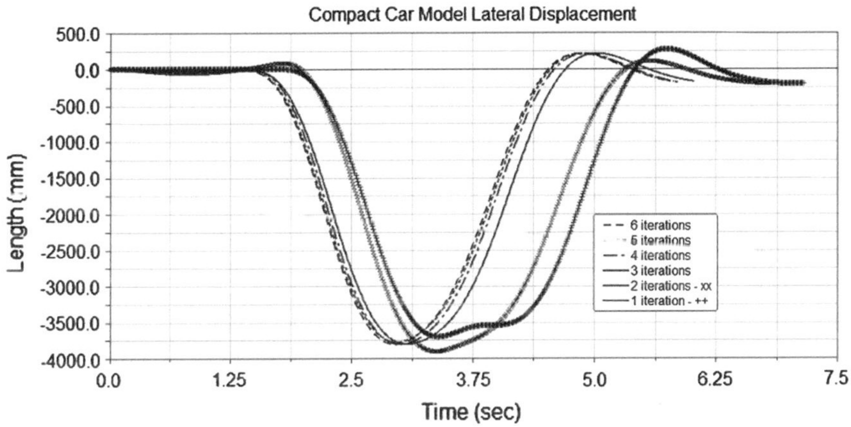

Fig. 6.6 Convergence of optimal solution for compact car lateral displacement, ISOLC

The start (constant) speed is increased and a new cycle begins. Depending on the user specified number of refinements, additional generations are run for the trajectory creation process. The event is completed when at least one combination of trajectory and maximum speed is identified which runs completely without a cone hit situation. The optimal is a weighted combination of highest speed and minimum time [15].

6.2.3 Offline Optimization Results

Simulations were run for three vehicle types to check the robustness of the algorithm; a high performance sports car, a sport-utility vehicle, and a compact car. Also, two cone layouts were examined; the ISO lane change layout and the Consumer Reports lane change layout. The Consumer Reports layout is similar to Fig. 6.4 but with only one pair of cones for the middle section.

The vehicle lateral displacement from the initial lane and the handwheel angle (steer demand) are used to illustrate convergence and for later comparison with a real driver. From Figs. 6.6, 6.7, 6.8, 6.9, it can be concluded that the optimization converges to a steady state lateral displacement and steering profile in six generations of the trajectory optimization for the ISO lane change maneuver and within 3 generations for the Consumer Reports lane change maneuver.

Table 6.1 shows the convergence of the optimizer for lap time and speed for each iteration of the ISO lane change maneuver. Note the unexpected change in maximum lateral acceleration from iteration 2 to 3.

A vehicle's characteristics, such as track width, wheel base, roll center height, and other design parameters will have an effect on optimal lane change performance. The results of the off-line optimization have been used to improve vehicle handling in the lane change maneuver through design parameter changes [15].

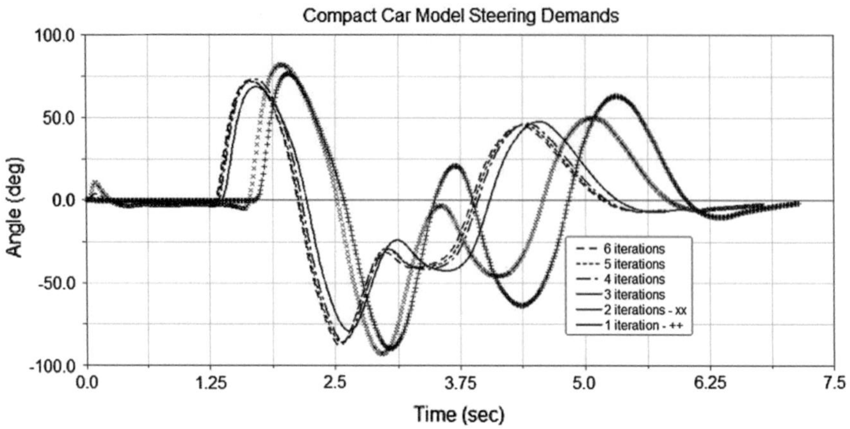

Fig. 6.7 Convergence of optimal solution for compact, steer demand, ISOLC

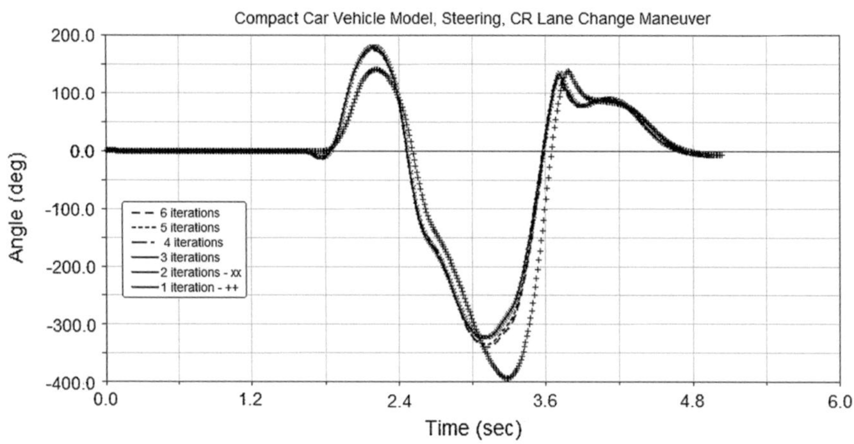

Fig. 6.8 Convergence of optimal solution for compact, steer demand, CR LC

6.3 Virtual Prototyping Environment for Subjective Evaluation

The previous sections have illustrated a method for evaluating the handling characteristics of a new vehicle to execute the lane change maneuver. The offline method of evaluating a vehicle is an objective measure of performance, but as discussed previously, vehicles are also evaluated for their subjective performance through a customer test. New technology for the subjective evaluation of a vehicle for handling characteristics during the early design stage was recently introduced [6]. This section describes this new technology for subjective evaluation of the handling characteristics of a vehicle.

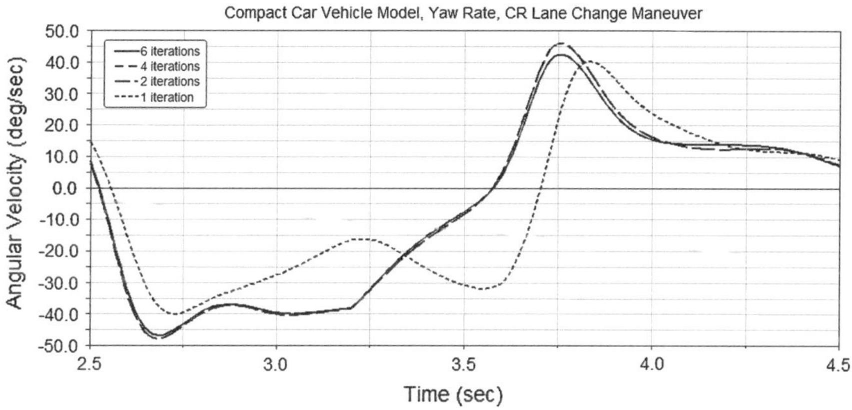

Fig. 6.9 Convergence of optimal solution for compact car yaw rate, CR LC

Table 6.1 Optimizer convergence results, ISLOC

Iteration	1	2	3	4	5	6
Lap time, seconds	7.14	6.78	6.03	5.87	5.82	5.78
Lateral accel. Max, G	0.761	0.786	0.752	0.79	0.793	0.802
Speed, avg, km/h	86.3	90.8	102.1	105.0	105.8	106.5

VI-DriveSim Dynamic is a high-fidelity, 'hardware and driver-in-the loop', full-motion vehicle simulator developed for the simulation of vehicle handling response. The simulator is the result of a collaborative effort between Multimatic, VI-grade, Concurrent, SimCoVR, and Ansible Motion [6]. Two complete systems have been built to date. Additionally, a complete new driving simulator design that also simulates vehicle ride quality in addition to handling quality has now been developed [13]

VI-DriveSim Dynamic is an integration of real-time vehicle dynamics simulation software, a six-degree-of-freedom (6DOF) motion platform, high-definition graphics, state-of-the-art motion cueing and a highly configurable driver interface. (see Figs. 6.10 and 6.11).

VI-CarRealTime provides real-time, high-fidelity, fully dynamic vehicle motion simulation. VI-DriveSim is used to integrate the analysis engine with an immersive graphics and high resolution visualization program. The motion platform utilizes an innovative compact mechanism to provide decoupled lateral, longitudinal, and yaw motion with large displacement capacity. The pitch, roll, and heave axes are less important for handling dynamics and are coupled.

The resultant lateral, longitudinal, and yaw motions provide the driver's vestibular system with appropriate onset 'cues'. The result is a superior response compared to a hexapod. The visual cues are projected onto a 160°, 1.5 m high, 4 m diameter cylindrical screen, which moves in X–Y, while the projectors move in X–Y and yaw, allowing for a large field of vision and a highly immersive viewing experience.

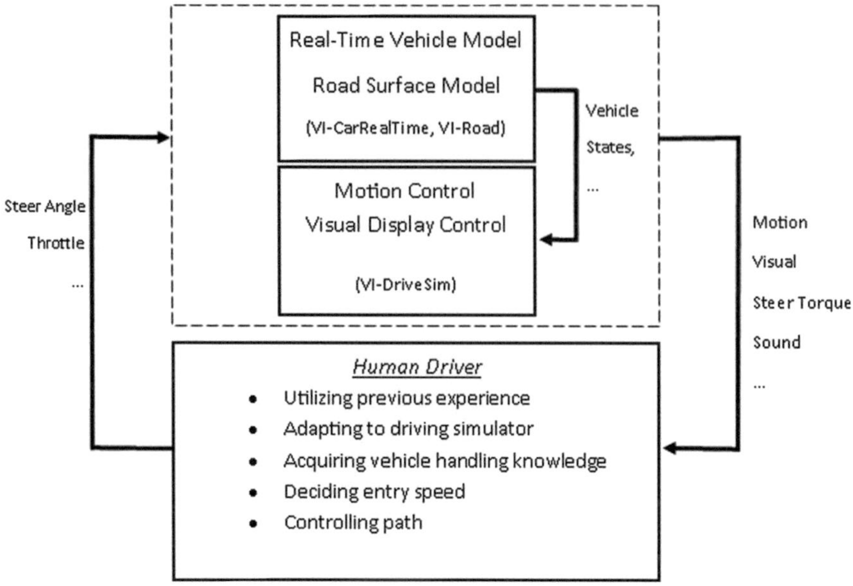

Fig. 6.10 Driving simulator functional block diagram

To achieve the most realistic simulation, particular attention was also given to the steering feedback.

Multimatic Technical Centre and their development driver have run hundreds of real and virtual laps of the company's lidar scanned and meshed test track, Calabogie Motorsports Park, Ontario, with the company's fully characterized parameter or P-Car. The traces in the plot (Fig. 6.12) show excellent correlation, but more importantly, their driver has repeatedly been able to demonstrate that even minor mechanical and aerodynamic changes made on the P-Car can be discerned on the simulator, and vice versa.

6.3.1 Driver Maneuvers in a Controlled Experiment

Prior to the offline optimization study, an experiment was conducted using a professional driver-engineer from the manufacturer of the same compact car used in the optimization study. This driver has a great deal of driving experience with the compact car as well as other vehicles on a physical test course.

The first step of the experiment was to familiarize the driver with the motion simulator environment and the vehicle model. The driver used a simulated skid pad to conduct two types of maneuvers to familiarize himself with the vehicle: constant radius cornering and the lane change maneuver. The driver results are for a double lane change maneuver with a slightly longer layout than the Consumer Reports cone

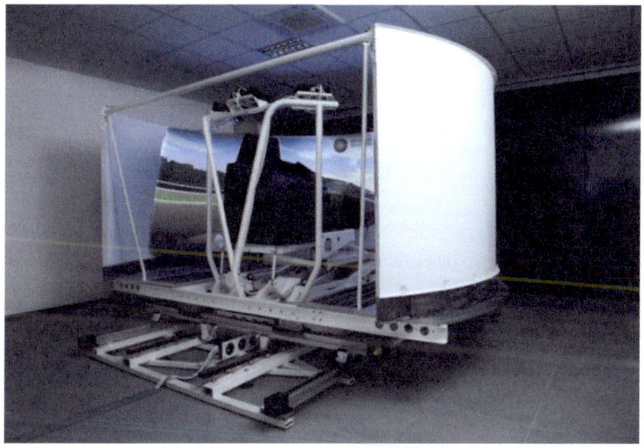

Fig. 6.11 The simulator's highly immersive visual experience

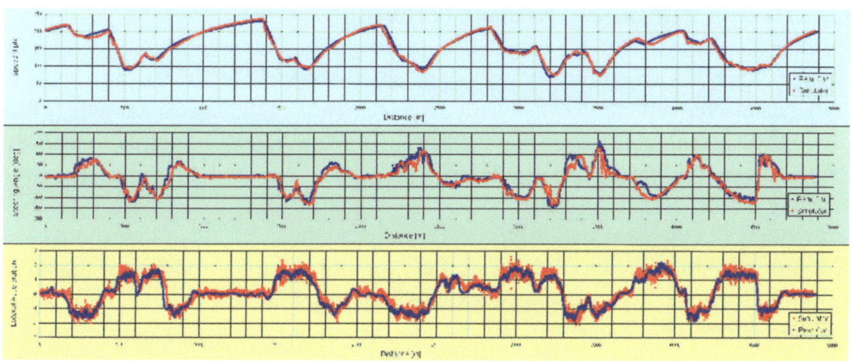

Fig. 6.12 Correlation between the P-Car (*blue*) and the simulation (*red*). The plots show speed (*top*), steering angle (*middle*), and lateral acceleration (*bottom*) for one lap at Calabogie Motorsports Park

spacing. This layout was more familiar to the driver but resulted in scaling differences between the results that are discussed later.

As part of the experiment, changes to the vehicle setup were made to determine if the driver had sensitivity to the change in the vehicle performance. The following changes were made, each acknowledged by the driver:

- Gross vehicle weight change (Added 800 lbm)
- Roll Stiffness Distribution Change by adding 250 N-m/deg rear roll stiffness.
- Reducing the rear lateral force compliance steer by 0.04 deg
- Faster Steering from 16.3 to 15.0
- Rear Toe to Zero (From .38 deg in).

The focus of this chapter is comparing the driver results on the simulator with the optimization results, and not on the assessment of the driving simulator as a subjective vehicle design tool. However, the ultimate intent is to use the driving simulator as a

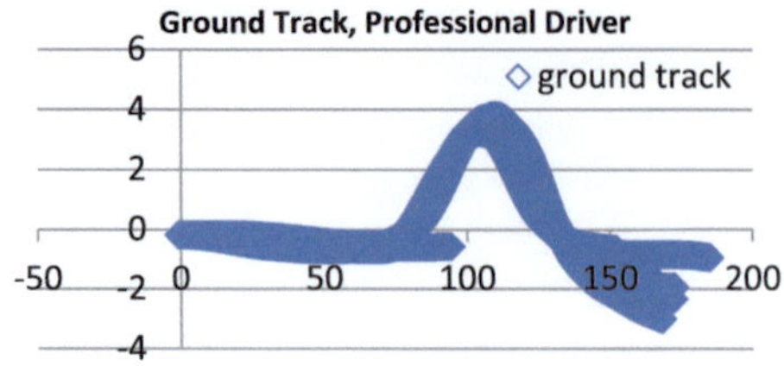

Fig. 6.13 Repeated ground track, professional driver, lane change

supplement to physical proving ground testing using a simulated vehicle model that is closer to the final design than a physical prototype.

6.4 Driving Simulator Results (Online)

Figure 6.13 shows ground track results that are obtained from the driving simulator as raw data. The vertical axis is lateral displacement and the horizontal axis is longitudinal displacement, both in meters. Because the driver has the ability to reset the simulation at any time, he can decide to continue the event until the end or start over, speeding up the learning process.

Figure 6.14 compares the results of the online simulation and offline simulation. In addition to slight layout differences, the professional driver modulated the throttle during the maneuver whereas the simulation was found to hold velocity better, resulting in differences in scaling of the yaw rates. For now, we are considering the general shape of the response to determine if the professional driver learns a similar optimal path and control as the genetic and driver control algorithm.

Figure 6.15 compares the steering angle results of the online simulation, and offline simulation. Again, differences between the offline and online experiment account for differences in scaling of the yaw rates between the two simulations. For now, we are considering the general shape of the steer angle response to determine differences in learning between the professional driver and the genetic algorithm.

6.4.1 Imposing Constraints on Simulated Driver Controls

In an attempt to better match the actual driver performance, two adjustments were made to the driver controls. The first adjustment was to increase the preview distance. The preview distance increases the distance to fit a connecting contour, which was expected to give a smoother control. As shown in Fig. 6.16, this adjustment results in lower and smoother yaw rates at a penalty of a longer transit time.

The second adjustment was to increase the driver anticipation effect, used to compensate for slower vehicle response due to inertia and compliance. An increase in anticipation effect moves the starting reference point of the connecting contour by

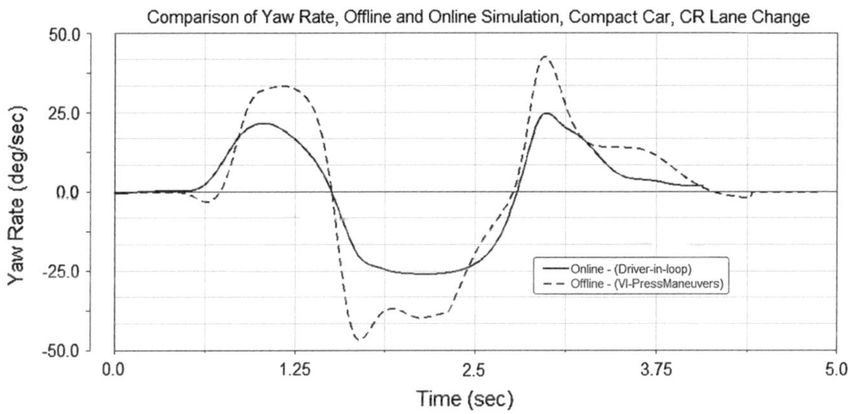

Fig. 6.14 Comparison of yaw rate, online and offline simulation

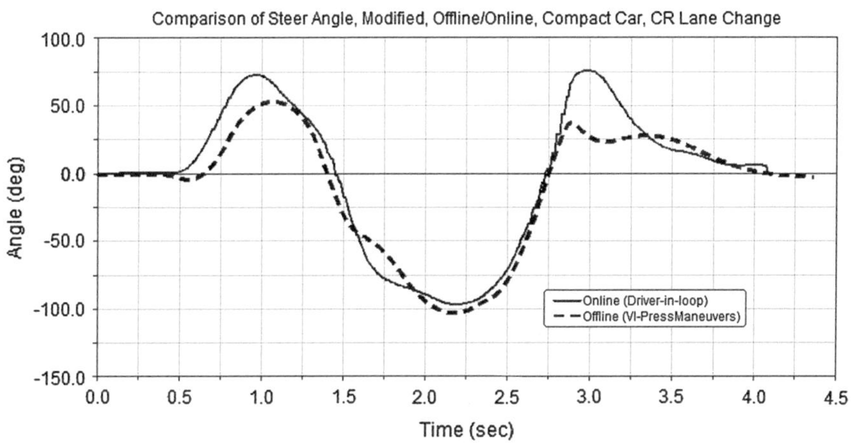

Fig. 6.15 Comparison of steering angle, online and offline simulation

an amount (AC · velocity). The results of this change are shown in Fig. 6.17. This adjustment shows promise as it results in faster maneuver times and in lower driver control requirements.

6.5 Conclusions

This paper has compared an objective method of using simulation-based optimization to determine the "best" automobile design for the lane change maneuver with a subjective method which utilizes a driving simulator.

The objective method is based on running a series of simulations with a vehicle model and determining both the driver steering control and a set of trajectory points

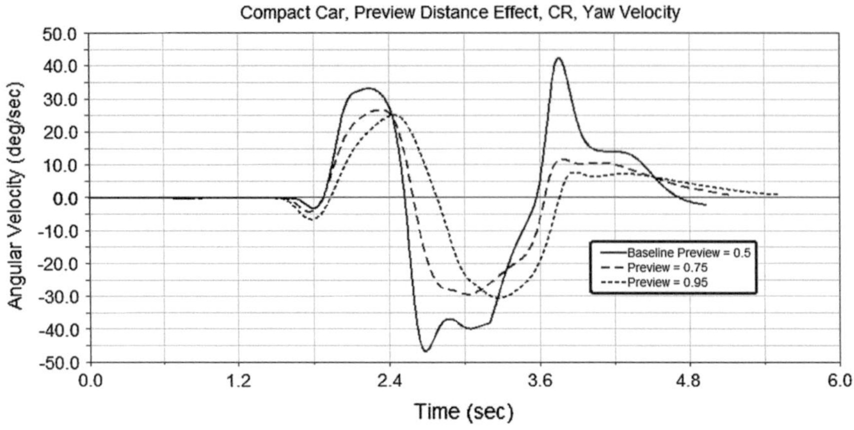

Fig. 6.16 Effect of preview distance changes

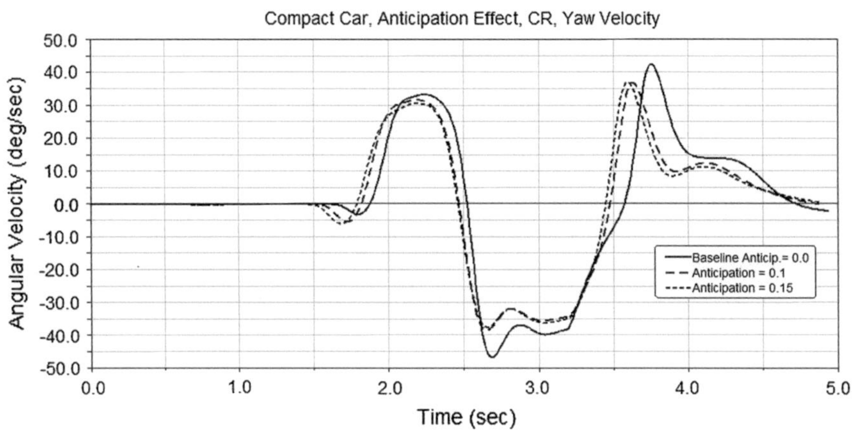

Fig. 6.17 Effect of anticipatory compensation changes

to determine the minimum time maneuver. The subjective method uses the same vehicle model running in a driver-in-the-loop simulation to allow an expert driver to give an assessment of the vehicle.

The results of the two methods were compared, primarily with the intention of improving the objective methods but also to illustrate a new vehicle design process that utilizes both objective and subjective metrics early in the design process, before the creation of physical prototypes. Work is in progress for a systematic comparison using a thoroughly validated vehicle model on the same lane change course for both offline and online simulations. Also, a mathematical formalism is under development for comparing objective and subjective results.

Acknowledgments I would like to thank Davide Bacchet of VI-grade Srl for genetic algorithm design discussions and Peter Gibbons and Lars Ogilvie of Multimatic Technical Centre for driving simulator results.

References

1. Anon. ADAMS/Car User Manual. MSC Software Corporation
2. Anon. VI-CarRealTime User Manual. VI-grade GmbH
3. Bulirsch R, Vögel M, von Stryk O, Chucholowski C, Wolter TM (2003) An optimal control approach to real-time vehicle guidance. In: Jäger W, Krebs H-J (eds) Mathematics—key technology for the future. Springer, Berlin, pp 84–102
4. Estefan JA (2007) Survey of model-based systems engineering (MBSE) methodologies. Incose MBSE Focus Group 25
5. Frezza R, Saccon A, Minen D, Ortmann C (2004) Smart driver: a research project for closed loop vehicle simulation in msc. adams. In: 3rd international symposium on multi-body dynamics: monitoring & simulation techniques (MBD-MST). Loughborough University, Loughborough, Leicestershire, UK
6. Gibbons P (2012) Multimatic for vehicle dynamics international magazine. Annual Showcase 2012 Issue, Product Profile-Innovative Simulator
7. Hatipoglu C, Ozguner U, Unyelioglu KA (1995) On optimal design of a lane change controller. In: Proceedings of the intelligent vehicles' 95 symposium. IEEE, pp 436–441
8. Kolb DA (1984) Experiential learning: experience as the source of learning and development, vol 1. Prentice-Hall, Englewood Cliffs, NJ
9. MacAdam CC (1981) Application of an optimal preview control for simulation of closed-loop automobile driving. IEEE Trans Syst Man and Cybern 11(6):393–399
10. Minen D, Bacchet D (2008) Advanced simulation technology for closed loop limit vehicle handling performance. In: 8th world congress on computational mechanics, Venice
11. Motoyama K, Yamanaka T, Hoshino H, McNally PJ, Yu Q, Shiratori M (2000) A study of automobile suspension design using optimization technique. In: AIAA-2000-4748, 8th AIAA/USAF/NASA/ISSMO symposium on multidisciplinary analysis and optimization
12. Schudt JA, Kodali R, Shah M, Babiak G (2011) Virtual road load data acquisition in practice at general motors. Technical Report 2011–01-0025, SAE Technical Paper
13. VI Grade Press Release (2013) Ferrari adopts revolutionary driving technology from saginomiya. www.vi-grade.com
14. Wang GG (2002) Definition and review of virtual prototyping. J Comput Inf Sci Eng (Trans ASME) 2(3):232–236
15. Wilhelm H, Audi AG (2007) Maximum performance simulation while avoiding cones. In: 1st VI-grade user conference. Marburg, Germany. www.vi-grade.com

Chapter 7
Real-Time Near-Optimal Feedback Control of Aggressive Vehicle Maneuvers

Panagiotis Tsiotras and Ricardo Sanz Diaz

Abstract Optimal control theory Patrick J. can be used to generate aggressive maneuvers for vehicles under a variety of conditions using minimal assumptions. Although optimal control provides a very powerful framework for generating aggressive maneuvers utilizing fully nonlinear vehicle and tire models, its use in practice is hindered by the lack of guarantees of convergence, and by the typically long time to generate a solution, which makes this approach unsuitable for real-time implementation unless the problem obeys certain convexity and/or linearity properties. In this chapter, we investigate the use of statistical interpolation (e.g., kriging) in order to synthesize *on-the-fly* near-optimal feedback control laws from pre-computed optimal solutions. We apply this methodology to the challenging scenario of generating a minimum-time yaw rotation maneuver of a speeding vehicle in order to change its posture prior to a collision with another vehicle, in an effort to remedy the effects of a head-on collision. It is shown that this approach offers a potentially appealing option for real-time, near-optimal, robust trajectory generation.

7.1 Introduction

An enormous amount of work has been devoted during the past three decades to the development of active safety systems for passenger automobiles. This effort has led to the development of a plethora of active safety systems, such as ABS, TCS, ESP, RCS, AFS and others [2, 11, 34, 39], many of which are now standard equipment in

P. Tsiotras (✉)
School of Aerospace Engineering, Georgia Institute of Technology,
Atlanta, GA 30332-0150, USA
e-mail: tsiotras@gatech.edu

R. S. Diaz
Department of Aerospace Engineering, Universidad Politécnica de Valencia, Valencia, Spain
e-mail: risanda@etsid.upv.es

H. Waschl et al. (eds.), *Optimization and Optimal Control in Automotive Systems*,
Lecture Notes in Control and Information Sciences 455,
DOI: 10.1007/978-3-319-05371-4_7,

production vehicles. The main goal of all these systems is to help the driver avoid, or prevent, the so-called "abnormal" driving scenarios (skidding, sliding, excessive under/oversteer, etc). In these conditions, nonlinear effects dominate the vehicle dynamics, and the tire friction is very close to (or exceeds) the adhesion limit(s). Driving at the boundary of the adhesion limits of the tires leads to a reduced operational stability margin for the driver. The main goal of most current active safety systems is therefore to restrict the operational envelope of the vehicle and the tires inside a linear, well-defined, stable regime. This is, however, an overly conservative approach. Enhanced stability comes at the cost of decreased maneuverability. There are many realistic scenarios where the occurrence (or the post-effects) of a collision can be alleviated by allowing (or even inducing) the vehicle to operate in its nonlinear regime *in a controlled manner*.

The previous observations naturally lead one to investigate algorithms that exploit the increased vehicle maneuverability brought about by operating the vehicle in nonlinear and/or unstable regimes. By extending the region of validity of the future generation of active safety systems one expects to increase their performance. In our previous work [6, 7, 35–38] we have investigated the mathematical modeling of vehicles operating in nonlinear and/or unstable regimes, and have demonstrated the potential benefits of such an approach to achieve collision avoidance and mitigation beyond what is possible with current active safety systems.

This point of view represents a philosophical departure from current practice, and differs significantly in scope from standard active safety system design for passenger vehicles. As a result—and understandably so—it brings along with it a slew of unanswered questions; among them, the key question is how to generate the necessary control actions (at the short time scales required) that are needed to perform such extreme maneuvers. Indeed, most drivers—except perhaps expert professional, stunt and race drivers—would have great difficulty initiating an aggressive maneuver and controlling the vehicle throughout the whole maneuver duration.

Optimal control is a powerful framework that has been used successfully in many engineering applications to generate feasible trajectories subject to constraints and complicated system dynamics. The field of numerical optimal control has experienced enormous advances during the recent years, to the point that we now have reliable numerical algorithms to generate optimal trajectories for a variety of practical engineering problems [4]. Despite these advances, the current state-of-the-art in numerical optimal control mainly focuses on generating only open-loop optimal controllers. Furthermore, and unless the underline problem (dynamics, cost) obeys certain convexity and/or linearity conditions, current trajectory optimizers do not allow the computation of optimal trajectories in *real-time*, at least for applications similar to the one we have in mind in this chapter, where the time allotted to solve the problem is in the order of a few milliseconds. One example where fast computation has become possible owing to the current advancement of embedded computing is the area of Model Predictive Control (MPC), where successive linearizations of the plant are used to generate a sequence of linear or convex optimization problems over a finite horizon that can be solved very efficiently on-line [10]. In general, however, optimal solutions for general nonlinear systems and general cost functions are notoriously sensitive to the provided initial guesses and, in the absence of timely re-planning, the robustness of these open-loop optimal control laws is questionable.

Consequently, several researchers have recently turned their attention to the generation of *near-optimal* trajectories using alternative methods, which bypass the exact on-line computations required for the solution of complicated, nonlinear optimal control problems, opting instead for approximate, near-optimal solutions. One typical approach uses interpolation over pre-computed optimal control actions for a variety of initial conditions. Naïve interpolation, however, does not ensure feasibility—let alone optimality—of the resulting interpolated trajectories. In [1], for instance, the authors used traditional interpolation over pre-computed optimal trajectories. However, this method turns out to be inaccurate and time-consuming. Another, more promising, approach is the one proposed in [14], where the optimal control problem is cast as one of *metamodeling*, in which the (unknown) map between control inputs/system response pairs is generated implicitly via a series of computer experiments. Specifically, the approach in [14] considers the solution to an optimal control problem obtained by numerical methods as the output of such a metamodel obtained by a series of off-line simulations. A vast number of publications about metamodeling of computer experiments can be found in the literature. Most of them are motivated by the low time-consuming optimization process, derived from having a metamodel of a given simulation.

In contrast to [1], the framework in [14] is based on rigorous interpolation between the off-line solutions (the "metamodel") using ideas from statistical interpolation theory via Gaussian processes, which in geostatistics it is also known as kriging [8, 16, 22, 33]. Kriging approximates a function observed at a set of discrete points with a convex combination of the observations so as to reduce the least mean-squared error (MSE), and is a special case of prediction using Gaussian processes [16, 22]. Although classical interpolation focuses on low-order polynomial regression, which is suitable for sensitivity analysis, kriging is an interpolation technique that provides better global predictions than classical methods [20, 33]. In this work, we use kriging to construct a (near-)optimal *feedback* controller from off-line computed extremal trajectories. Prior use of kriging has been focused mainly on simulation and metamodeling [18, 30, 32]. A brief overview of interpolation using Gaussian processes and kriging is given in Sect. 7.3.

We apply a technique similar to the one proposed in [14] to obtain near-optimal "feedback" controllers for the problem of minimum-time aggressive yaw maneuver generation for a high-speed vehicle impeding a collision with another vehicle at an intersection (T-Bone collision). Our results show that kriging interpolation is able to generate very accurate parameterized trajectories in real-time, and hence it may be a potential option for real-time, near-optimal trajectory generation under such extreme driving conditions, where the time constraints do not allow the computation of an exact optimal trajectory in a timely manner using current state of technology.

Prior similar work that uses parameterized trajectory generation includes [9], which developed an algorithm to generate a whole set of trajectories between two pre-computed solutions for two different initial conditions, and [31], where parameterized trajectories were generated using experimental demonstrations of the maneuver.

However, the control laws obtained in [9, 31] are open-loop and thus susceptible to uncertainties in the initial conditions and unknown model parameters. The advantage of the method described in this chapter is that the control is obtained as a function of the actual state, hence is a "feedback" control.

The chapter is structured as follows. In the next section the problem to be investigated is introduced, along with the dynamical model of the vehicle and the tire friction dynamics. Next, the optimal control problem is formulated, which is solved over a discrete grid of initial conditions. This series of generated solutions at several discrete points is stored in memory, and is used in Sect. 7.4 to generate a feedback control by interpolating between the stored solutions on-line using kriging. For the benefit of the uninformed reader, a brief summary of kriging theory as used in this chapter is given in Sect. 7.3. In Sect. 7.4.2 we present numerical results from the application of the proposed approach to the problem of T-Bone collision mitigation at an intersection between two speeding vehicles, as a demonstration of the possibilities enabled by the proposed approach for optimal on-line controller generation.

7.2 Aggressive Yaw Maneuver of a Speeding Vehicle

7.2.1 Problem Statement

One of the most lethal collisions between two speeding vehicles is the so-called "T-bone" collision (Fig. 7.1), which occurs when one of the vehicles drives into the side of the other vehicle [29]. The vehicle suffering the frontal impact is often referred to as the "bullet" vehicle, while the one suffering the side impact is said to have been "T-boned." If there is inadequate side impact protection, the occupants of a T-boned vehicle risk serious injury or even death.

Although the bullet vehicle is driving much faster, this collision scenario is specially dangerous for the driver or the side passenger of the target vehicle. This is owing to the fact that the requirements in terms of frontal crashworthiness of cars on the market nowadays is excellent [12]. Frontal-crash tests are carried out at velocities up to 64 km/h, with the result of the passengers cabin being almost intact. The suitable design and choice of materials of the front part of the vehicle allows large structural deformations and thus absorption of the residual energy during impact. Moreover, the installation of frontal airbags, mandatory in the US since September 1998, has resulted in a great decrease of deaths and injuries owing to frontal collisions. On the other hand, the side part of the chassis is structurally weak, and large deformations would result in lethal injuries for the occupants. Unfortunately, side airbags are currently available only in upscale or mid-range cars, although it is envisioned that they will also become standard safety equipment for all passenger vehicles in the future. Reference [13] offers a detailed study on occupant injuries during side impact crashes. As expected, the most frequent source of severe injuries is the contact between the chest and the door panel.

Fig. 7.1 T-Bone collision

Car manufactures are aware of the high risk involved in side collisions. Volvo, for instance, in 1991 introduced a special protection system against side collision named Side Impact Protection System (SIPS). Other car manufacturers have introduced similar passive safety systems.

In our previous work [6, 7] we investigated the possibility of mitigating the results from an unavoidable T-Bone collision by using an aggressive yaw maneuver for the incoming bullet vehicle. The proposed collision mitigation maneuver involves a rapid yaw rotation of the bullet vehicle at an approximately 90 deg angle that brings the longitudinal axes of the two vehicles into a nearly parallel alignment, in order to distribute the residual kinetic energy of the collision over a larger surface area, thus mitigating its effects. Although this represents a worst-case scenario, where the target vehicle does not respond (a more optimal strategy would involve a rapid yaw maneuver of the target vehicle as well), our initial study focuses only on the case when the bullet vehicle is actively maneuvered during the pre-collision phase. The generalization to the case when both vehicles collaboratively try to avoid the collision will probably involve some vehicle-to-vehicle (V2V) communication and it is left for future investigation. Henceforth, we thus only consider the problem when only one (the bullet) vehicle is actively controlled. This problem was posed in [6, 7] as a time-optimal control problem, and it was solved using pseudospectral methods [25]. In the next two sections we briefly summarize the problem definition and its numerical solution.

7.2.2 Vehicle and Tire Model

The model used in this chapter is the so-called "bicycle model" [26], augmented with wheel dynamics. The nomenclature and conventions regarding this model are shown in Fig. 7.2. The state is given by $\mathbf{x} = [u, v, r, \psi, \omega_f, \omega_r]^{\mathsf{T}}$, where u and v are, respectively, the body-fixed longitudinal and lateral velocities, r is the vehicle yaw rate, ψ is the vehicle heading, and $\omega_f \geq 0$ and $\omega_r \geq 0$ are the angular speeds of the

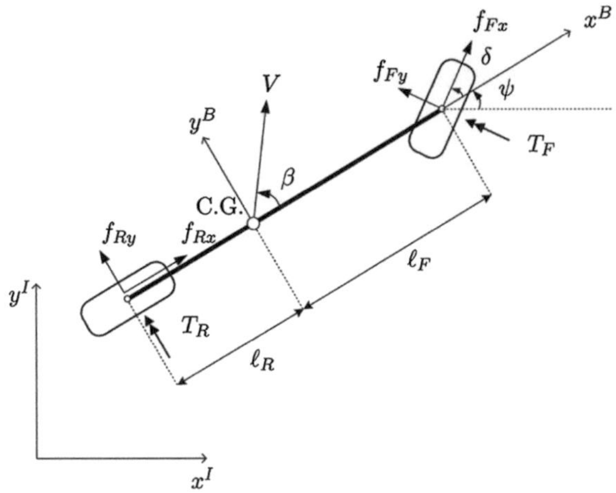

Fig. 7.2 Schematic of bicycle model

front and rear wheels, respectively. The system is controlled by $\mathbf{u} = [\delta, T_b, T_{hb}]^\mathsf{T}$, where δ is the steering angle and T_b, T_{hb} denote the torques generated by the footbrake and handbrake, respectively.

The equations of motion of the vehicle can be written as shown in (7.1)–(7.4)

$$\dot{u} = \frac{1}{m}(F_{xf}\cos\delta - F_{yf}\sin\delta + F_{xr}) + vr, \tag{7.1}$$

$$\dot{v} = \frac{1}{m}(F_{xf}\sin\delta + F_{yf}\cos\delta + F_{yr}) - ur, \tag{7.2}$$

$$\dot{r} = \frac{1}{I_z}\big(\ell_f(F_{xf}\sin\delta + F_{yf}\cos\delta) - \ell_r F_{yr}\big), \tag{7.3}$$

$$\dot{\psi} = r, \tag{7.4}$$

along with the wheel dynamics

$$\dot{\omega}_f = \frac{1}{I_w}(T_{bf} - F_{xf}R), \tag{7.5}$$

$$\dot{\omega}_r = \frac{1}{I_w}(T_{br} - F_{xr}R), \tag{7.6}$$

where m, I_z are, respectively, the mass and yaw moment of inertia of the vehicle, I_w is the rotational inertia of each wheel about its axis, R is the effective tire radius, and ℓ_f, ℓ_r are, respectively, the distances of the front and rear axles from the vehicle center of mass. In (7.1)–(7.6) F_{ij} $(i = x,\ y;\ j = f,\ r)$ denote the longitudinal and lateral force components developed by the tires, defined in a tire-fixed reference frame.

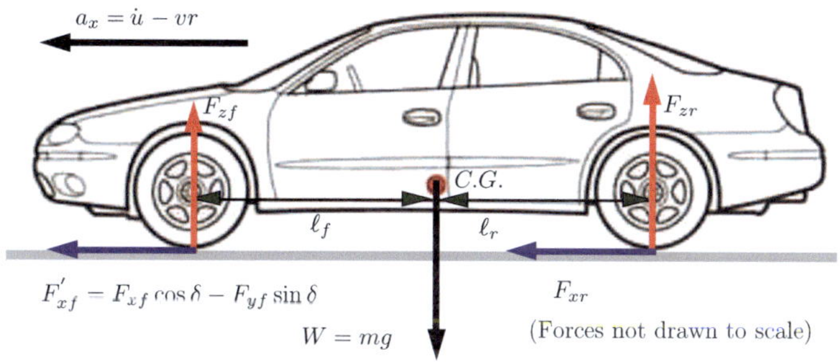

Fig. 7.3 Longitudinal load transfer force distribution

These forces depend on the normal loads on the front and rear axles, F_{zf} and F_{zr}, given by

$$F_{zf} = \frac{mg\ell_r - hmg\mu_{xr}}{\ell_f + \ell_r + h(\mu_{xf}\cos\delta - \mu_{yf}\sin\delta - \mu_{xr})}, \tag{7.7}$$

$$F_{zr} = \frac{mg\ell_f + hmg(\mu_{xf}\cos\delta - \mu_{yf}\sin\delta)}{\ell_f + \ell_r + h(\mu_{xf}\cos\delta - \mu_{yf}\sin\delta - \mu_{xr})} \tag{7.8}$$

where h is the distance of the vehicle center of mass from the ground (see Fig. 7.3), and where

$$\mu_j = D\sin(C\arctan(Bs_j)), \quad \mu_{ij} = -(s_{ij}/s_j)\mu_j, \quad i = x, y;\ j = f, r, \tag{7.9}$$

for some constants C, B and D. Expression (7.9) is a simplified version of the well-known Pacejka "Magic Formula" (MF) [24] for the tire friction modeling, and combines the longitudinal and lateral motion, thus intrinsically incorporating the non-linear effect of the lateral/longitudinal coupling also known as the "friction circle" (see Fig. 7.4), according to which, the constraint $F_{x,j}^2 + F_{y,j}^2 \le F_{\max,j}^2 = (\mu_j F_{z,j})^2$ ($j = f, r$) couples the allowable values of longitudinal and lateral tire friction forces.

Incorporating the friction circle constraint is necessary for the correct modeling of the dynamics occurring during the aggressive maneuvers we consider in this work.

In Eq. (7.9) s_{ij} denote the tire longitudinal and lateral slip ratios, given by

$$s_{xj} = \frac{V_{xj} - \omega_j R}{\omega_j R} = \frac{V_{xj}}{\omega_j R} - 1, \quad s_{yj} = (1 + s_{xj})\frac{V_{yj}}{V_{xj}}, \qquad j = f, r, \tag{7.10}$$

where the longitudinal and lateral velocity components, defined in the tire-fixed reference frame, are given by

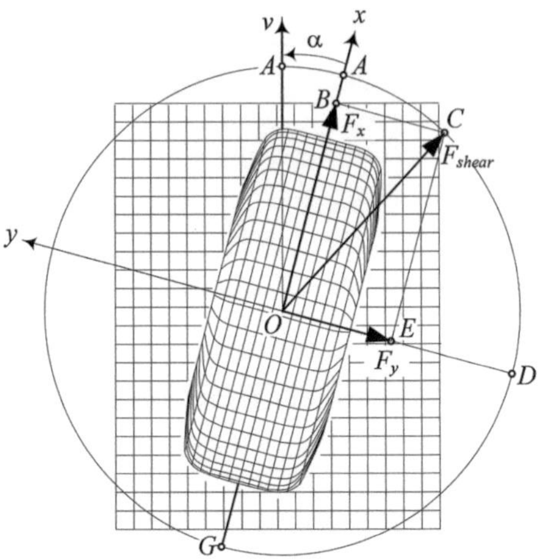

Fig. 7.4 The friction circle concept (from [19])

$$V_{xf} = u\cos\delta + v\sin\delta + r\ell_f \sin\delta, \quad V_{yf} = -u\sin\delta + v\cos\delta + r\ell_f\cos\delta, \tag{7.11}$$

$$V_{xr} = u, \qquad V_{yr} = v - r\ell_r, \tag{7.12}$$

and s denotes the total slip, computed as $s_j = (s_{xj}^2 + s_{yj}^2)^{\frac{1}{2}}$, $(j = f, r)$. Finally, the tire forces in (7.1)–(7.6) are computed by $F_{ij} = F_{zj}\mu_{ij}$, $(i = x, y;\ j = f, r)$.

Following current vehicle technology, it is assumed that handbrake torque is only applied on the rear axle and the footbrake torque is distributed to both axles by a factor γ_b, according to $T_{bf}/T_{br} = (1-\gamma_b)/\gamma_b$, so that $T_{bf} = -(1-\gamma_b)T_b$ and $T_{br} = -\gamma_b T_b - T_{hb}$. It is further assumed that the controls are bounded in magnitude between upper and lower bounds as follows

$$\delta_{\min} \le \delta \le \delta_{\max}, \quad 0 \le T_b \le T_{b,\max}, \quad 0 \le T_{hb} \le T_{hb,\max}, \tag{7.13}$$

which define the allowable control constraint set, $\mathbf{u} \in \mathscr{U} \subset \mathbb{R}^3$. For more details on the vehicle and tire model used in this work, the reader is referred to [7, 35, 38].

7.2.3 Optimal Control Formulation

Assuming that the vehicle is initially moving on a straight line along the positive x direction with velocity $V_0 = u(0)$, our main goal is to find the control input history $\mathbf{u}(t)$ to bring the posture of the vehicle to $\psi(t_f) = 90°$ as fast as possible. Without loss of generality, it will be assumed that the angular velocity of the front and rear wheels is such that a no-slip condition is satisfied, i.e., $\omega_f(0) = \omega_r(0) = V_0/R$.

We therefore wish to solve the following optimal control problem

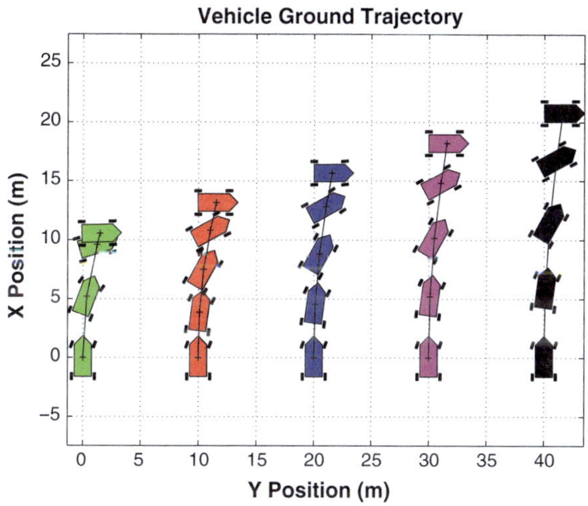

Fig. 7.5 Pre-computed solutions for different initial vehicle velocities (dry asphalt case, corresponding to $\mu = 0.8$)

$$\min_{\mathbf{u}\in\mathscr{U}} J = \int_0^{t_f} \mathrm{d}t, \tag{7.14}$$

$$\text{s.t.} \quad \dot{\mathbf{x}} = f(\mathbf{x}, \mathbf{u}), \tag{7.15}$$

$$\mathbf{x}(0) = [V_0,\ 0,\ 0,\ 0,\ V_0/R,\ V_0/R]^{\mathsf{T}}, \tag{7.16}$$

$$\psi(t_f) = \pi/2, \tag{7.17}$$

where $f(\mathbf{x}, \mathbf{u})$ is given by the right-hand side of (7.1)–(7.6) and, besides ψ, the rest of final states are free.

This problem can be solved using a variety of numerical methods [3–5, 17, 23, 27, 28]. In this work, we have used the package GPOPS based on pseudospectral methods to solve the previous optimal control problem [25]. The problem was solved for a variety of initial conditions. A typical maneuver obtained by the solution of the optimal control problem is shown in Fig. 7.5. For more details, the interested reader is referred to [7]. Table 7.1 summarizes the vehicle model data and tire parameters used in the numerical examples of Sect. 7.4.2.

In the sequel, we focus on generating optimal solutions for different values of initial conditions by interpolating between the pre-computed optimal trajectories. The interpolation method we use is based on representing the input (initial conditions) and output [control commands obtained from the numerical solution of the optimal control problem (7.14)–(7.17)] as a realization of a (hidden) Gaussian process. The goal is then to find the unknown parameters of this Gaussian process in order to predict the optimal control inputs for different problem parameters. Although we only present the results for different initial conditions, the approach can be easily

Table 7.1 Vehicle and tire data used in the numerical simulations

Variable	Value	Unit	Variable	Value	Unit
m	1,245	Kg	B	7	–
I_z	1,200	Kg m^2	C	1.4	–
I_w	1.8	Kg m^2	$\delta_{\max} = -\delta_{\min}$	45	deg
ℓ_f	1.1	m	$T_{b,\max}$	3,000	Nm
ℓ_r	1.3	m	$T_{hb,\max}$	1,000	Nm
h	0.58	m	γ_b	0.4	–
R	0.29	m	g	9.81	m/s^2

generalized to the case of different vehicle or road parameters, as long as we have enough data points within the range of the parameters of interest.

7.3 Statistical Interpolation Using Gaussian Processes

7.3.1 Basic Theory

The basic idea behind statistical interpolation is that the actual values for all possible observations are a *realization* from an underlying stochastic process [16]. It is essentially an interpolation technique over random data fields and it provides accurate interpolation even if there is no a priori trend. Kriging is a common term referred to the case when the underlying statistical process is Gaussian. The basic idea that differentiates kriging from the traditional Generalized Least Squares (GLS) approach is the assumption that, given a point where a prediction is to be made, points closer to this new point should have a larger weight, i.e., they should have more influence on the prediction than points that are further away. This implies that the interpolation weights are not constant, but rather they must be specifically computed at each new location.

A kriging interpolation model has the following features:

(a) It is unbiased, i.e., the expected value of the error is zero.
(b) It is optimal, in the sense that minimizes the variance of the error.
(c) It provides exact interpolation, i.e., the predicted output values at the already observed points are equal to the observations.
(d) It is computationally very efficient, hence on-line implementation is feasible.

Below we briefly summarize the basic ingredients of the approach. The discussion in this section is taken mainly from [15]. In order to understand how statistical prediction works, let us consider a set of given locations $\mathbf{X} = [\mathbf{x}_1 \ \dots \ \mathbf{x}_N] \in \mathbb{R}^{n \times N}$ with $\mathbf{x}_i \in \mathbb{R}^n$, where an unknown function $y : \mathbb{R}^n \to \mathbb{R}$ is observed. A simple regression model is to assume that

$$y(\mathbf{x}) = \sum_{k=1}^{r} \beta_k f_k(\mathbf{x}) + z = f(\mathbf{x})^\mathsf{T}\beta + z, \tag{7.18}$$

for some basis functions (regressors) $f(\mathbf{x}) = [f_1(\mathbf{x}) \ldots f_r(\mathbf{x})]^\mathsf{T}$, where $\beta = [\beta_1 \ldots \beta_r]^\mathsf{T} \in \mathbb{R}^r$ is the vector of regression coefficients, and $z \in \mathbb{R}$ is the observation error. Let now $\mathbf{y} = [y(\mathbf{x}_1) \ldots y(\mathbf{x}_N)]^\mathsf{T} = [y_1 \ldots y_N]^\mathsf{T} \in \mathbb{R}^N$ be the vector of observations. The generalized regression model given the data $(\mathbf{y}, \mathbf{x}_1, \ldots, \mathbf{x}_N)$ follows easily from (7.18)

$$\mathbf{y} = F(\mathbf{X})\beta + \mathbf{z}, \tag{7.19}$$

where $\mathbf{z} = [z_1 \ldots z_N]^\mathsf{T} \in \mathbb{R}^N$ is the vector of observation errors, and $F(\mathbf{X}) \in \mathbb{R}^{N\times r}$ is the matrix of regressors, given by

$$F(\mathbf{X}) = [f(\mathbf{x}_1)^\mathsf{T} \ldots f(\mathbf{x}_N)^\mathsf{T}]^\mathsf{T} = \begin{bmatrix} f_1(\mathbf{x}_1) & f_2(\mathbf{x}_1) & \ldots & f_r(\mathbf{x}_1) \\ f_1(\mathbf{x}_2) & f_2(\mathbf{x}_2) & \ldots & f_r(\mathbf{x}_2) \\ \vdots & \vdots & \ddots & \vdots \\ f_1(\mathbf{x}_N) & f_2(\mathbf{x}_N) & \ldots & f_r(\mathbf{x}_N) \end{bmatrix}. \tag{7.20}$$

In statistical prediction the errors $\mathbf{z}$ in (7.19) are modeled as a stationary covariance stochastic process[1] having the properties

$$\mathbb{E}[\mathbf{z}] = 0, \tag{7.21}$$

$$\mathrm{cov}[\mathbf{z}] = \mathbb{E}[\mathbf{z}\mathbf{z}^\mathsf{T}] = \mathbf{C} = \sigma^2\mathbf{R}, \tag{7.22}$$

where $\mathbf{C}, \mathbf{R} \in \mathbb{R}^{N\times N}$ are the covariance and correlation matrices, respectively, defined by

$$\mathbb{E}[z_i z_j] = \mathbf{C}_{ij} = \sigma^2\mathbf{R}_{ij}(\mathbf{x}_i, \mathbf{x}_j), \quad i, j = 1, \ldots, N. \tag{7.23}$$

where $\mathbf{R}_{ij}(\mathbf{x}_i, \mathbf{x}_j)$ are stationary correlation functions to be defined later.

Suppose now that we want to predict the value $y(\mathbf{x}_0)$ at the new location $\mathbf{x}_0 \in \mathrm{co}(\mathbf{x}_1, \mathbf{x}_2, \ldots, \mathbf{x}_N)$, where $\mathrm{co}(\cdot)$ denotes convex hull. From (7.18), the predicted value of $y(\mathbf{x}_0)$ is then given by

$$y(\mathbf{x}_0) = f(\mathbf{x}_0)^\mathsf{T}\beta + z_0, \tag{7.24}$$

where the scalar z_0 represents the prediction error. Here is where kriging and GLS differ. The later assumes that both the sample disturbances in (7.18) and the predictor disturbance in (7.24) are independent, that is, $\mathrm{cov}[\mathbf{z}, z_0] = 0$. However, in view of the interdependence of disturbances in the samples ($\mathbf{C}$ has non-zero off-diagonal elements), it seems more reasonable to assume that [15]

$$\mathbb{E}[z_0] = 0, \tag{7.25}$$

[1] A stationary covariance process has constant mean and variance and the covariance matrix depends only on the distance between the corresponding inputs.

$$\mathrm{cov}[z_0] = \mathbb{E}[z_0^2] = \sigma^2, \tag{7.26}$$

$$\mathrm{cov}[\mathbf{z}, z_0] = \sigma^2 r(\mathbf{x}_0), \tag{7.27}$$

where $r(\mathbf{x}_0) \in \mathbb{R}^N$ is the vector of correlations between $\mathbf{z}$ and z_0.

Assuming now that the optimal linear predictor of (7.24) can be written in terms of the observed values, one obtains

$$\hat{y}(\mathbf{x}_0) = \sum_{i=1}^{N} w_i y_i = \mathbf{w}^\mathsf{T}\mathbf{y}, \tag{7.28}$$

where $\mathbf{w} = [w_1 \ \dots \ w_N]^\mathsf{T} \in \mathbb{R}^N$ is the column vector of weights. The residual error of the approximation is given by

$$\varepsilon(\mathbf{x}_0) = \hat{y}(\mathbf{x}_0) - y(\mathbf{x}_0) = \sum_{i=1}^{N} w_i y_i - y(\mathbf{x}_0). \tag{7.29}$$

In order to determine the optimal weights $\mathbf{w}$ kriging imposes the conditions [15, 20]

$$\min_{\mathbf{w}} \ \mathrm{var}[\varepsilon(\mathbf{x}_0)] \qquad \text{s.t.} \quad \mathbb{E}[\varepsilon(\mathbf{x}_0)] = 0, \tag{7.30}$$

to obtain the Best Linear Unbiased Predictor (BLUP). In some texts [20, 21] the criterion involves the minimization of the mean square error instead. It turns out that both criteria are equivalent if the estimator is unbiased.

The minimization problem in (7.30) can be re-written as a quadratic programming (QP) problem in the form

$$\begin{aligned} &\min_{\mathbf{w}} \ \mathrm{var}[\varepsilon(\mathbf{x}_0)] = \min_{\mathbf{w}} \sigma^2(1 + \mathbf{w}^\mathsf{T}\mathbf{R}\mathbf{w} - 2\mathbf{w}^\mathsf{T} r(\mathbf{x}_0)), \\ &\text{subject to} \quad F(\mathbf{X})^\mathsf{T}\mathbf{w} - f(\mathbf{x}_0) = 0, \end{aligned} \tag{7.31}$$

whose solution is readily obtained as follows

$$\mathbf{w}^* = \mathbf{R}^{-1}(r(\mathbf{x}_0) - F(\mathbf{X})\lambda^*), \tag{7.32}$$

$$\lambda^* = \left(F(\mathbf{X})^\mathsf{T}\mathbf{R}^{-1}F(\mathbf{X})\right)^{-1}\left(F(\mathbf{X})^\mathsf{T}\mathbf{R}^{-1}r(\mathbf{x}_0) - f(\mathbf{x}_0)\right). \tag{7.33}$$

Using the previous expressions, one may finally express the best linear unbiased predictor of (7.28) as

$$\hat{y}(\mathbf{x}_0) = \mathbf{R}^{-1}\left[r(\mathbf{x}_0) - F(\mathbf{X})\left(F(\mathbf{X})^\mathsf{T}\mathbf{R}^{-1}F(\mathbf{X})\right)^{-1}\left(F(\mathbf{X})^\mathsf{T}\mathbf{R}^{-1}r(\mathbf{x}_0) - f(\mathbf{x}_0)\right)\right]\mathbf{y}. \tag{7.34}$$

A deeper insight in the predictor can be obtained by expressing (7.34) as

$$\hat{y}(\mathbf{x}_0) = f(\mathbf{x}_0)^{\mathsf{T}}\beta^* + r(\mathbf{x}_0)\gamma^*, \tag{7.35}$$

where

$$\beta^* = (F(\mathbf{X})^{\mathsf{T}}\mathbf{R}^{-1}F(\mathbf{X}))^{-1}F(\mathbf{X})^{\mathsf{T}}\mathbf{R}^{-1}\mathbf{y}, \qquad \gamma^* = \mathbf{R}^{-1}(\mathbf{y} - F(\mathbf{X})\beta^*). \tag{7.36}$$

The term β^* is the GLS solution to the regression problem $\mathbf{y} \approx F(\mathbf{X})\beta$, also known as Aitken's GLS estimator [15]. From (7.35) it can be seen that, if independence of the disturbances is considered,that is, $r(\mathbf{x}_0) = 0$, then the solution becomes equivalent to GLS. Another important point is that β^* and γ^* are fixed for a given set of design data $\mathbf{x}_1, \mathbf{x}_2, \ldots, \mathbf{x}_N$ and $\mathbf{y}$. Thus the computational effort required to calculate the value of the interpolated function at one point involves only the computation of two vectors (by evaluating the regression basis functions and the correlation function) and two simple products.

As mentioned previously, (7.35) is an exact interpolator, in the sense that it returns the observed value at the design points. This can be easily shown from (7.35) by choosing $\mathbf{x}_0 = \mathbf{x}_i$. Then $r(\mathbf{x}_i)$ is just the ith column of the correlation matrix $\mathbf{R}$. Hence $\mathbf{R}^{-1}r(\mathbf{x}_i) = e_i$ where e_i is the ith column of the identity matrix. It follows that

$$\begin{aligned}\hat{y}(\mathbf{x}_i) &= f(\mathbf{x}_i)^{\mathsf{T}}\beta^* + r(\mathbf{x}_i)\mathbf{R}^{-1}\left(\mathbf{y} - F(\mathbf{X})\beta^*\right) \\ &= f(\mathbf{x}_i)^{\mathsf{T}}\beta^* + e_i(\mathbf{y} - F(\mathbf{X})\beta^*) \\ &= f(\mathbf{x}_i)^{\mathsf{T}}\beta^* + y_i - f(\mathbf{x}_i)^{\mathsf{T}}\beta^* = y_i.\end{aligned} \tag{7.37}$$

7.3.2 Choice of Correlation Functions

It is important to emphasize that the accuracy of the method is highly dependent on the choice of correlation functions in (7.23) and (7.27), since they determine the influence of the observed values in the surrounding locations. These are not known a priori, however, and they have to be estimated from the data. In order to find a way to approximate the correlation functions, it is customary to assume that they can be expressed as

$$\mathbf{R}_{ij}(\theta; \mathbf{x}_i, \mathbf{x}_j) = \prod_{k=1}^{n} \rho(\theta; \mathbf{x}_i^{(k)}, \mathbf{x}_j^{(k)}) = \prod_{k=1}^{n} \rho(\theta; |\mathbf{x}_i^{(k)} - \mathbf{x}_j^{(k)}|), \tag{7.38}$$

for some parameter θ and $\mathbf{x}_i, \mathbf{x}_j \in \mathbb{R}^n$ with $\mathbf{x}^{(k)}$ denoting the kth component of the vector $\mathbf{x}$. The expression (7.38) implies that multi-dimensional correlations are expressed as a product of n one-dimensional correlation functions. Spatial correlation functions depend on both the parameter θ and the distance between the considered points $\ell = |\mathbf{x}_i^{(k)} - \mathbf{x}_j^{(k)}|$. In order to result in proper correlation functions $\mathbf{R}_{ij}$, the coordinate correlation function ρ must satisfy $0 \leq \rho(\theta; \ell) \leq 1$ for all $\ell \geq 0$. Furthermore, it must satisfy $\rho(\theta; 0) = 1$ and $\lim_{\ell \to \infty} \rho(\theta; \ell) = 0$, encoding the fact that far-away points have weaker or no correlation, whereas coincident points yield maximum correlation.

The parameter θ determines how fast the correlation function goes to zero. This parameter can be obtained using Maximum Likelihood Estimation (MLE). Figure 7.6 shows the effect of θ is the response surface for the function $f(x, y) = x^2 + y^3$. For simplicity, a constant polynomial is selected as the general trend. Different kriging models are built for different values of θ. The resulting metamodel representation and the observation points for the decreasing values of θ are also shown in Fig. 7.6.

The spatial evolution according to the distance from the origin and the influence of the parameter θ, for different correlation functions, is shown in Fig. 7.7. As it is customary in practice, the state variables are normalized so that have unit length. Consequently, the normalized support ($|d| = \ell$) of ρ is this figure is $0 \leq |d| \lesssim 2$.

7.4 Application to On-line Aggressive Vehicle Maneuver Generation

7.4.1 Feedback Controller Synthesis

Using the method outlined in Sect. 7.2.3, a set of trajectories was computed offline using five equidistant initial conditions corresponding to vehicle initial speeds $V_0 = [40,\ 48,\ 56,\ 64,\ 72]$ km/h. The pre-computed open-loop optimal trajectories for three of the cases considered are shown in Fig. 7.8.

We are interested in obtaining a controller able to perform the maneuver described in Sect. 7.2.1 in a (near-)optimal manner for *any* initial velocity in the interval $40\ \text{km/h} \leq V_0 \leq 72\ \text{km/h}$. To this end, we use the interpolation expressions derived in Sect. 7.3.1, specifically, Eq. (7.35). A separate interpolation model is needed for each variable we want to interpolate. In this case we have a total of four interpolating metamodels: three for the control signals and one more for the optimal final time. A uniform discretization of the optimal trajectories provides the input data $\mathbf{X} = [\mathbf{x}_1 \ldots \mathbf{x}_N]$, whereas the control commands δ, T_b, T_{bh}, and the final time t_f comprise the vector of the observed variables $\mathbf{y}$ at the same time instances. Given now a state $\mathbf{x}_0 = (u(t_k), v(t_k), \psi(t_k), \omega_f(t_k), \omega_f(t_k))$ of the vehicle trajectory at some time t_k, we obtain the required control inputs as a function of the current state as follows

$$\delta(t_k) = \hat{y}_1(\mathbf{x}_0) = \kappa_1(u(t_k), v(t_k), \psi(t_k), \omega_f(t_k), \omega_f(t_k)), \tag{7.39}$$

$$T_b(t_k) = \hat{y}_2(\mathbf{x}_0) = \kappa_2(u(t_k), v(t_k), \psi(t_k), \omega_f(t_k), \omega_f(t_k)), \tag{7.40}$$

$$T_{bh}(t_k) = \hat{y}_3(\mathbf{x}_0) = \kappa_3(u(t_k), v(t_k), \psi(t_k), \omega_f(t_k), \omega_f(t_k)), \tag{7.41}$$

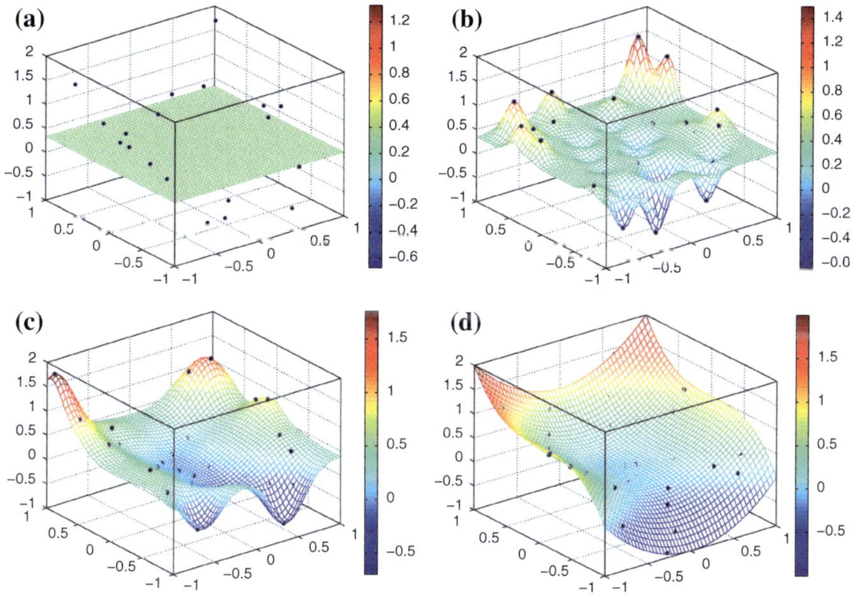

Fig. 7.6 Influence of the parameter θ in the response surface. **a** $\theta = 1{,}000$, **b** $\theta = 20$, **c** $\theta = 5$ **d** θ obtained from MLE

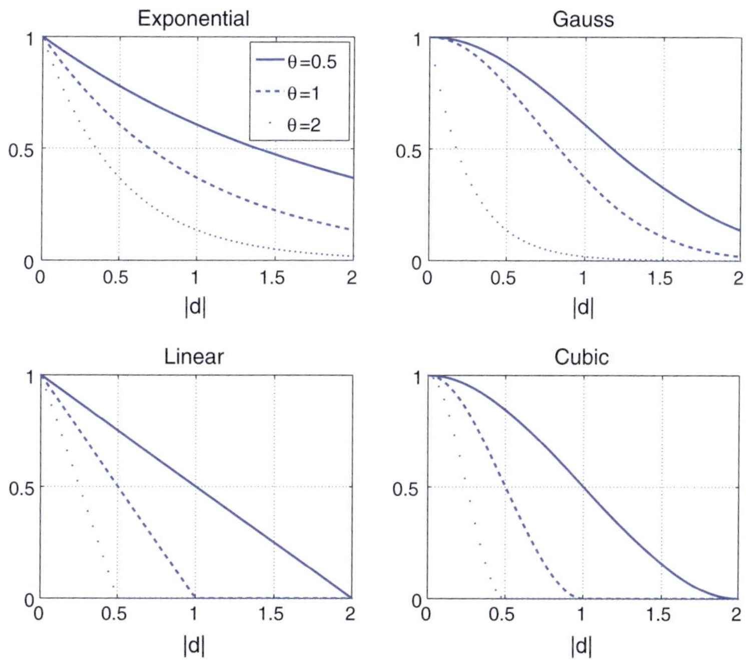

Fig. 7.7 Different possible choices for correlation functions $\rho(\theta; |d|)$

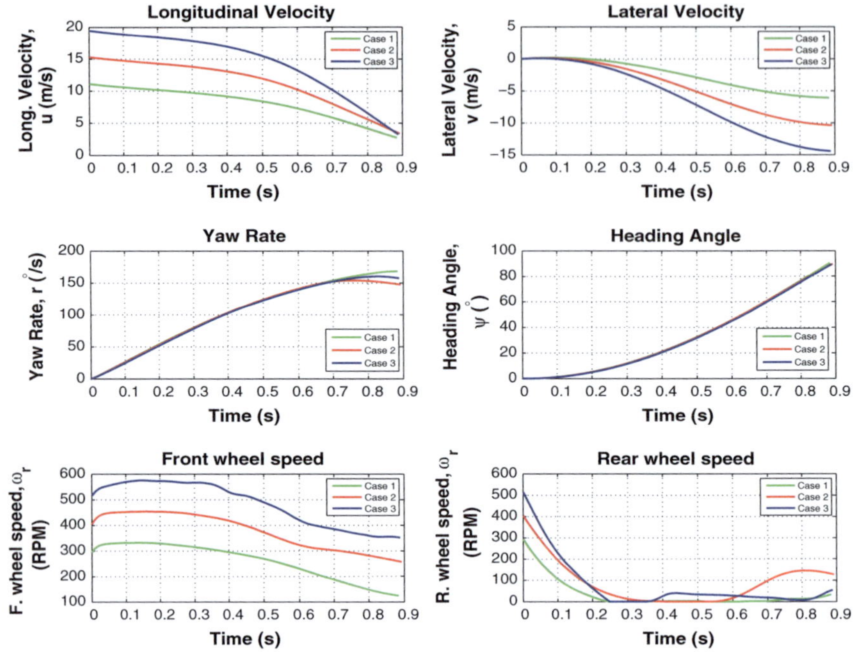

Fig. 7.8 Optimal open-loop state trajectories generated with GPOPS

where, for notational convenience, we have introduced, rather informally, the interpolating functions κ_i $(i = 1, 2, 3, 4)$ to denote the right-hand-side of (7.35). Similarly, the optimal time to perform the maneuver from the current state is given by

$$t_f(t_k) = \hat{y}_4(\mathbf{x}_0) = \kappa_4(u(t_k), v(t)k), \psi(t_k), \omega_f(t_k), \omega_f(t_k)). \tag{7.42}$$

Note from (7.39) to (7.41) that the approach yields, at each instant of time, a control action that depends on the current state, that is, the resulting control has a feedback structure. In essence, we have developed a tool for controller *synthesis* where the open-loop optimal controllers are combined to a single feedback strategy. The difference with standard approaches is that this synthesis is not performed analytically, but rather numerically, via an implicit interpolation of the pre-computed open-loop control laws.

For all computations we have used the DACE toolbox for Matlab [21]. Both the correlation functions and the allowable values for the parameter θ were determined by trial and error. Constant and first order polynomials were sufficient for this problem, along with cubic correlation functions (see Fig. 7.7). The optimal value of θ was obtained using MSE, as explained in Sect. 7.3.2.

7.4.2 Numerical Results

The family of near-optimal controls is shown in Fig. 7.9a–c. The red lines highlight the pre-computed solutions used to obtain the interpolating metamodel.

These results show that the controller obtained using the proposed statistical interpolation technique generates near-optimal solutions for the whole range of initial velocities considered. In all simulations the trajectories reach the final constraint, $\psi = 90°$, as required. Furthermore, notice in Fig. 7.9a–c how the interpolated solutions match the pre-computed ones at the trial sites. This is a consequence of the exact interpolation property of the interpolation scheme, shown in (7.37). Notice also that the solutions vary smoothly along the whole range of initial velocities.

The average to compute a single interpolation of all three controls was 1.2 ms (or a rate of 800 Hz) on a Intel Pentium Core 2 Duo processor running at 2.4 GHz with 4 GB of RAM. This rate is considered fast enough for real-time controller implementation.

The parameter θ that affects the behavior of the correlation functions is shown in Fig. 7.10 for the case of the footbrake command. For very large values of θ the solution tends to the GLS solution ($r(\mathbf{x}_0) = 0$ except at the observation points). The oscillatory behavior for large values of θ observed in these figures is owing to the fact that we have chosen a zero order polynomial for the footbrake, in which case the interpolating terms tend to a superposition of impulse functions at the observed points (see Fig. 7.7). As the value of θ is reduced, each observation increases its "region of influence" over a larger area of the space, thus "averaging out" the contributions from neighboring observation points.

It is also of interest to explore the positive attributes that arise from having a controller in feedback form [see again (7.39–(7.41)]. Although there is no analytic expression for the feedback controller, the control action is obtained as a function of the current state. Feedback controllers are more desirable than open-loop controllers since they can account for sudden changes in the state, unmodelled uncertainties, etc. In order to evaluate the benefits of having a controller in a feedback form, a simulation with a disturbance representing a 30 % reduction in the yaw rate at $t = 0.6t_f$ was carried out. The comparison was performed at one of the trial locations where the interpolated solution matches the pre-computed one, so the comparison is fair. The initial velocity was chosen as $V_0 = 56$ km/h. Figure 7.11 shows how the interpolated control changes when the disturbance is applied and how the system is finally guided to the final constraint despite the abrupt change in the state at $t = 0.6t_f$.

7.5 Conclusions

The future generation of active safety systems for passenger vehicles will have to take advantage of the nonlinearities of the vehicle and tire friction dynamics in order to safely implement more aggressive obstacle avoidance maneuvers in the case of an impeding accident. Unfortunately, generating optimally such aggressive maneuvers—at

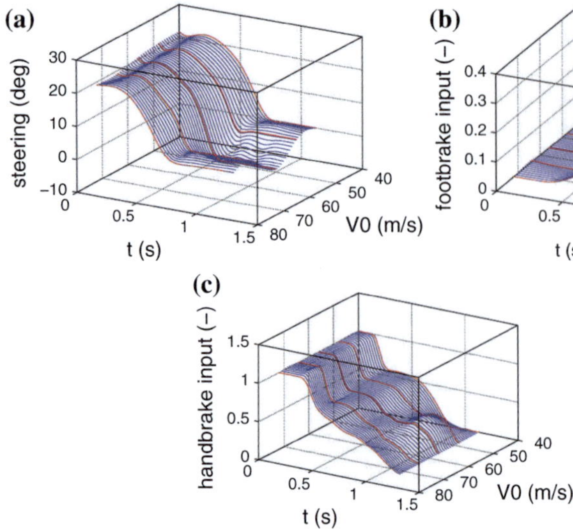

Fig. 7.9 Interpolated optimal control histories. **a** Steering command. **b** Footbrake command. **c** Handbrake command

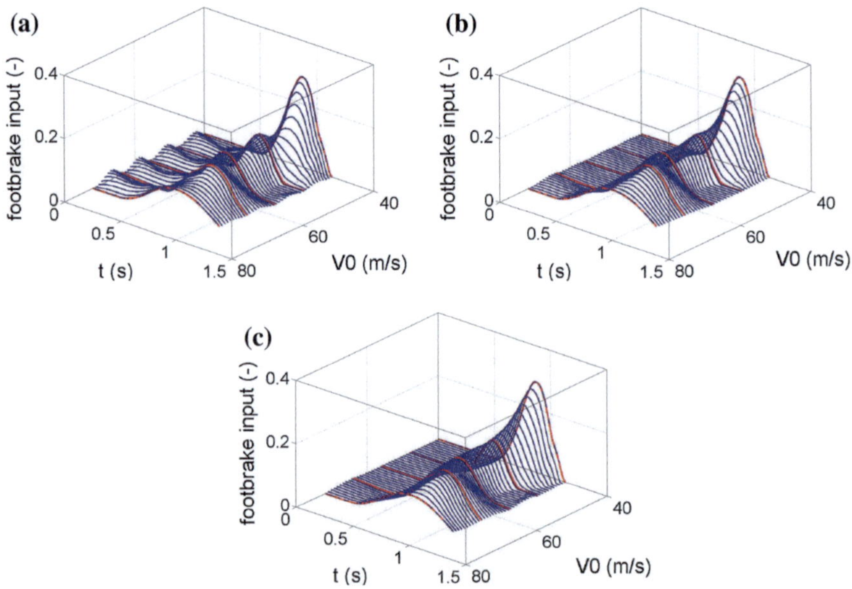

Fig. 7.10 Effect of correlation parameter θ on the optimal footbrake command response surface. **a** Large value of θ. **b** Medium value of θ. **c** Low value of θ

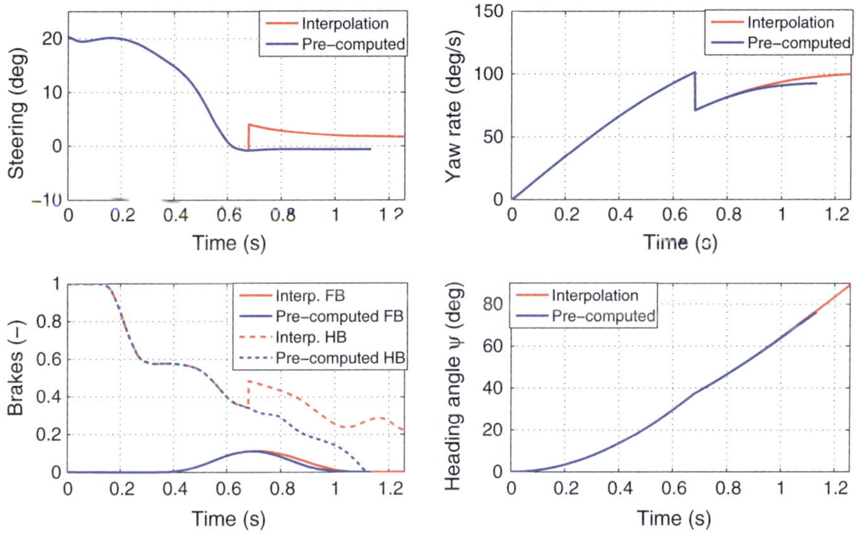

Fig. 7.11 State evolution comparison under disturbance

the time scales required along with convergence guarantees—is still an elusive goal with current trajectory optimizers. In this chapter we investigate the use of a statistical interpolation technique based on Gaussian processes (e.g., kriging) to generate near-optimal trajectories, along the corresponding control actions, from a set of off-line pre-computed optimal trajectories. The resulting approach essentially generates a metamodel of the action-response map based on the pre-computed optimal control solutions. The resulting interpolation model emulates an optimal feedback controller, as long as the initial conditions are contained in the convex hull of the off-line test locations.

Our numerical results show that the resulting controller has excellent performance, always guiding the system to the exact terminal constraint. Furthermore, the controller is extremely fast to compute, since it is based on simple algebraic manipulations and hence it is beneficial for all similar situations where decisions must be taken within extremely short deadlines.

Acknowledgments Partial support for the work presented in this chapter has been provided by NSF through award no. CMMI-1234286 and ARO via MURI award no. W911NF-11-1-0046.

References

1. Adams J (2012) An interpolation approach to optimal trajectory planning for helicopter unmanned aerial vehicles. Master's thesis, Naval Postgraduate School

2. Assadian F, Hancock M (2005) A comparison of yaw stability control strategies for the active differential. In: Proceedings of IEEE international symposium on industrial electronics ISIE 2005, vol 1, pp 373–378. doi:10.1109/ISIE.2005.1528939
3. Becerra VM (2011) PSOPT optimal control solver user manual
4. Betts JT (1998) Survey of numerical methods for trajectory optimization. AIAA J Guidance Control Dyn 21(2):193–207
5. Betts JT, Huffman WP (1997) Sparse optimal control software SOCS. Mathematics and engineering analysis technical document mealr-085, Boeing Information and Support Services, The Boeing Company, Seattle
6. Chakraborty I, Tsiotras P (2011) Mitigation of unavoidable T-bone colisions at intersections through aggressive maneuvering. In: Proceedings of the 50th IEEE conference on decision and control and European control conference, Orlando, FL, pp 3264–3269. doi:10.1109/CDC.2011.6161241
7. Chakraborty I, Tsiotras P, Diaz RS (2013) Time-optimal vehicle posture control to mitigate unavoidable collisions using conventional control inputs. In: American control conference, Washington, DC, pp 2165–2170
8. Cressie N (1990) The origins of kriging. Math Geol 22(3):239–252
9. Dever C, Mettler B, Feron E, Popovic J, McConley M (2006) Nonlinear trajectory generation for autonomous vehicles via parameterized maneuver classes. J Guidance Control Dyn 29(2):289–302
10. del Re L, Allgöwer FF, Glielmo L, Guardiola C, Kolmanovsky I (2010) Automotive model predictive control: models, methods and applications. Lecture notes in control and information sciences. Springer, Berlin
11. Di Cairano S, Tseng HE (2010) Driver-assist steering by active front steering and differential braking: design, implementation and experimental evaluation of a switched model predictive control approach. In: Proceedings of 49th IEEE conference on decision and control, pp 2886–2891. doi:10.1109/CDC.2010.5716954
12. Doe R (2012) European new car assessment programme, Results (www.euroncap.com), Brussels Belgium
13. Fildes B, Lane J, Lenard J, Vulvan A (1994) Passenger cars and ocupant injury: side impact crashes. Monash University, Accident Research Center, Report No. 134, Canberra, Australia
14. Ghosh P, Conway B (2012) Near-optimal feedback strategies for optimal control and pursuit-evasion games: a spatial statistical approach. In: AIAA/AAS astrodynamics specialist conference, Minneapolis, MN, AIAA Paper 2012–4590. doi:10.2514/6.2012-4590
15. Goldberger A (1962) Best linear unbiased prediction in the generalized linear regression model. J Am Stat Assoc 57(298):369–375
16. Handcock MS, Stein ML (1993) A Bayesian analysis of kriging. Technometrics 35(4):403–410
17. Hargraves CR, Paris SW (1987) Direct tracjectory optimization using nonlinear programming and collocation. AIAA J Guidance Control Dyn 10(4):338–342
18. Huang D, Allen T, Notz W, Miller R (2006) Sequential kriging optimization using multiple-fidelity evaluations. Struct Multi Optim 32(5):369–382
19. Jazar RN (2008) Vehicle dynamics: theory and application. Springer, New York
20. Kleijnen J (2009) Kriging metamodeling in simulation: a review. Eur J Oper Res 192(3):707–716
21. Lophaven SN, Nielsen HB, Søndergaard J (2002) DACE: a MATLAB kriging toolbox, Technical report, Technical University of Denmark
22. MacKay DJC (1998) Introduction to Gaussian processes. NATO ASI Ser F Comput Syst Sci 168:133–166
23. Oberle H, Grimm W (1985) BNDSCO: a program for the numerical solution of optimal control problems. English translation of DFVLR-Mitt. 85–05
24. Pacejka H, Bakker E, Nyborg L (1987) Tyre modelling for use in vehicle dynamics studies. SAE paper 870421
25. Rao AV, Benson D, Darby CL, Mahon B, Francolin C, Patterson M, Sanders I, Huntington GT (2011) User's manual for GPOPS version 4.x: a MATLAB software for solving multiple-phase

optimal control problems using hp-adaptive pseudospectral methods. http://www.gpops.org/gpopsManual.pdf
26. Riekert P, Schunck TE (1940) Zür Fahrmechanik des gummibereiften Kraftfahrzeugs. Arch Appl Mech 11(3):210–224
27. Ross IM (2003) User's manual for DIDO: a MATLAB application package for solving optimal control problems. NPS technical report MAE-03-005, Naval Postgraduate School, Monterey, CA
28. Schwartz AL (1996) Theory and implementation of numerical methods based on Runge-Kutta integration for solving optimal control problems. Ph.D. thesis, Berkeley, University of California
29. Side Impacts: Few second chances. http://www.southafrica.co.za/2011/02/10/side-impacts-few-second-chances/
30. Simpson T, Martin J, Booker A, Giunta A, Haftka R, Renaud J, Kleijnen J (2005) Use of kriging models to approximate deterministic computer models. AIAA J 43(4):853–863
31. Tang J, Singh A, Goehausen N, Abbeel P (2010) Parameterized maneuver learning for autonomous helicopter flight. In: 2010 IEEE international conference on robotics and automation (ICRA), IEEE, pp 1142–1148
32. Simpson TW, Mauery TM, Korte JJ, Mistree F (2001) Kriging models for global approximation in simulation-based multidisciplinary design optimization. AIAA J 39(12):2233–2241
33. Van Beers W, Kleijnen J (2004) Kriging interpolation in simulation: a survey. In: Proceedings of the 2004 winter simulation conference, vol 1. IEEE
34. van Zanten AT (2002) Evolution of electronic control system for improving the vehicle dynamic behavior. In: Advanced vehicle control conference (AVEC), Hiroshima, Japan
35. Velenis E, Frazzoli E, Tsiotras P (2010) Steady-state cornering equilibria and stabilization for a vehicle during extreme operating conditions. Int J Veh Auton Syst 8(2–4):217–241. doi:10.1504/IJVAS.2010.035797
36. Velenis E, Katzourakis D, Frazzoli E, Tsiotras P, Happee R (2011) Steady-state drifting stabilization of RWD vehicles. Control Eng Pract 19(11):1363–1376. doi:10.1016/j.conengprac.2011.07.010
37. Velenis E, Tsiotras P, Lu J (2007) Modeling aggressive maneuvers on loose surfaces: the cases of trail-braking and pendulum-turn. In: European control conference, Kos, Greece, pp 1233–1240
38. Velenis E, Tsiotras P, Lu J (2008) Optimality properties and driver input parameterization for trail-braking cornering. Eur J Control 14(4):308–320. doi:10.3166/EJC.14.308-320
39. Yamamoto M (1991) Active control strategy for improved handling and stability. SAE Trans 100(6):1638–1648

Chapter 8
Applications of Computational Optimal Control to Vehicle Dynamics

Joško Deur, Mirko Corić, Josip Kasać, Francis Assadian and Davor Hrovat

Abstract Modern vehicle dynamic control systems are based on new types of actuators, such as active steering and active differentials, in order to improve the overall handling performance including stability, responsiveness, and agility. Numerical techniques of off-line optimization of vehicle dynamics control variables can conveniently be used to facilitate decisions on optimal actuator configurations and provide guidance for design of realistic, on-line controllers. This chapter overviews the previous authors' results of assessment of various vehicle dynamics actuator configurations based on application of a back propagation through time (BPTT) conjugate gradient optimization algorithm. It is then focused on detailed optimization of active front and rear steering control variables for various maneuvers and design specifications, where a nonlinear programming-based optimization tool is used.

8.1 Introduction

The conventional vehicle dynamics control systems are based on using wheel brakes as actuators [13, 15]. By braking solely or predominantly one of the wheels, a proper amount of active yaw torque is generated and at the same time the vehicle is decelerated, which contributes to the vehicle handling stability and responsiveness (i.e. oversteer and understeer compensation, respectively). In recent years the active chassis systems are being equipped with different types of additional actuators such

J. Deur (✉) · M. Corić · J. Kasać
Automotive Control Group at the Faculty of Mechanical Engineering and Naval Architecture, University of Zagreb, Zagreb, Croatia
e-mail: josko.deur@fsb.hr

F. Assadian
Department of Automotive Engineering, Cranfield University, Milton Keynes, UK

D. Hrovat
Research and Advanced Engineering, Ford Motor Company, Dearborn, MI, USA

H. Waschl et al. (eds.), *Optimization and Optimal Control in Automotive Systems*, 131
Lecture Notes in Control and Information Sciences 455,
DOI: 10.1007/978-3-319-05371-4_8,

as active steering or active differentials [7, 8, 10], which can provide superior handling performance without being intrusive to the driver (no NVH disturbance unlike for hydraulic brakes) and without affecting agility (minimal reduction of the vehicle velocity).

In the early stage of active chassis hardware specification/selection, optimization techniques can be used for assessment of various single- or multi-actuator chassis configurations. More specifically, an open-loop optimal control algorithm is applied to a proper nonlinear vehicle dynamics model to find time response of control variables that minimize the specified cost function (e.g. minimal trajectory following error) under various equality and inequality constraints (e.g. limited control input authority or state variable magnitude). At the same time, the optimization results can be used as a benchmark for achievable performance of production-oriented (on-line) control systems (e.g. model predictive control systems [5]) as well as to give good insights that can facilitate control system design and calibration.

Although different control variable optimization algorithms have been used in various automotive power train studies (e.g. dynamic programming for optimization of hybrid electric power trains [2, 6] and nonlinear programming for turbocharger power assist system optimization [12]), their application to vehicle dynamics control field have been relatively scarce (with exceptions of, for instance, [9, 16] where handling analyses were conducted). The authors' recent chapter [4] have dealt with assessment of various vehicle dynamics actuator configurations, based on the application of a back propagation through time optimization (BPTT) algorithm [11]. Those assessment results are outlined in this chapter, as a background for presenting a more recent active front and rear steering optimization study based on a fundamentally different nonlinear programming method implemented through TOMLAB optimization platform [14].

8.2 Overview of Previous Optimization and Assessment Results

8.2.1 Optimization Algorithm

The general optimization problem is to find a control vector input $\mathbf{u}(t), 0 \leq t \leq tf$, which minimizes the Bolza-type cost function $J_0 = \Theta(\widehat{\mathbf{x}}(t_f) + \int_0^{t_f} F_0(\widehat{\mathbf{x}}(t), \mathbf{u}(t))dt$, subject to nonlinear plant state equations $\dot{\widehat{\mathbf{x}}} = \phi(\widehat{\mathbf{x}}(t), \mathbf{u}(t))dt, \widehat{\mathbf{x}}(0) = \widehat{\mathbf{x}}_0$, and subject to the final conditions on the state vector $b(\widehat{\mathbf{x}}(t_f)) = 0$, and subject to the control and state vector inequality constraints $g(\widehat{\mathbf{x}}(t), \mathbf{u}(t)) \geq 0$ and equality constraints $h(\widehat{\mathbf{x}}(t), \mathbf{u}(t)) = 0$.

In the previous work [4], the optimal control vector input $\mathbf{u}(t)$ was found by using an iterative gradient descent algorithm with respect to control vector: $\mathbf{u}^{(l+1)}(t) = \mathbf{u}^{(l)}(t) - \eta \frac{\partial J}{\partial \mathbf{u}^{(l)}(t)}$, where the total cost function J included the basic (user-defined) function J_0 extended with the penalty terms for the final conditions, and the equality and inequality constraints. The gradient $\frac{\partial J}{\partial \mathbf{u}}$ was calculated by utilizing the BPTT algorithm based on the plant model discretized in time using the Adams method [11]. The Jacobians were calculated numerically. The learning rate η was made time variant using the Dai-Yuan conjugate gradient method, in order to improve the algorithm convergence rate.

8.2.2 Vehicle Model

A 10-Degree of Freedom (DoF) vehicle dynamics model [8] with a full "magic" formula tire model was implemented within the optimization program. The model includes six state variables, which are related to the longitudinal (U), lateral (V), and heave (W) velocities, and roll (p), pitch (q), and yaw (r) rates. The remaining four state variables correspond to the rotational speeds of each wheel. In addition, six state variables related to suspension submodel and vehicle trajectory in the inertial (X–Y) coordinate system are used (roll, pitch, and yaw angles, heave displacement, and vehicle X,Y position), as well as state variable(s) related to first-order lag term(s) describing the actuator dynamics. Thus, the overall nonlinear dynamics model includes from 17 to 19 state variables and from 1 to 3 control variables, depending on the number of actuators used in particular vehicle configuration.

Due to the paper length constraints, only the main model state equations, related to the lateral velocity and yaw rate DOFs, are given:

$$M(\dot{V} + Ur) = F_{yf} + F_{yr} \tag{8.1}$$

$$I_{zz}\dot{r} \cong bF_{yf} - cF_{yr} - \frac{w}{2}(F_{x1} + F_{x2} + F_{x3} + F_{x4}) \tag{8.2}$$

where M is the vehicle mass, I_{zz} is the vehicle moment of inertia around the vertical axis, b and c are the distances between the vehicle center of gravity and the front and rear axles, respectively, w is the vehicle track (the distance between the left and right wheel axes). F_x and F_y denote the longitudinal and lateral tire forces, respectively, after being transformed from the tire to vehicle coordinate system. Here, the index $i = 1, \ldots, 4$ represent the tire number starting from the front-left tire towards the rear-left tire in the clockwise direction, and the indices f and r refer to front and rear axle, with $F_{yf} = F_{y1} + F_{y2}$ and $F_{yr} = F_{y3} + F_{y4}$. The sign conventions are such that the yaw rate r is positive for counter-clockwise turning, where the wheels 1 and 3 are inner wheels. For the sake of simplicity of presentation, the yaw rate state Eq. (8.2) omits a roll cross-coupling term that is of secondary importance.

The tire model describes nonlinear, combined-slip tire static curves as functions of the tire normal force $F_{zi}, i = 1, \ldots, 4$, calculated from a nonlinear suspension model, and the tire sideslip angle α_i and the longitudinal slip η_i, given by:

$$\alpha_i = \delta_i - \arctan\left(\frac{V_i}{U_i}\right) \tag{8.3}$$

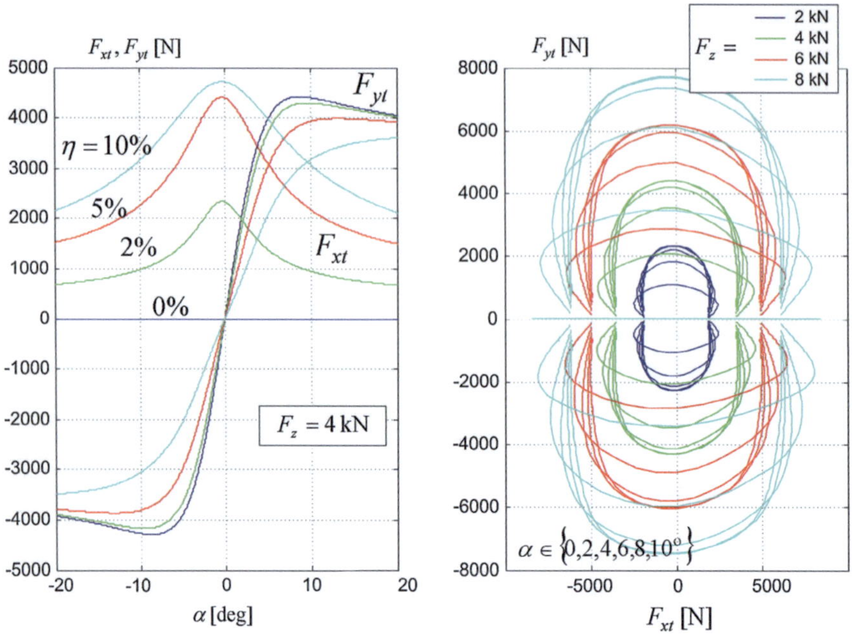

Fig. 8.1 Tire model static curves

$$\eta_i = \frac{R\omega_i - U_i}{U_i} \tag{8.4}$$

where ω_i are the wheel rotational speeds, δ_i are the road wheel angles, R is the effective tire radius, and U_i and V_i are the longitudinal and lateral wheel center velocity components, respectively:

$$U_i = U + (-1)^i \frac{w}{2} r, \quad V_{1,2} = V + br, \quad V_{3,4} = V - cr \tag{8.5}$$

The modeled tire static curves are shown in Fig. 8.1 The obtained tire forces F_{xti} and F_{yti} are transformed into the vehicle coordinate system by using the wheel steering input δ_i, thus obtaining the forces F_{xi} and F_{yi} that are fed into the state-space model (see Eqs. 8.1 and 8.2).

The considered actuators include active front and rear steering (AFS and ARS), active rear, front and, central differentials of limited slip and torque vectoring types (ALSRD, TVRD, ALSFD, TVFD, ALSCD, and TVCD). The vehicle (a mid-size sedan) is of rear wheel drive (RWD) type, except in the cases of using the front or central differentials, when a FWD or 4WD configuration is considered, respectively. The active differential torque transfer is subject to the following constraints [3, 10]: (1) the ALSD cannot transfer the torque to faster wheel, and (2) the TVD can transfer

Table 8.1 List of double lane change maneuver variants

No.	*Ti* (Nm)	μ	Specifics
1	0	0.6	Initial understeer, then oversteer
2	350	1	Understeer under torque
3	250	0.6	Initial understeer, then instability

the torque to faster wheel (or faster axle for TVCD) only if the faster-slower wheel relative speed difference is smaller than the Allowable Wheel Speed Difference factor (AWSD = 28.6 % for the given TVDs). These constraints were conveniently implemented through the vehicle model rather than via inequality constraints.

8.2.3 Formulation of Optimization Problem

The optimization study was conducted for three characteristic types of double lane change maneuver (DLC; Table 8.1). The objective was to find the control inputs for a given actuator configuration, which minimized the path following error given by the cost function subintegral function $F_0(\widehat{\mathbf{x}}(t), \mathbf{u}(t)) = K_h(Y - Y_R(X))^2$, where $Y_R(X_R)$ is the DLC reference trajectory. This objective is subject to inequality constraints (limits) on control input variables and sideslip angle $\beta = arctan(V/U)$ (20° and 15°, respectively), as well as on the active front steering angle rate (to avoid largely affecting the perceived driver torque feedback).

Maneuver 1 is defined as follows. The front steering input δ_f is first optimized to follow the given DLC reference path for the tire-road friction coefficient $\mu = 1$ and zero driveline input torque ($T_i = 0$). The optimized variable δ_f is then used as an open-loop driver input (again with $T_i = 0$) for optimizations of vehicle dynamics control inputs for the reference trajectory achieved in the first step, but reduced friction coefficient $\mu = 0.6$. Due to the already saturated lateral tire forces (for $\mu = 1$), the friction reduction results in tendency of passive vehicle to initially understeer and then strongly oversteer (see 'no control' response in Fig. 8.2a). Maneuver 2 utilizes the same front steering input δ_f and the reference trajectory, but it is applied under the accelerating conditions ($T_i = 350$ Nm) and the nominal road condition $\mu = 1$. Obviously, in the presence of accelerating torque and unchanged driver steering input, the passive vehicle is prone to understeer, particularly in the second part of response (Fig. 8.2b). Maneuver 3 is similar to Maneuver 1. The main difference is that the optimization relates to a non-zero driveline torque ($T_i = 250$ Nm). According to tire friction ellipse (Fig. 8.1), this reduces the rear (driven) tire lateral forces, thus making the passive vehicle unstable (the sideslip angle β starts diverging around the peak trajectory point, see Sect. 8.3). Also, the initial understeer is more emphasized than in Maneuver 1 (cf. Fig. 8.2a and c), because of the growth of vehicle velocity and corresponding rise of achievable (stationary) turning radius

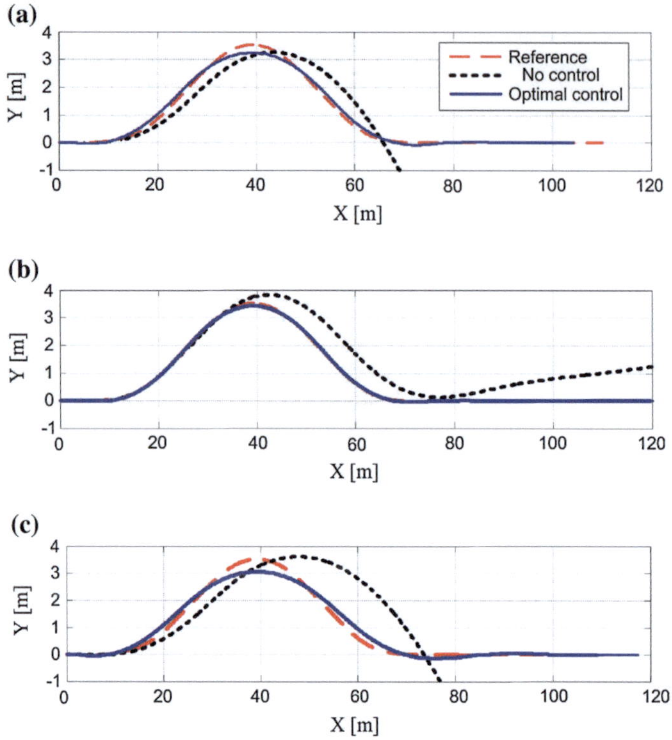

Fig. 8.2 Trajectory following optimization results for three DLC maneuvers and ARS actuator

$R_t = U^2/a_{y,sat}$ for the given (saturated) lateral acceleration $a_{y,sat} = \mu g$ (see Eq. 8.1 and note that $a_y = \dot{V} + rU$ and $(F_{yf} + F_{yr})_{sat} = Mg\mu$).

8.2.4 Optimization and Assessment Results

The optimization results for different single- and multi-actuator configurations and the three DLC maneuvers are given in Table 8.2. The trajectory following accuracy is characterized by the root-mean-square (RMS) error of $Y(X) - Y_R(X)$. The vehicle stabilization is reflected in the ability of suppressing the magnitude of sideslip angle β. The agility is better if the traveled distance $X(t_f)$ over the constant maneuver time period is larger. Based on these indices, the different actuator configurations have been ranked, with the final ranking/assessment results also included in Table 8.2. The assessment results are briefly discussed below, with the reference to [4] for more detailed elaboration.

According to the RMS data in Table 8.2 and also the trajectory plots in Fig. 8.2, Maneuver 3 is the most difficult in terms of trajectory following accuracy. This is

Table 8.2 Comparative optimization results for various active-steering and active-differential actuator configurations and maneuvers (using BPTT algorithm)

No.	Actuator configuration	Maneuver 1			Maneuver 2			Maneuver 3			Single actuator rank	Multi-actuator rank
		RMS (m)	$\beta\|_{max}$ (deg)	$X(t_f)$ (m)	RMS (m)	$\beta\|_{max}$ (deg)	$X(t_f)$ (m)	RMS (m)	$\beta\|_{max}$ (deg)	$X(t_f)$ (m)		
1	ARS	0.116	4.565	104.0	0.038	4.403	122.7	0.190	4.307	117.5	1	1
2	AFS	0.146	5.502	104.8	0.045	4.817	122.9	0.234	5.354	118.9	2	5
3	TVRD	0.140	4.774	103.1	0.086	3.478	121.8	0.326	4.774	117.5	4	9
4	ALSRD	0.290	4.835	105.7	0.243	3.436	124.1	5.467	18.3	111.2	7	12
5	TVFD	0.124	5.747	102.8	0.112	3.552	123.4	0.269	5.822	117.4	3	8
6	ALSFD	0.258	4.822	105.7	0.158	2.604	124.8	0.432	4.734	119.7	6	11
7	TVCD	0.358	5.410	102.9	0.130	2.967	119.9	0.410	5.036	116.7	5	11
8	ALSCD	2.404	7.295	104.6	0.604	3.457	124.2	0.557	6.123	119.2	8	13
9	AFS & TVRD	0.149	4.901	104.7	0.043	4.834	122.4	0.220	5.064	117.8	**oo****	3
10	AFS & ALSRD	0.160	4.733	105.3	0.064	3.782	123.6	0.248	4.346	119.3	**o**	6
11	TVRD & TVCD	0.165	5.211	102.6	0.069	4.064	119.7	0.243	6.108	115.1	**	7
12	ALSRD & TVCD	0.283	5.002	104.0	0.130	3.165	121.0	0.375	5.010	116.9	*	10
13	TVRD & TVFD	0.111	7.263	101.8	0.075	4.660	122.0	0.186	7.244	115.4	***	2
14	TVRD & TVCD & TVFD	0.101	6.854	100.6	0.080	3.886	121.2	0.192	6.423	115.0	*	4

* = improvement by introducing more actuators in terms of performance
o = improvement by introducing more actuators in terms of smaller control effort (better **AFS** torque feedback)

because it is hard to stabilize the accelerating passive vehicle on the low-μ surface (see Table 8.1), and also to compensate for the initial understeer in the case of reduced μ. Maneuver 2 is the least difficult, because the friction coefficient is the highest.

The active rear steering (ARS) provides the best overall performance, owing to a high control authority of both yaw rate and lateral velocity/acceleration dynamics (the control input $\delta_r = \delta_{3,4}$ can influence both state Eqs. 8.1 and 8.2, through the rear lateral forces $Fy3, 4$). Although quite effective, the individual rear or front torque vectoring differentials (TVRD or TVFD) are inferior to the ARS actuator. This is because they directly influence only the yaw rate dynamics through generating difference in left and right longitudinal tire forces (see Eq. 8.2), and because the longitudinal force generation is limited by a low normal load of inner tires. The combined front TVD and rear TVD configuration provides excel-lent performance, owing to the balanced front/rear axle torque distribution (for this 4WD configuration) and increased TVD control authority (both axles generate the yaw torque). However, when compared to the ARS actuator, the TVRD+TVFD configuration is characterized by somewhat reduced agility (due to the certain decelerating action of TVDs [3]) and increased sideslip magnitudes (due to strong yaw torque generation). The central TVD can rarely contribute to the performance boost of the TVFD+TVRD configuration.

The active front steering (AFS) is also among the best performers, but it is consistently worse than the ARS actuator due to a lower control authority (see Sect. 8.3). The AFS+TVRD and TVRD+TVCD controls have also been found to be viable configurations, where the former provides a good trade off between the performance and limited AFS control effort, and the latter improves the TVRD performance for the throttle-on Maneuvers 2 and 3 due to its 4WD feature and additional understeer and oversteer compensation imposed by TVCD. However, they do not offer as significant improvement over the individual controls as the TVRD+TVFD configuration.

The TVCD is inferior to TVFD or TVRD for the coasting Maneuver 1, because it acts on the lateral tire forces indirectly—through spinning the front axle for oversteer compensation or rear axle for understeer compensation. On the other hand, it gives rather good performance for the accelerating Maneuvers 2 and 3, partly because of its 4WD feature. The front and rear active limited slip differentials (ALSFD and ALSRD) give worse performance than TVDs, which is explained by their limited, zero-AWSD-related control authority: (1) the oversteer compensation is constrained by the weakly-loaded inner wheel being spun above the outer wheel speed, and (2) the understeer compensation is unfeasible (the torque cannot be transferred to faster/outer wheel), unless the inner wheel tends to spin for an accelerating maneuver. The ALSCD barely gives any improvements over the passive vehicle.

The presented comparative study has demonstrated that the BPTT optimization algorithm can represent a useful tool for assessment of various vehicle dynamics actuator configurations. However, certain issues have been observed in the optimization results: (1) the optimization of some multi-actuator configurations give worse (RMS error) results than in the case of corresponding single-actuator configuration (see e.g. AFS+ALSRD configuration in Table 8.2), and (2) the optimization was inconsistent for the ALSRD actuator and Maneuver 3 (unrealistically high RMS error). The former can be explained by the fact that optimization ends in a local optimum for more complex-multi actuator configurations (especially those with actuators of different types). The latter can be caused by highly nonlinear clutch model used to describe the active differential dynamics. In order to avoid issues with multi-actuator configuration optimization, a more detailed ARS/AFS study has been conducted by using the TOMLAB optimization platform (see Sect. 8.3).

8.3 Detailed Optimization for Active Steering Configurations

8.3.1 Optimization Algorithm

TOMLAB optimization platform is Matlab interface package for modeling, optimization and optimal control [14]. It consists of modules for optimal control formulation (PROPT), automatic differentiation (MAD) and variety of problem solvers (SNOPT, KNITRO, CPLEX etc.). The user specifies state equations that are considered as equality constraints, initial and final conditions, constraints and cost function. The optimization approach is based on pseudospectral collocation method in which the optimal control problem is discretized and transformed into large-scale nonlinear programming (NLP) problem [1]. The solution takes polynomial form of nth order, where n is the number of collocation (time grid) points specified by the user. The number of grid points is significantly smaller than in the case of BPTT algorithm, where a dense grid was needed to explicitly calculate the control input values at each grid point. This difference may contribute to a lower local minima sensitivity of the NLP algorithm when compared to its BPTT counterpart. The polynomial solution needs to satisfy the discretized state equations and constraints in the collocation points. If solver claims that solution is optimal, it satisfies necessary, not sufficient, optimality conditions. In this work the large-scale sparse solver SNOPT is used, which is based on the sequential quadratic programming (SQP) method. As SNOPT uses nonlinear function gradients, calculated automatically by MAD, the problem must be smooth (first-order differentiable) around the optimal solution.

8.3.2 Active Rear Steering (ARS)

The ARS optimization results are shown in Fig. 8.3 for the three DLC maneuvers defined in Table 8.1, and a soft constraint imposed on the control input. The related performance indices are given in first two rows of Table 8.3. The sign conventions are such that negative ARS control input, the rear road wheel angle δ_r, contributes to increase of positive yaw rate r. That is, the ARS actuator generates oversteer (or compensate for understeer) if $sgn(\delta_r) \neq sgn(r)$, and the understeer is generated

Table 8.3 Comparative optimization results for various active-steering configurations and different maneuvers (using Tomlab tool)

Actuator configuration	Control input constraint	RMS (m)	$\mid \beta \mid_{max}$ (deg)	$X(t_f)$ (m)	$\mid \alpha \mid_{max}$ (deg)	η_{max} (%)	$\mid Mz \mid_{max}$ (kNm)
(a) *Maneuver 1*							
ARS	Yes	0.112	4.58	103.9	9.2	0.06	8.0
ARS	No	0.087	4.75	100.4	27.2	0.3	11.8
AFS	Yes	0.137	5.26	104.4	9.4	0.04	10.3
ARS & AFS	Yes	0.091	4.96	102.5	9.7	0.1	12.5
ARS & AFS	No	0.031	17.7	92.4	37.3	0.4	16.7
ARS & AFS	Weak	0.069	8.38	100.0	18.5	0.2	16.5
(b) *Maneuver 2*							
ARS	Yes	0.044	4.06	123.2	8.7	6.7	8.0
ARS	No	0.017	4.20	118.0	32.1	37.8	22.0
AFS	Yes	0.045	4.05	123.2	8.1	7.0	12.9
ARS & AFS	Yes	0.030	4.66	122.2	9.4	11.0	11.3
ARS & AFS	No	0.006	20.7	109.7	48.1	49.9	28.4
ARS & AFS	Weak	0.027	4.69	121.7	10.0	5.8	19.8
(c) *Maneuver 3*							
ARS	Yes	0.176	4.75	116.6	9.8	12.3	9.0
ARS	No	0.158	4.43	112.7	25.8	26.2	11.6
AFS	Yes	0.229	5.14	118.6	10.3	3.7	10.1
ARS & AFS	Yes	0.163	4.29	115.7	9.7	11.2	13.7
ARS & AFS	No	0.101	17.86	105.2	32.0	47.7	16.9
ARS & AFS	Weak	0.145	5.23	112.9	22.1	8.0	16.6

(i.e. oversteer is compensated for) if $sgn(\delta_r) = sgn(r)$. Therefore, the δ_r and r time responses in Fig. 8.3a (Maneuver 1) indicate that the control action is such that there are three characteristic, relatively short periods of understeer compensation (USC): $t \in [0, 0.3]$, [1, 1.25] and [2.5, 2.7]s. The first USC action steers the vehicle towards the reference path, with the note that the actual vehicle trajectory response is delayed with respect to control action due to the influence of vehicle dynamics (i.e. the ARS provides a preview control action during initially straight motion). The second USC intervention results in "cutting" the trajectory around its peak point, in order to minimize the trajectory following error in the absence of ability to accurately follow the reference trajectory peak on the reduced-μ surface. The third USC action is similar to the first one.

The USC actions result in characteristic peaks of the yaw torque, calculated as $M_z \cong bF_{yf} - cF_{yr}$ (cf. Eq. 8.2), which provides boost of yaw rate r to generate de-sired oversteer. Accordingly, each USC action provokes large rates of change of the sideslip angle β (see Eq. 8.1 and recall that $\beta = arctan(V/U)$). In order to effectively limit the sideslip angle β, and, thus, provide an ample stability margin (compared to the passive/uncontrolled vehicle), each The USC interval is followed by a relatively long interval of oversteer compensation (OSC). USC-related boosts

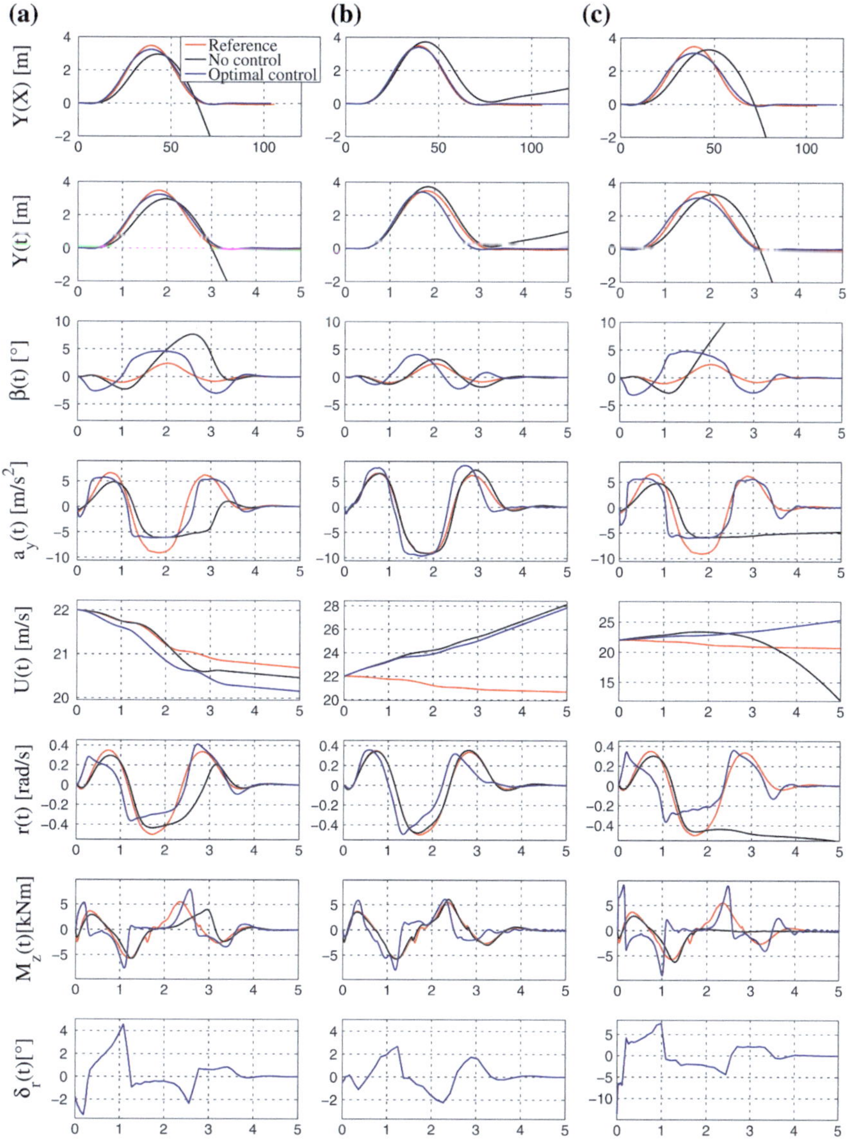

Fig. 8.3 Optimization results for ARS actuator and three DLC maneuvers

of side slip angle, i.e. its large deviation, result in decelerating force on the vehicle, thus somewhat affecting the agility (the end speed $U(t_f)$ is reduced compared to reference and uncontrolled vehicle, Fig. 8.3).

The lateral acceleration response ay tends to follow that of the reference vehicle (related to $\mu = 1$). However, since the lateral acceleration is limited to $0.6g$ for the reduced $\mu = 0.6$, accurate following of the lateral acceleration peaks around $1g$ is not possible, which is "compensated for" by wider and sharper edges of lateral acceleration. This gives a characteristic square-wave-like shape of a_y response. The tire sideslip angles α_i and lateral forces F_{yi} assume similar response shape (not shown in Fig. 8.3), particularly for rear tires ($i = 3, 4$) over which the ARS actuator has a direct control authority.

Since the high-μ Maneuver 2 is less critical than Maneuver 1 (Sect. 8.2), the ARS control input magnitudes are rather modest in this case (Fig. 8.3b). The dominant control action relates to 2nd USC during the interval $t \in [1, 1.4]$s (and related 2nd OSC for $t \in [1.5, 2.2]$s), which compensates for the emphasized understeer behavior after the trajectory peak point. Other response features are qualitatively similar as described above for Maneuver 1, including the effect of square-wave shaping of lateral acceleration (particularly, during the dominant 2nd USC action).

The results for Maneuver 3 (Fig. 8.3c and Table 8.3c) are qualitatively similar to those of Maneuver 1, with the main difference that the control action is strongest for this more critical maneuver. Due to inability of accelerated vehicle to follow the sharp trajectory around the peak point, the 1st USC action (around $t = 0.1$ s) is made stronger to overshoot the initial part of trajectory, i.e. to "cut" the trajectory, to a larger extent than in Maneuver 1. Accordingly, the lateral acceleration response widely saturates afterwards, as well, thus getting closer to the square-wave shape. The 2nd USC intervention (around $t = 1s$) is also stronger to provide a more emphasized "cutting" of the trajectory around its peak point.

Table 8.3 indicates that the trajectory following error can be notably reduced when the control input constraint is omitted. However, the control input δ_r then reaches excessive values (and has rather oscillatory behavior), the tire operates deeply in the unstable saturation region (large magnitudes of α_i), the vehicle side-slip angle magnitudes increase for the acceleration Maneuvers 2 and 3, and the longitudinal slip of inner rear/driven tire is large. Also, due to an extremely abrupt and high-magnitude nature of (control input) response, the obtained optimization results could not have been fully reproduced through post-optimization simulation (a certain trajectory drift occurred).

8.3.3 Active Front Steering (AFS)

The AFS optimization results, with the soft control input constraint included, are given in third rows of Table 8.3. These results, as well as previous ARS results, are in a good agreement with the BPTT optimization results (cf. Tables 8.2 and 8.3).

Although the magnitudes of the control inputs $\Delta\delta_f$ and δ_r are comparable (see Fig. 8.4), the AFS control performance is notably worse than the ARS one (Table 8.3). This is explained by the fact that the driver already utilizes the front steering input to a large extent, while the rear steering control action can effectively boost the unused transient performance of rear tires. More specifically, the driver's front steering effort $\delta_f = \delta_{1,2}$ is directly transferred to the front tire sideslip angles $\alpha_{1,2}$ and, thus, to

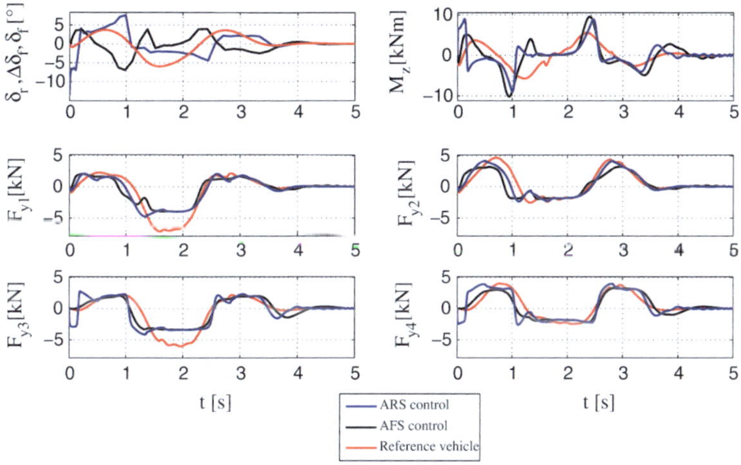

Fig. 8.4 Comparative ARS and AFS optimization results for Maneuver 3

the front lateral forces $F_{y1,2}$, while the rear tire sideslip angles $\alpha_{3,4}$ (and the related forces $F_{y3,4}$, are built up indirectly through a lagged response of the lateral velocity V (or the sideslip angle β) and the yaw rate r (see Eq. 8.3).

On the other hand, the ARS control action $\delta_r = \delta_{3,4}$ can directly influence $\alpha_{3,4}$ and $F_{y3,4}$, thus significantly improving the lateral vehicle responsiveness. This is illustrated by the comparative ARS and AFS responses of lateral forces F_{yi} (Fig. 8.4, Maneuver 3), which shows the following (with particular attention to the 1st USC event around t = 0.1 s): (1) the AFS action boosts the front lateral forces during the transient interval, but it has a limited authority over the rear lateral forces, (2) the ARS action makes the rear lateral forces response very fast and timely to assist the AFS action in boosting the overall yaw torque and lateral acceleration response.

8.3.4 Four Wheel Steering (4WS = ARS & AFS)

When the ARS is combined with AFS in a 4WS configuration, the trajectory following RMS error can be reduced by 7–30 % depending on maneuver (Table 8.3, regular input constraint). Here, the more critical the maneuver, the lower is the RMS error reduction. Figure 8.5a shows the corresponding time responses for Maneuver 1. The supporting AFS action is manifested in a strong 1st OSC intervention around t = 0.3 s, which has a counter-steering meaning, and supplemental 2nd and 3rd USC intervention around 1.1 and 2.65 s, respectively. The counter-steering interventions allows for a stronger 1st USC action by the ARS actuator, thus boosting the critical first peak of (oversteer) yaw torque M_z and sideslip angle β when compared with

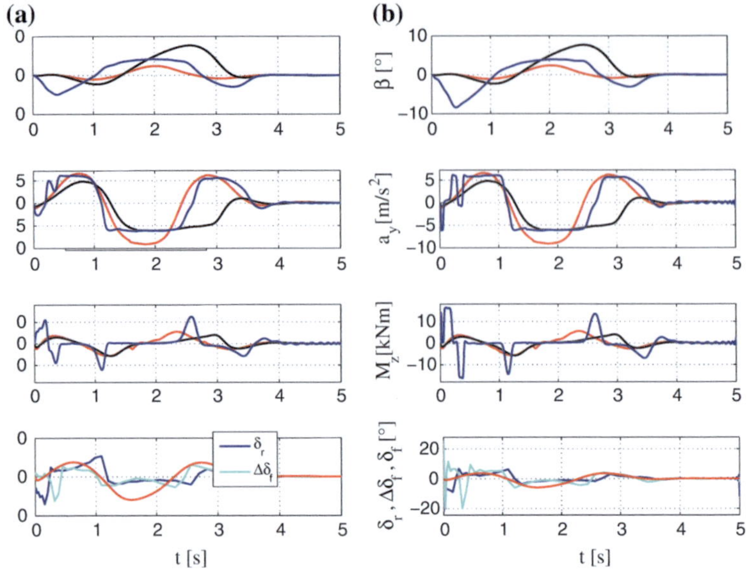

Fig. 8.5 Comparative 4WS optimization results for cases of regular (**a**) and weak (**b**) constraint on ARS and AFS control inputs, and Maneuver 1

ARS control only (cf. Figs. 8.3a and 8.5a and see also Table 8.3a). The β-boost affects the agility to some extent (Table 8.3a).

The 4WS can reduce the trajectory following RMS error by 40–65 % if the control input constraint is omitted (Table 8.3). However, this results in an unacceptable high-magnitude oscillatory behavior of control inputs and other variables, as already explained in Sect. 8.3.1. Also, the magnitudes of sideslip angle β and longitudinal slip η are very high (a "drifting" vehicle with spinning inner rear wheel).

If a weak control input constraint is used, the RMS error can be reduced by 11–24 % when compared to the regular input constraint (Table 8.3), with a generally acceptable, well-damped response (Fig. 8.5b), suppressed magnitudes of sideslip angle β, and a characteristic, strong and impulsive response of the yaw torque M_z. However, the AFS control input magnitudes are high, particularly the counter-steering one, which may not be convenient from the standpoint of perceived driver torque feedback. Also, the agility is somewhat deteriorated due to the increased β-peaks, particularly after the 1st USC phase.

8.4 Conclusion

The presented study has demonstrated that advanced algorithms of computational optimal control can represent a useful tool for assessment of various vehicle dynamics actuator configurations and gain the insights into optimal control actions of those actuators. The back propagation through time (BPTT) conjugate-gradient algorithm and nonlinear programming (NLP)-based TOMLAB tool have given comparable optimization results for single-actuator configurations, while the NLP approach has proven to be more effective in multi-actuator configurations.

Among various single-actuator configurations the active rear steering (ARS) actuator gives the best overall performance, because of its ability to directly influence the, otherwise unused/lagged, dynamic potential of rear tire lateral forces. The active front steering (AFS) actuator can effectively assist the ARS actuator, but the related performance boost is strongly related to an excessive AFS counter-steering action that can affect the perceived driver torque feedback. When considering the active differential application, it has been found out that the combined front and rear torque vectoring differential (TVD) configuration has the ability of reaching the ARS control performance. The AFS & TVRD and TVRD & TVCD configurations have also been found to provide good performance. The front or rear TVDs are inferior to ARS, but they may approach the AFS performance. The active limited slip differentials (ALSDs) are inferior to TVDs, particularly for maneuvers with emphasized understeer behavior.

References

1. Betts JT (2001) Practical methods for optimal control using nonlinear programming. SIAM, Philadelphia
2. Cipek M, Corić M, Skugor B, Kasać J, Deur J (2013) Dynamic programming-based optimization of control variables of an extended range electric vehicle. SAE paper No. 2013–01-1481
3. Deur J, Ivanović V, Hancock M, Assadian F (2010) Modeling and analysis of active differential dynamics. ASME J Dyn Syst Measur Control 132:061501/1–13
4. Deur J, Kasać J, Hancock M, Barber P (2011) A study of optimization-based assessment of global chassis control actuator configurations. In: Proceedings of IAVSD 2011
5. Falcone P, Tseng EH, Borelli F, Asgari J, Hrovat D (2008) Mpc-based yaw and lateral stabilization via active front steering and braking. Veh Syst Dyn 46:611–628
6. Guzzella L, Sciarretta A (2007) Vehicle propulsion systems. In: Introduction to modeling and optimization. Springer, Berlin
7. Hac A, Doman D, Oppenheimer M (2006) Unified control of brake- and steer-by-wire systems using optimal control allocation methods. SAE paper No. 2006–01-0924
8. Hancock M (2006) Vehicle handling control using active differentials. Ph.D. thesis, University of Loughborough, UK
9. Horiuchi S (2011) Evaluation of chassis control method through optimization based controllability region computation. In: Proceedings of IAVSD 2011
10. Hrovat D, Tseng HE, Lu J, Deur J, Assadian F, Borrelli F, Falcone P (2011) Vehicle control. In: Levine WS (ed) The control handbook: control system applications, 2nd edn. CRC Press, Boca Raton
11. Kasać J, Deur J, Novaković B, Kolmanovsky I, Assadian F (2012) A conjugate gradient-based bptt-like optimal control algorithm with vehicle dynamics control application. IEEE Trans Control Syst Technol 19:1587–1595
12. Kolmanovsky I, Stefanopoulou A (2001) Optimal control techniques for assessing feasibility and defining subsystem level requirements: an automotive case study. IEEE Trans Control Syst Technol 9(3):524–534. doi:10.1109/87.918904

13. Pilutti T, Ulsoy G, Hrovat D (2000) Vehicle steering system and method for controlling vehicle direction through differential braking of left and right road wheels. U.S. patent No. 6021367
14. PROPT-Matlab Optimal Control Software. User's guide (2010)
15. van Zanten AT (2000) Bosch ESP systems: 5 years of experience. SAE paper No. 2000-01-1633
16. Velenis E, Tsiotras P (2005) Minimum time vs maximum exit velocity path optimization during cornering. IEEE ISIE 2005:355–360

Chapter 9
Stochastic Fuel Efficient Optimal Control of Vehicle Speed

Kevin McDonough, Ilya Kolmanovsky, Dimitar Filev, Steve Szwabowski, Diana Yanakiev and John Michelini

Abstract Stochastic dynamic programming (SDP) is applied to generate control policies that adjust vehicle speed to improve average fuel economy without degrading, significantly, the average travel speed. The SDP policies take into account statistical patterns in traffic speed and road topography. Specific problems of fuel efficient in-traffic driving and fuel efficient lead vehicle following are considered, and it is shown how these problems can be treated within an SDP framework. Simulation results are summarized to quantify fuel economy improvements, and experimental results are reported for the fuel efficient lead vehicle following case. The properties of vehicle speed trajectories induced by SDP policies are examined.

9.1 Introduction

Fuel efficient driving has emerged as one of the pathways to increasing fuel economy of passenger cars. Fuel economy can be improved by driving at efficient speeds, accelerating quickly, but smoothly, coasting, and burn and coast (also known as

---

K. McDonough (✉) · I. Kolmanovsky
Department of Aerospace Engineering, The University of Michigan, Ann Arbor, MI, USA
e-mail: kevinmcd@umich.edu

I. Kolmanovsky
e-mail: ilya@umich.edu

D. Filev · S. Szwabowski · D. Yanakiev · J. Michelini
Ford Research and Advanced Engineering, Dearborn, MI, USA
e-mail: dfilev@ford.com

S. Szwabowski
e-mail: sszwabow@ford.com

D. Yanakiev
e-mail: dyanakie@ford.com

J. Michelini
e-mail: jmichel1@ford.com

H. Waschl et al. (eds.), *Optimization and Optimal Control in Automotive Systems*, Lecture Notes in Control and Information Sciences 455,
DOI: 10.1007/978-3-319-05371-4_9,

Pulse-and-Glide), see [22]. "Pulse-and-Glide" policies are particularly interesting in view of their connections with periodic optimal control [7, 19, 20].

References [3, 6, 8, 10, 13, 21] describe the use of deterministic dynamic programming and model predictive control to improve fuel economy. Systems that control vehicle speed based on electronic horizon have been introduced for commercial heavy vehicle applications [5]. References [2, 14, 18] focus on stability and the design of adaptive cruise control (ACC) systems.

Our approach to fuel efficient speed control [11, 12, 16, 17] in this and in previous publications is based on the perspective that the vehicle operating environment (e.g., speed of vehicle(s) in front of our vehicle in traffic) is inherently stochastic due to the uncertain nature of traffic. Even though for a known route, the road topography is known, the road grade can also be treated as stochastic if the route being traveled is not a priori known or to avoid relying on detailed topographic information. We model the statistics of transitions in traffic speed and road grade by Markov Chain models. With the knowledge of these statistics (e.g., specific to a certain geographic region where the driver operates the vehicle), vehicle speed and following distance can be optimized for best-on-average fuel economy and travel time performance using Stochastic Dynamic Programming (SDP).

Such an SDP policy prescribes the optimal vehicle speed as a function of current values of road grade, reference speed ("local" traffic speed, lead vehicle speed or our vehicle speed) and distance to preceding vehicle for best-on-average performance in terms of fuel consumption and travel time in a given geographic region. The SDP policy is generated off-line using value iterations or linear programming (along with appropriate approximations and simplifications) and is stored for on-line use. This approach opens up a possibility of implementing fuel efficient vehicle speed control at lower cost and complexity versus alternatives that use electronic horizon.

The previous solutions to fuel efficient vehicle speed control, with possible exception of [3, 21] which focused on the application of stochastic MPC, did not treat the vehicle operating environment as stochastic. The present chapter summarizes and unifies our results for fuel efficient in-traffic driving and fuel efficient following of a vehicle driving at a constant speed, provides further details, discussion, analysis, and includes previously unreported experimental results for the fuel efficient vehicle following case. Because this chapter contains a summary of previous work, it encompasses different simulations and vehicle experiments that track the evolution of our study. Throughout this evolution, while the general approach remained the same, tuning, settings, and models have changed because these modifications were beneficial, e.g., to improve experimental vehicle drivability.

9.2 Modeling for SDP Policy Generation

The models used for SDP control policy generation include the representations for vehicle dynamics, the stochastic models of road grade and traffic speed, and the cost function and its separate components. The models and the optimization are all based on spatial discretization. That is, given a distance segment of length Δs, and values of road grade, reference speed, host vehicle speed, and relative distance at the beginning of the current segment, the models produce the values of these variables at the start of the next distance segment.

9.2.1 Longitudinal Vehicle Dynamics

A simplified model for longitudinal vehicle dynamics has been adopted for control policy generation. The controlled vehicle is referred to as the host vehicle, and its velocity is denoted by v. The reference speed, v_{ref}, represents either the local traffic speed in the in-traffic driving scenario or the speed of the preceding (lead) vehicle in the lead vehicle following scenario. Finally, the relative distance between the vehicles is denoted by ρ. By v_{ref}^+, v^+, and ρ^+ we denote reference vehicle speed, host vehicle speed, and relative distance values one distance segment ahead.

The host vehicle speed is updated according to the following dynamic equation,

$$v^+ = v_{ref} + u, \tag{9.1}$$

where u is a control input corresponding to an offset in the host vehicle speed relative to traffic speed at the beginning of the next traffic segment. This model has been used to generate SDP policies for our simulation case studies of fuel efficient in-traffic driving and lead vehicle following.

For in-vehicle experiments, we replaced (9.1) with an alternative model,

$$v^+ = v + u, \tag{9.2}$$

where u is a control input corresponding to an offset in the host vehicle speed at the next distance segment relative to the host vehicle speed at the previous distance segment. The use of model (9.2) resulted in better drivability of the experimental host vehicle.

An approximate model for relative distance dynamics is defined by

$$\rho^+ = \rho + (v^+ - v)\Delta T, \tag{9.3}$$

where

$$\Delta T = \frac{2\Delta s}{v_{ref}^+ + v_{ref}}, \tag{9.4}$$

is (approximately) the time to travel the road segment of length Δs. We use $\Delta s = 30$ m to generate control policies in our simulation case studies. To provide faster control updates and improve experimental vehicle safety, we reduced Δs to 15 m when generating control policies for the vehicle experiments.

In both the experimental vehicle testing and the computer simulations, we provided a set point speed, v^+, to existing speed controllers and relied upon these existing controllers to execute the necessary changes in speed.

9.2.2 Stochastic Models of Reference Speed and Road Grade

Transitions in reference speed and road grade are modeled using Markov Chains. Transition probabilities are defined for changes in reference speed and road grade over a road segment of length Δs,

$$P(v_{ref}^+|v_{ref}), \quad P(\theta^+|\theta), \tag{9.5}$$

with values of v_{ref} and θ quantized to values on a discrete grid, $v_{ref} \in \{0, 1, \ldots, 36\}$ m/s, $\theta \in \{-6, -5, \ldots, 5, 6\}\,\%$.

The identification of the transition probabilities has been performed from 16 data sets collected on-board of an experimental vehicle. The vehicle has been driven along a route on and around M-39 (mixed highway and city driving), in northbound and southbound directions, during different times of day (in rush hour traffic and off-peak hours traffic) and with the driver emulating two different driving styles ('smooth' and 'aggressive'). Transition probability models have been identified using

$$P(x_j \mid x_i) \approx \frac{N_{x_i,x_j}}{M_{x_i}}, \tag{9.6}$$

where $P(x_j|x_i)$ is the approximate probability of transition of a variable $x \in \{v_{ref}, \theta\}$ from a discrete state x_i to state x_j, N_{x_i,x_j} is the number of transitions from state x_i to state x_j observed in the data, and M_{x_i} is the total number of transitions out of state x_i. Figure 9.1 illustrates the transition probability matrices learned based on one of the data sets. Note the near diagonal character of the model suggesting that continuing with the same reference speed and grade, i.e., $v_{ref}^+ = v_{ref}$ and $\theta^+ = \theta$, is most likely. On the right side of Fig. 9.1 we observe more off-diagonal entries at lower vehicle speeds as it takes longer in time for the vehicle to travel the distance segment and thus more time is available to accelerate to different speeds.

Assuming that θ and v_{ref} are independent, it follows that

$$P(\theta^+, v_{ref}^+|\theta, v_{ref}) = P(v_{ref}^+|v_{ref})P(\theta^+|\theta). \tag{9.7}$$

Remark 1 For the in-traffic driving, we ultimately advocate a multi-model approach where several transition probability models (TPMs) and SDP policies are developed for different traffic types (e.g., rush hour traffic versus off-peak hour traffic) and road types. A Kullback-Liebler (KL) divergence can be used to establish similarity between currently observed transition probabilities and transition probability models

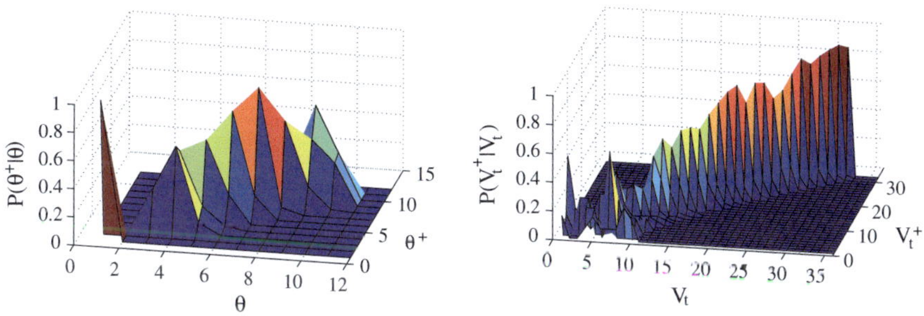

Fig. 9.1 A surface plot of grade (*left*) and reference vehicle speed (*right*) transition probabilities

that have been stored and for which the corresponding SDP policy is available. The SDP policy for the best matching TPM is then switched to for the on-board use. References [9, 15] discuss an evolving models approach where multiple models are generated to cover the drives of interest. However, [9, 15] do not generate SDP policies.

9.2.3 Cost Function Constituents

The SDP problem uses an incremental cost function of the general form,

$$R = \bar{W}_f + \lambda \bar{T}_t + \phi(\rho), \tag{9.8}$$

where the constituents are the expected fuel consumption, $\bar{W}_f$, over the segment of length Δs, the expected segment travel time, $\bar{T}_t$, and the distance constraint violation penalty, $\phi(\rho)$. The subsequent formulations of the SDP problem for the in-traffic driving use (9.8) with $\phi(\rho) = 0$. The constituents of the cost function are now discussed.

9.2.3.1 Fuel Consumption

The general model for the fuel flow is of the form,

$$W_f = W_f(v, v^+, \theta, \theta^+), \tag{9.9}$$

where v and θ are the vehicle speed and the road grade, respectively, at the beginning of the current road segment of length Δs, while v^+ and θ^+ are the vehicle speed and road grade, respectively, at the beginning of the next road segment of length Δs.

For the simulation case studies, the model (9.9) has been developed based on a single hidden layer neural network and has the form,

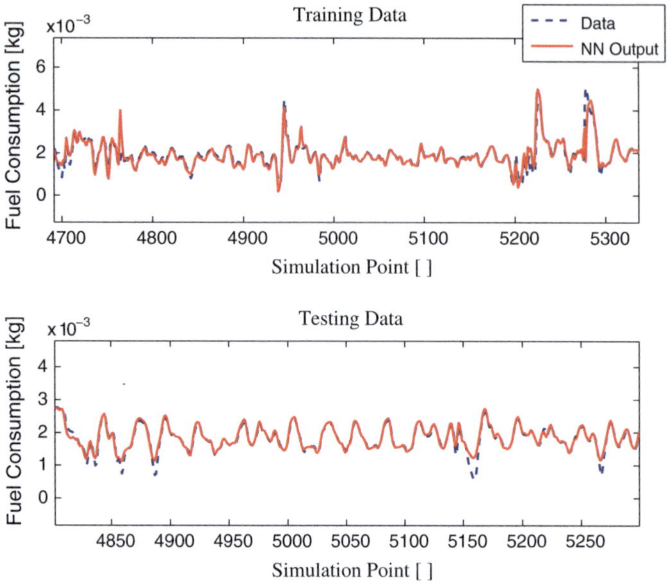

Fig. 9.2 Example of the training and testing data for the neural network model of the fuel flow based on CARSIM data. Note only a small sample of the 7,000 data points was selected in each plot for clarity

$$W_f = \sigma_2(w_2\sigma_1(w_1 u_{nn} + b_1) + b_2), \tag{9.10}$$

where σ_1 and σ_2 are the hyperbolic and linear activation functions, respectively; w_1 and w_2 are the corresponding vectors of weights; b_1 and b_2 are the corresponding vectors of biases; and $u_{nn} = (v, v^+, v^+ - v, \theta, \theta^+)^T$ is the model input vector. Using the *trainbr* function in MATLAB neural network toolbox, the neural network was trained using Bayesian regularization back propagation applied to 16 vehicle drives replicated in CARSIM environment. Roughly 14,000 data points were used to generate the neural network −7,000 to train and 7,000 to test. See Fig. 9.2.

For the experimental vehicle testing, a physics-based fuel consumption model representative of the 2007 Ford Edge experimental vehicle has been used. Note that both models account for gear shifts and torque converter lock/unlock events.

The expected value of fuel consumption used in the incremental cost function (9.8) has the form

$$\bar{W}_f = E_{v^+,\theta^+}[W_f(v, v^+, \theta, \theta^+)]. \tag{9.11}$$

9.2.3.2 Travel Time

The travel time of a road segment of length Δs can be approximated as

$$T_t(v, v^+) = \frac{2\Delta s}{v + v^+}. \tag{9.12}$$

The expected value of travel time used in the incremental cost function (9.8) has the form,

$$\bar{T}_t = E_{v^+}[T_t(v, v^+)]. \tag{9.13}$$

9.2.3.3 Distance Constraint Penalty

The penalty function for the distance constraint violation in (9.8) has one of the following forms. The first form,

$$\phi(\rho) = \begin{cases} \kappa & \text{if } \rho > \rho_{max}, \\ \kappa & \text{if } \rho < \rho_{min}, \\ 0 & \text{else}, \end{cases} \tag{9.14}$$

was used for the simulations while the second form,

$$\phi(\rho) = \begin{cases} (\kappa e^{\rho - \rho_{max}} - \kappa) & \text{if } \rho > \rho_{max}, \\ (\kappa e^{\rho_{min} - \rho} - \kappa) & \text{if } \rho < \rho_{min}, \\ 0 & \text{else}, \end{cases} \tag{9.15}$$

was used for in-vehicle experiments. In the above, $[\rho_{min}, \rho_{max}]$ is the desired interval in which the distance is to be maintained and $\kappa > 0$ is a parameter. The value of $\kappa = 10$ has been used to generate SDP policies for fuel efficient vehicle following for both simulations and experiments. Values of $\rho_{min} = 3$ m and $\rho_{max} = 10$ m have been used for the simulations with (9.14) while values of $\rho_{min} = 5$ m and $\rho_{max} = 15$ m have been used for the vehicle experiments using (9.15). A switch was made from (9.14) to (9.15) in order to eliminate discontinuities within the cost function.

We note that the minimum distance constraint is imposed to prevent host vehicle colliding with another vehicle in front, while the maximum distance constraint is imposed to improve driving comfort and reduce the impact of the host vehicle actions on the flow of traffic around it. We do not attempt to take advantage of drag reduction in close following even though in principle such a possibility can be realized by prescribing a small value for ρ_{max}. We also do not attempt to use asymmetric penalty factors in (9.14) or (9.15) and equally weight both relative distance extremes as both extremes are important for the safety of the host vehicle and surrounding traffic.

9.3 Stochastic Dynamic Programming

SDP is used to generate best-on-average control policies. For the incremental cost (9.8), the stochastic optimal control problem that SDP solves is of the form

$$J = E\left[\sum_{k=0}^{\infty} q^k R(v(k), \theta(k), \rho(k))\right] \to \min_{u \in U}, \tag{9.16}$$

subject to the model in Sect. 9.2 where $0 \leq q < 1$ is a discount factor introduced to guarantee that the cost is finite and the set U denotes the set of feasible control values. In (9.16), the value of $q = 0.96$ in order to approximate the average cost.

We use the value iteration approach [1] to solve the SDP problem. Under the standard assumptions, the following iterations converge, as $n \to \infty$, to the value function $V^*(v_{ref}, v, \theta, \rho)$,

$$\begin{aligned} V_{n+1}(v_{ref}, v, \theta, \rho) = min_{u \in U} \quad & Q_n(v_{ref}, v, \theta, \rho, u), \\ & Q_n(v_{ref}, v, \theta, \rho, u) = \\ R(v_{ref}, v, \theta) + \sum \theta^+, v_{ref}^+ \quad & q V_n(v_{ref}^+, v^+, \theta^+, \rho^+) P(v_{ref}^+, \theta^+ \mid v_{ref}, \theta), \\ & V_0(v_{ref}, v, \theta, \rho) = 0. \end{aligned} \tag{9.17}$$

The value iterations are performed numerically using standard gridding and linear interpolation techniques.

9.4 Simulation Case Studies

To quantify the fuel economy benefits of the SDP approach, simulation case studies were implemented in a simulation environment based on CARSIM and Matlab/Simulink. We used CARSIM to emulate a *traffic analog* vehicle that replicates the data collection drives of our experimental vehicle. We also implement in CARSIM the host vehicle responding to control inputs generated by SDP policy executed in Simulink. The SDP policy was generated to prescribe the offset relative to reference speed in (9.1),

$$u \in \{-3, -2, -1, 0, 1, 2, 3\} \text{ m/s}.$$

The metrics used to evaluate the results are the percent mpg improvement and percent difference in average vehicle speed. They are defined by

$$\%mpg_{imp} = \frac{mpg_h - mpg_t}{mpg_t} \times 100, \quad \%v_{diff} = \frac{\bar{v}_h - \bar{v}_t}{\bar{v}_t} \times 100, \tag{9.18}$$

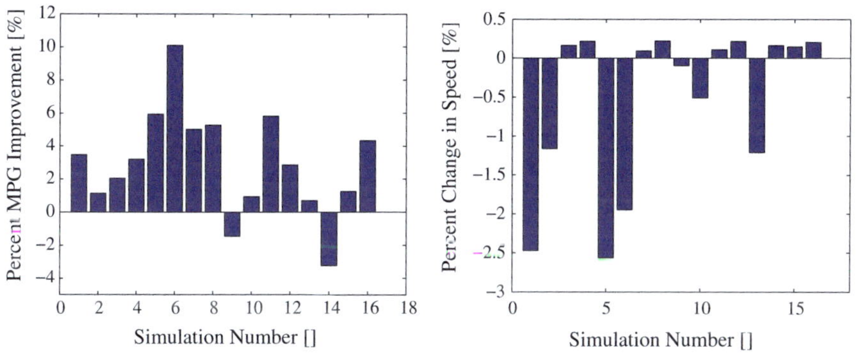

Fig. 9.3 *Left* the percent fuel economy increase for the in-traffic driving case study in the non-porous traffic case. *Right* the percent change in average speed

where $mpg_t, \bar{v}_t$ are, respectively, the fuel economy and the average speed of the traffic analog vehicle, while $mpg_h, \bar{v}_h$ are, respectively, the fuel economy and the average speed of the host vehicle. Positive values in $\%mpg_{imp}$ indicate the fuel economy improvements of the host vehicle and positive values in $\%v_{diff}$ faster average speed of the host vehicle sersus the traffic analog.

9.4.1 *In-traffic Driving*

For the in-traffic driving, $v_{ref} = v_t$. The value of the weight $\lambda = 0.002$ in (9.8) was chosen after some tuning. We assumed $\phi(\rho) \equiv 0$ in (9.8) at the stage of generation of SDP policies. As a result, the SDP policy became independent of the relative distance, ρ, which simplified its offline computation and online implementation. When evaluating this SDP policy, two case studies of non-porous traffic and porous traffic were considered.

9.4.1.1 Non-porous Traffic

In the non-porous traffic case, the host vehicle was not able to pass the traffic analog vehicle and the brakes would be activated in the simulation to ensure there was no collision. The non-porous traffic represented the worst case for the evaluation of SDP policy. The results based on our 16 drives are summarized in Fig. 9.3. For the non-porous traffic case, the average improvement in mpg is 2.97 % with an average drop in average speed of 0.78 %. Slight positive values of the average speed difference on the right of Fig. 9.3 for some of the drives indicate that the host and travel analog vehicles ended up closer together at the end of the simulation than when the simulation was started.

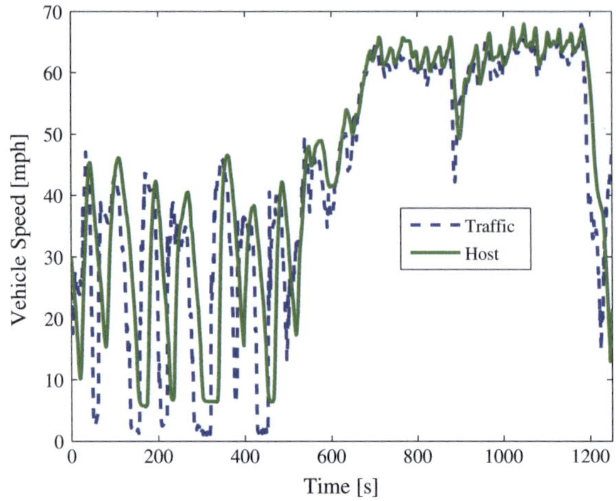

Fig. 9.4 The time histories of traffic analog (*dotted*) and host vehicle (*solid*) speeds for the simulation 6 in the non-porous traffic case

Figure 9.4 compares the vehicle speed time history of the traffic analog vehicle and host vehicle for drive 6 where the largest fuel economy improvement has been observed. The speed of the host vehicle with the SDP policy has fewer and smaller abrupt changes and an overall "smoothed-out" character, especially for large accelerations and decelerations. This conclusion is also confirmed by examining the vehicle speed traces for other drives.

9.4.1.2 Porous Traffic

In the porous traffic case, the host vehicle is able to follow the SDP policy freely passing or being passed by the traffic analog vehicle if needed. This case is consistent with the cost function used for SDP policy generation and represents the best case scenario. The results based on our 16 drives are shown in Fig. 9.5. For the porous traffic case, the SDP policy achieved simultaneous improvement in mpg (average improvement is 5.67 %) and an increase in average speed (average increase is 5.38 %).

Note that in the non-porous traffic case, the average speed for many of the simulations is lower than the traffic analog. Though one can argue that some of the fuel economy improvement can, in principle, be due to this reduction, it is clear from the porous traffic case results show that simultaneous increases in average speed and fuel economy are also feasible.

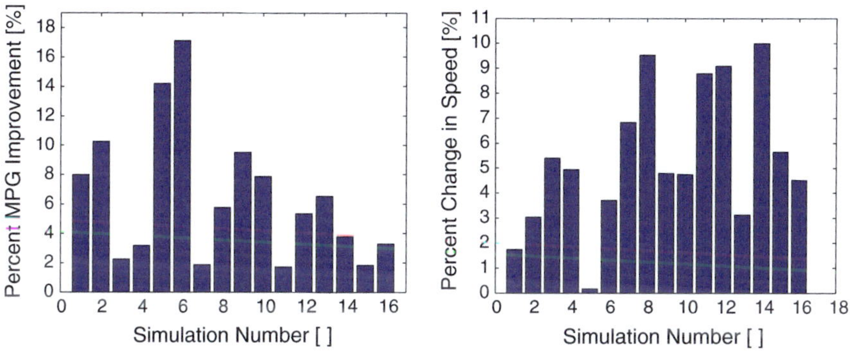

Fig. 9.5 *Left* The percent fuel economy increase for the in-traffic driving case study in the porous traffic case. *Right* The percent change in average speed

Table 9.1 Fuel economy improvement broken down between traffic conditions and traffic type

Traffic condition	Traffic type	Percent improvement in mpg
Rush hour	Porous	8.98
Rush hour	Non-porous	2.20
Off-peak	Porous	2.39
Off-peak	Non-porous	3.73

9.4.1.3 Rush Hour Versus Off-peak Traffic Conditions

Table 9.1 compares the average results over rush hour simulations (drives 1, 2, 5, 6, 9, 10, 13, 14) and off-peak traffic conditions (drives 3, 4, 7, 8, 11, 12, 15, 16). In both the porous and non-porous cases, the fuel savings, as compared to the lead vehicle, come from smoothing the accelerations and a reduction in the amount of vehicle braking. This observation is consistent with the results that in the non-porous case the host vehicle performs better than traffic during off-peak times while in the porous case the host vehicle does better during rush-hour times.

9.4.2 Optimal Vehicle Following

We consider next the vehicle following scenario, where the host vehicle follows, without passing, the lead vehicle that is driven at a constant speed, v_{ref}, where $v_{ref} \in \{45, 50, 55, 60\}$ mph. For this case, separate SDP policies with $\phi(\rho) \neq 0$ in (9.15) and $\lambda = 0.0012$ in (9.8) were generated for each of four values of v_{ref}. Furthermore, the cases of uniformly zero grade along the route and non-zero grade modeled stochastically, as discussed in Sect. 9.2.2, were considered separately. The SDP policies are functions of ρ, v, and θ (in the non-zero grade case) and prescribe

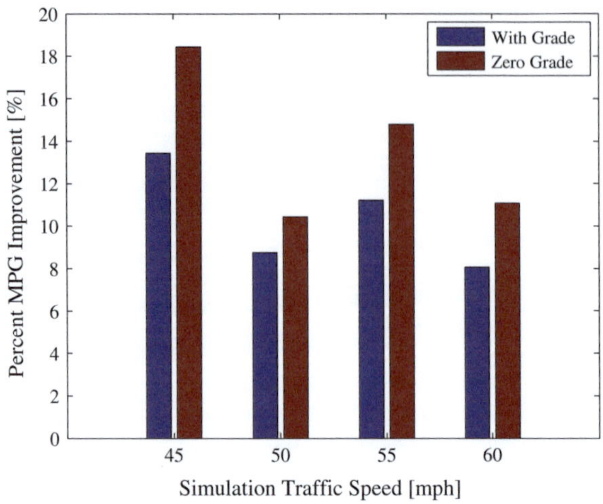

Fig. 9.6 Fuel economy improvement with SDP policies. The *blue* (*first*) set of bars are for the non-zero grade policies and the *red* (*second*) set of bars are for zero grade

the offset u in (9.1). Note that the zero grade policies are basically deterministic policies. The zero grade policies were simulated assuming zero grade while the non-zero grade policies were simulated with the grade profile recorded over the same route as in Sect. 9.2.2.

The fuel economy is improved with SDP policies, see Fig. 9.6. This improvement is attributed to Pulse-and-Glide behavior (oscillation) of the host vehicle speed induced by the SDP policies, see Figs. 9.7 and 9.8. The Pulse-and-Glide behavior has been observed in both non-zero grade (Fig. 9.7) and zero grade (Fig. 9.8) simulated drives. The sharp acceleration slope and more gradual deceleration (coasting) slope in non-zero grade simulation become more regular in the zero grade case. The fuel flow during sharp acceleration (pulse) exceeds fuel flow of the lead vehicle, but is less during the deceleration (coasting) phase. This results in a lower overall fuel consumption, seen in the right side of both Figs. 9.7 and 9.8 of the host vehicle compared to the lead vehicle. Due to the distance constraint, average vehicle speeds of the host and lead vehicles are nearly the same.

The fuel improvement mechanism of Pulse-and-Glide differs from that of acceleration smoothing. The acceleration smoothing limits the magnitude of the accelerations based on prediction of the lead vehicle velocity. It keeps the host vehicle from braking as much as traffic, potentially also avoiding unnecessary downshifts and torque converter unlocks. With Pulse-and-Glide, the lead vehicle is assumed to be at a constant and known speed, and the control policy takes advantage of differences in efficiency between operating points in the engine map, and of efficiently executed transients, to reduce fuel consumption.

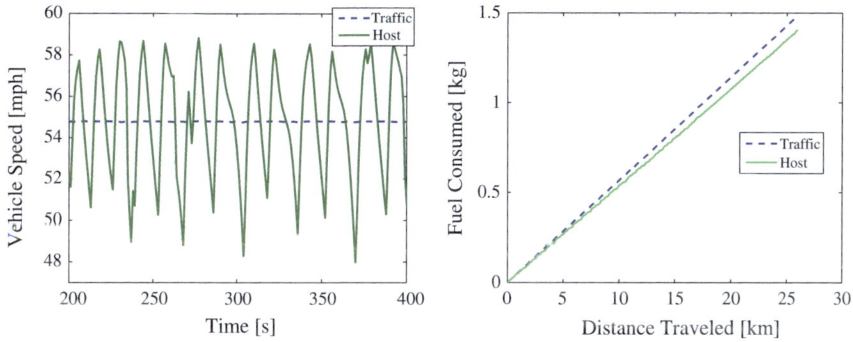

Fig. 9.7 *Left* time history of lead vehicle speed (*dashed*) and host vehicle (*solid*) for a part of the drive with non-zero grade. *Right* fuel consumption versus distance traveled of lead vehicle speed (*dashed*) and host vehicle (*solid*)

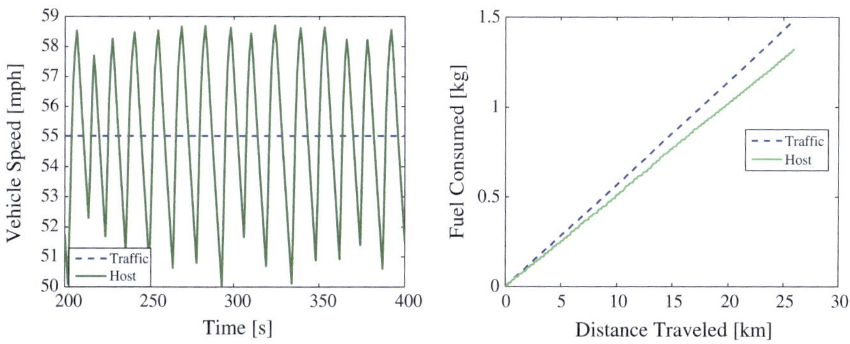

Fig. 9.8 *Left* time history of lead vehicle speed (*dashed*) and host vehicle (*solid*) for a part of the drive with zero grade. *Right* fuel consumption versus distance traveled of lead vehicle speed (*dashed*) and host vehicle (*solid*)

9.5 Vehicle Experiments

The experimental testing of fuel efficient vehicle following was performed in a 2007 Ford Edge. The tests were performed along the M-39 highway portion of the same route used for data collection in Sect. 9.2.2. We bypassed the vehicle speed set-point fed to the vehicle cruise controller with the output of SDP policy running on a dSPACE RTI 1005 board. The testing was restricted to vehicle speed following scenario, where the lead vehicle driving at a constant speed was implemented as a virtual vehicle (in software) while the host vehicle was the experimental vehicle. This implementation permitted testing with a single available vehicle which did not have a radar nor ACC. The vehicle was driven over the same route twice whenever testing was performed, once with the regular cruise control and then with the SDP policy. This was done in order to reduce errors due to day-to-day fluctuations in fuel economy.

After preliminary testing, to improve driving comfort we regenerated the SDP policies to prescribe the offset to the current host vehicle speed (see (9.2)) rather than to the lead vehicle speed, v_{ref}, and restricted the offset values to a small range $u \in \{-2, 0, 2\}$ mph to prevent large speed change requests. We additionally constrained u by adding to U the constraint $50 \leq v + u \leq 60$. We used $\lambda = 0.0012$ in the cost (9.8) as this value improved vehicle drivability.

Over 12 drives on M-39 (half in northbound and half in southbound direction) with $v_{ref} = 54$ mph, the average fuel economy improvement is 4.51 % with a maximum of 11.58 % and a minimum of -3.28 %. The maximum and minimum were observed on the same day during which there was a strong wind that may have skewed the results. Without these outliers, the average fuel economy improvement is 5.38 %.

Figure 9.9 shows the results of vehicle experiments for one of the drives. An oscillating pattern, representative of Pulse-and-Glide, clearly emerges in the experimental vehicle speed. The difference in the character of the response is attributed to changes in SDP problem formulation that we made to improve vehicle drivability and to the slow and asymmetric (between acceleration and deceleration) response of the nominal cruise controller to the set-point provided by the SDP policy. Also due to the nominal cruise controller response (unmodelled at the stage of generating SDP control policies), the average speed of the host vehicle ended up below the lead vehicle average speed. The distance ρ computed by a limited integrator in our implementation was thus saturated at the upper bound, ρ_{max}, during the tests.

The average speed difference over all 12 drives is -1.95 mph. To account for the increased fuel economy due to a lower average speed, test drives were performed using the standard cruise control set to 50, 52, and 54 mph in order to build a relationship between a reduction in average speed and increase in fuel economy. The resulting relationship is, on average, a 0.49 % reduction in fuel consumption for every percent drop in average speed. This results in a corrected fuel consumption improvement of 2.74 % when considering all of the tests and an improvement of 3.61 % when not considering the outliers.

9.6 Concluding Remarks

Fuel economy can be improved by optimal control of vehicle speed. One approach to developing such a speed controller, that does not use an electronic horizon of preview information, is through the stochastic modeling of traffic and terrain, and through the application of stochastic optimal control. The chapter provided a demonstration of this approach and showed that stochastic dynamic programming based policies can improve fuel economy through the mechanisms of smoother accelerations in in-traffic driving scenarios and through Pulse-and-Glide when following another vehicle moving at a constant (or nearly) constant speed. While these fuel economy improvement mechanisms have been discussed in previous literature, it is interesting

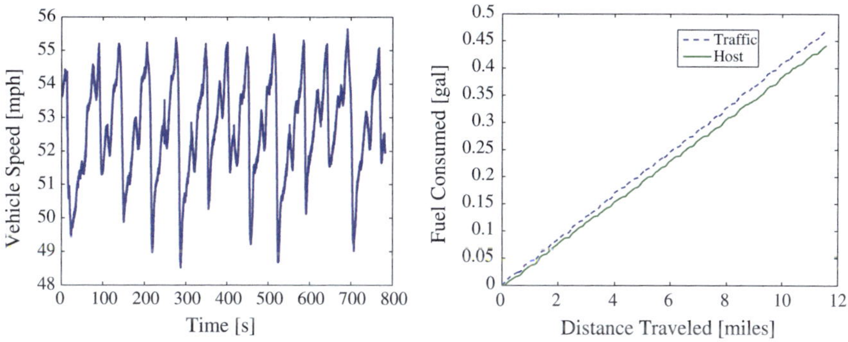

Fig. 9.9 *Left* time history of vehicle speed from vehicle testing. *Right* cumulative flow of the host vehicle (*solid*) versus lead vehicle (*dashed*)

that they also emerge in the process of solving a stochastic optimal control problem. The simulation and experimental results in the chapter provide a quantification of the benefits that can be achieved with this approach.

For the in-traffic driving, the smoothing of accelerations can be viewed as a way to eliminate aggressive driver behaviors that are detrimental to fuel economy. In the case of vehicle following, the lead car is not exhibiting any aggressive behaviors and thus the optimization exploits a Pulse-and-Glide mechanism. Consequently, the results of our case studies are not contradictory.

Much room remains for further research. This includes further vehicle experiments and benefit quantification in relation to existing Adaptive Cruise Control (ACC) systems. The issue of acceptability of Pulse-and-Glide type policies also requires further study in terms of customer and traffic flow impact. Our experience in the experimental vehicle suggests that driving comfort can be improved to a level that can be acceptable to some customers, and the impact on traffic flow can be minimized by a judicious choice of the cost and constraints.

We note that we have not pursued the development of a policy for in-traffic driving with a distance constraint because such a policy depends on more state variables and due to the curse of dimensionality is more complex to generate and store. The development of such a comprehensive policy is left to future work.

Acknowledgments The authors would like to thank Ford-University of Michigan Alliance for supporting this project.

References

1. Altman E (1999) Constrained Markov decision processes. Chapman and Hall/CRC, London/Boca Raton
2. Bareket Z, Fancher PS, Peng H, Lee K, Assaf CA (2003) Methodology for assessing adaptive cruise control behavior. IEEE Trans Intell Transp Syst 4(3):123–131

3. Bichi M, Ripaccioli G, Di Cairano S, Bernardini D, Bemporad A, Kolmanovsky IV (2010) Stochastic model predictive control with driver behavior learning for improved powertrain control. In: Proceedings of the 49th IEEE conference on decision and control, pp 6077–6082
4. Filev DP, Kolmanovsky I (2010) Markov chain modeling approaches for on board applications. In: Proceedings of American control conference (ACC), pp 4139–4145
5. Freightliner RunSmart Predictive Cruise Control, http://www.autoblog.com/2009/03/22/freightliner-debuts-runsmart-predictive-cruise-control/
6. Gausemeir S, Jaker K, Trachtler A (2010) Multi-objective optimization of a vehicle velocity profile by means of dynamic programming. In: Proceedings of 6th IFAC symposium advances in automotive control, Munich, July 2010
7. Gilbert EG (1976) Vehicle cruise: improved fuel economy by periodic control. Automatica 12(2):159–166
8. Hellstrom E, Aslund J, Nielsen L (2010) Design of an efficient algorithm for fuel-optimal look-ahead control. IFAC J Control Eng Pract 18:1318–1327
9. Hoekstra A, McDonough K, Filev D, Szwabowski S, Kolmanovsky IV (2013) Evolving Markov chain models of driving conditions using onboard learning. In: Proceedings of IEEE international conference on cybernetics. Lausanne, Switzerland
10. Kamal M, Mukai M, Murata J, Kawabe T (2009) Development of ecological driving system using model predictive control. In: Proceedings of ICCAS-SICE, pp 3549–3554
11. Kolmanovsky IV, Filev DP (2009) Stochastic optimal control of systems with soft constraints and opportunities for automotive applications. In: Proceedings of 2009 IEEE multi-conference on systems and control, pp 1265–1270
12. Kolmanovsky IV, Filev D (2010) Terrain and traffic optimized vehicle speed control. In: Advances in automotive control, 6th IFAC symposium, pp 378–383
13. Li S, Li K, Rajamani R, Wang J (2011) Model predictive vehicular adaptive cruise control. IEEE Trans Control Syst Technol 19:556–566
14. Liang CY, Peng H (1999) Optimal adaptive cruise control with guaranteed string stability. Veh Syst Dyn 32(4–5):313–330
15. McDonough K, Kolmanovsky IV, Filev D, Yanakiev D, Szwabowski S, Michelini J, Abou-Nasr M (2011) Modeling of vehicle driving conditions using transition probability models. In: Proceedings of IEEE multi-conference on systems and control, pp 544–549
16. McDonough K, Kolmanovsky IV, Filev D, Yanakiev D, Szwabowski S, Michelini J (2012) Stochastic dynamic programming control policies for fuel efficient in-traffic driving. In: Proceedings of American control conference, pp 3986–3991
17. McDonough K, Kolmanovsky IV, Filev D, Yanakiev D, Szwabowski S, Michelini J (2013) Stochastic dynamic programming control policies for fuel efficient vehicle following. In: Proceedings of American control conference
18. Naus G, Vugts R, Ploeg J, van de Molengraft R, Steinbuch M (2010) Cooperative adaptive cruise control, design and experiments. In: American Control Conference (ACC), IEEE, pp 6145–6150, June 2010
19. Sciarretta A, Guzzella L (2005) Fuel-optimal control of rendezvous maneuvers for passenger cars. Automatisierungstechnik 53:244–250
20. Stoicescu AP (1995) On fuel-optimal velocity control of a motor vehicle. Int J Veh Des 17(2/3):229–256
21. Stanger T, del Re L (2013) A model predictive cooperative adaptive cruise control approach. In: Proceedings of American control conference (ACC), pp 1374–1379
22. Wikipedia, Fuel economy-maximizing behaviors. http://en.wikipedia.org/wiki/Fuel_economy-maximizing_behaviors

Chapter 10
Predictive Cooperative Adaptive Cruise Control: Fuel Consumption Benefits and Implementability

Dominik Lang, Thomas Stanger, Roman Schmied and Luigi del Re

Abstract Impressive improvements of efficiency and safety of vehicles have been achieved over the last decade, but increasing traffic density and drivers' age accentuate the need of further improvements. The contributions summarized in this chapter argue that a substantial additional fuel benefit can be achieved by extending the well introduced Adaptive Cruise Control in a predictive sense, e.g. taking into account a predicted behavior of other traffic components. This chapter starts by discussing results on the potential benefits in the ideal case (full information, no limits on computing power) and then examines how much of the potential benefits is retained if approximate solutions are used to cope with a realistic situation, with limited information and computing power. Two setups are considered: vehicles exchanging a small set of simple data over a V2V link and the case of mixed traffic, in which some vehicles will not provide any information, but the information must be obtained by a probabilistic estimator. The outcome of these considerations is that the approach is able to provide—statistically—a substantial fuel consumption benefit without affecting negatively the driveability or the driver comfort like other methods, e.g. platooning, would.

D. Lang · T. Stanger · R. Schmied · L. del Re (✉)
Institute for Design and Control of Mechatronical Systems, Johannes Kepler University Linz, Altenberger Straße 69, 4040 Linz, Austria
e-mail: dominik.lang@jku.at

T. Stanger
e-mail: thomas.stanger@jku.at

R. Schmied
e-mail: roman.schmied@jku.at

L. del Re
e-mail: luigi.delre@jku.at

H. Waschl et al. (eds.), *Optimization and Optimal Control in Automotive Systems*, Lecture Notes in Control and Information Sciences 455,
DOI: 10.1007/978-3-319-05371-4_10,

10.1 Introduction

Classical vehicle design aims to fulfill the needs of a wide variety of potential drivers in a wide variety of situations. While emissions and fuel consumption are determined under very well defined test conditions, actual consumptions and emissions arise from the operation of generic drivers under generic conditions. Different driver styles are known to affect very strongly the fuel consumption [1]. Advanced Driving Assistance Systems (ADAS) and autonomous vehicles can do more than support a different driving style, i.e. take a fuel optimal decision in real time considering several boundary conditions. This chapter shows that this idea—which can be seen as a seamless extension of normal adaptive cruise control (ACC)—offers a high fuel benefit potential. It can be realized with different degrees of complexity and can be used both in a connected traffic or—with some performance reduction—in a mixed traffic environment with high predictability and comfort.

The driving force behind the development of ADAS has clearly been safety, as in the case of anti-lock brake systems (ABS), first developed in 1929 for airplanes, followed by traction control, electronic stability programs (ESP) and brake assistants and many more (see [2] for an overview). Some ADAS have an impact on energy efficiency, in particular the Cruise Control (CC), which enforces a constant driving speed, and ACC [3–6], which additionally enforces a safe headway.

Recently, the focus has been extended to Cooperative ACC (CACC) [7–10] made possible by modern communication possibilities. Still, the focus is mainly safety, at the level of single vehicles, or fuel consumption, at the traffic level, by preventing jams, scheduling gear shifts etc. Some works have also addressed fuel efficiency, see. e.g. in [7] for a Model Predictive Control (MPC) solution or [3] for a Pulse and Glide (PnG) algorithm for a continuous variable transmission. Stochastic traffic models have been examined in [11] and applied to stochastic DP [12, 13].

This chapter focuses on the minimization of fuel consumption in moderate, non-congested traffic typical of a commuter travel. Computations and measurements are done on the basis of the available test car and engine (BMW320d and N47).

10.2 Problem Statement

We consider the speed control of a vehicle with a target speed v_{ref} in a multi vehicle scenario in which it has to cope with other vehicles with an own speed profile, in particular with a preceding one which acts as a disturbance, as shown in Fig. 10.1. We shall assume the controlled car to be equipped with some distance measurement device, e.g. radar. For simplicity, we shall not consider lane or gear changes. In this case, the related problem can be stated intuitively as

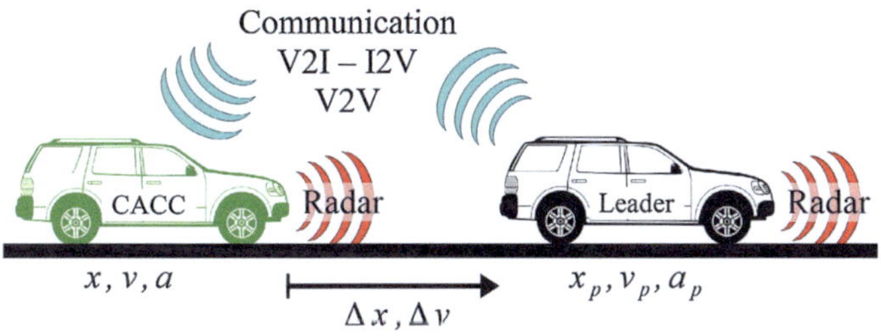

Fig. 10.1 Considered scenario, V2V communication is advantageous but optional

$$\min_{a(t)} Q_{f,T} = \min_{a(t)} \frac{1}{T} \int_0^T q_f\,(v(t), a(t))\ \mathrm{d}t, \tag{10.1}$$

where q_f is the current fuel consumption depending on the vehicle's speed $v(t)$ and acceleration $a(t)$, T the observation time, all other conditions assumed to be equal (path, temperature, pressure, humidity etc.).

10.2.1 Casting the Problem into the Mathematical Form

The basic dynamic equations of two vehicles moving within a string are given in the simplest form by

$$\begin{aligned} \Delta\dot{x} &= v_p - v \\ \dot{v} &= a \end{aligned} \tag{10.2}$$

where v_p is the velocity of a disturbing, preceding vehicle, a the acceleration and $\Delta x > 0$ the inter-vehicle distance.

Fuel consumption of a Diesel engine can be approximated by a static map depending on the engine torque and the rotational engine speed or accordingly on vehicle acceleration and vehicle speed, as shown in Fig. 10.2. The transient fuel consumption, not covered by the static map,adds typically about 4 % [14]. Boundary conditions can be derived by additional requirements.

A minimum distance is needed to prevent rear-end crashes. In [15] a historical review on car following models and their underlying spacing policies is given and in [16] collision avoidance capabilities of spacing policies are discussed.The most common spacing policy used, e.g. in [10], is linearly depending on the vehicle speed v

$$\Delta x \geq \Delta x_{min,0} + h_{min} v. \tag{10.3}$$

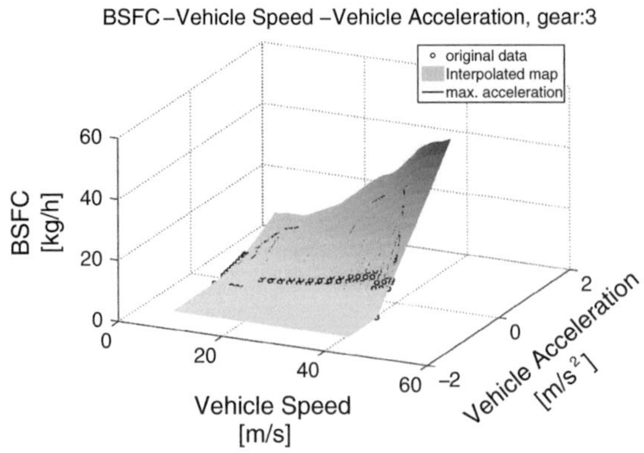

Fig. 10.2 Interpolated map of static fuel consumption, fixed gear, BSFC means brake-specific-fuel-consumption, q_f

where $\Delta x_{min,0}$ denotes the minimum inter-vehicle distance at stand still and h_{min} is the time-headway. *Maximum distance* is not a typical boundary condition, as, if the distance is large enough, both ACC and CACC would work as a standard CC. However, in congested traffic it seems advisable to limit this quantity to prevent loss of road capacity (and also to reduce frequency of third vehicle entering in the space between the actual vehicle and the predecessor). A maximum inter-vehicle distance is enforced by the constraint

$$\Delta x \leq \Delta x_{max,0} + h_{max} v, \quad v_P < v_{ref} \tag{10.4a}$$

$$v = v_{ref}, \quad v_P \geq v_{ref} \tag{10.4b}$$

considering the case that the speed of the preceding vehicle exceeds the reference speed, $\Delta x_{max,0}$ being the maximum distance at stand still and h_{max} the time headway coresponding to the maximum inter-vehicle distance.

Additional boundary conditions have to be included to respect system inherent limitations like maximum accelerating or decelerating power as well as maximum and minimum speed of the vehicle

$$a_{min}(v) \leq a \leq a_{max}(v) \tag{10.5a}$$

$$v_{min} \leq v \leq v_{max} \tag{10.5b}$$

where a_{max} and a_{min} are the speed dependent maximum and minimum acceleration and v_{max} and v_{min} denote the velocity limits of the vehicle.

Combining all this leads to the nonlinear optimization problem

$$\min_{a(t)} \int_0^T \left(q_f(v(t), a(t))\right) \mathrm{d}t \tag{10.6a}$$

Table 10.1 Potential fuel benefits with respect to a preceding vehicle assuming perfect knowledge of the future

Floating margin (m)	10	20	30	40	50	60	70	80
Potential fuel benefits (%)	9.2	14.7	18.3	20.2	22.1	23.9	24.4	25.7

$$
\begin{aligned}
\text{s.t.}\quad \Delta\dot{x}(t) &= v_p(t) - v(t)\\
\dot{v}(t) &= a(t)\\
\Delta x_{min,0} + h_{min}v(t) &\le \Delta x(t) \le \Delta x_{max,0} + h_{max}v(t)\\
a_{min}(v(t)) &\le a(t) \le a_{max}(v(t))\\
v_{min} &\le v(t) \le v_{max}
\end{aligned}
\tag{10.6b}
$$

Several other constraints can be introduced to ensure ride comfort and the acceptance by a human driver, for example limitations of acceleration or jerk.

10.3 Assessment of Potential

This nonlinear optimization problem cannot be solved in real time, but an approximate off-line numerical evaluation is possible in order to ascertain the potential of the method, providing the problem is simplified by discretization in time (see [22] for details). For this evaluation, we assume the preceding vehicle to follow a trajectory as in the FTP-75 (between 640 and 1,140 s) shifted by 24 km/h to better approximate a commuter traffic situation (assuming no jams). The behavior of the controlled vehicle 1 is shown in Fig. 10.3, in which the bottom subplot depicts the speed profile of the preceding vehicle, as well as the fuel optimal speed profile while the top graph illustrates the corresponding inter-vehicle distance. The optimal behavior corresponds to the expectations, the controlled vehicle uses the information on the future speed profile of the preceding vehicle to smooth its own speed profile by floating between a minimum and maximum bound. A floating margin of 20 m leads to a fuel consumption reduction of approximately 14.7 % with respect to the preceding vehicle. As portrayed in Table 10.1, the results show a strong dependency on the allowed floating margin.[1] The setup can be extended to a string of vehicles.

Figure 10.4 portrays the speed profiles of a string consisting of vehicles. As expected, a further reduction in acceleration and deceleration maneuvers towards the rear of the string is clearly visible. The effect on fuel economy is shown in Table 10.2.

[1] maximum–minimum allowed distance.

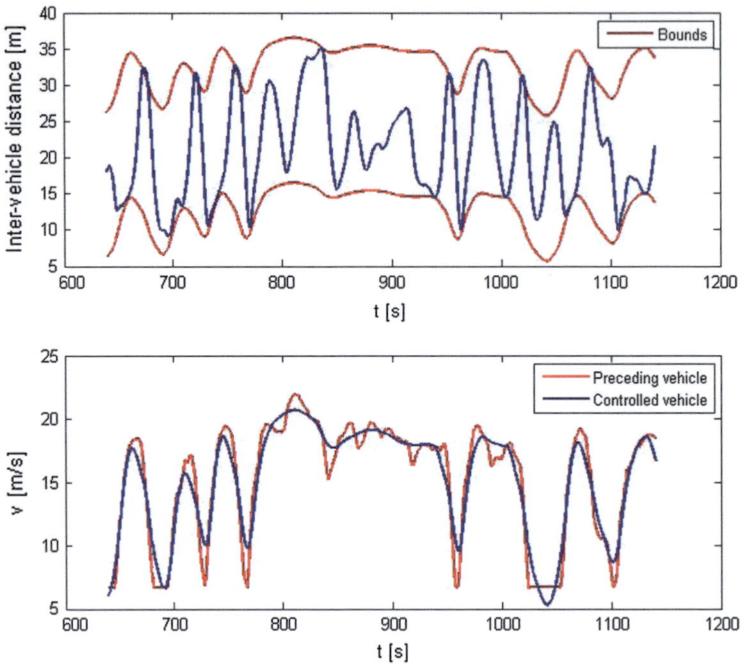

Fig. 10.3 The typical behavior of the optimized following vehicle. The optimized speed profile of the succeeding vehicle is essentially smoothed and is floating between minimum and maximum inter-vehicle distance

Table 10.2 Potential fuel benefits with respect to a predecessor in a string of vehicles assuming perfect knowledge of the future (floating margin = 20 m)

Vehicle position in string	2	3	4	5	6	7
Potential fuel benefits (%)	14.7	23.4	28	31.3	32.9	33.4

10.4 Nonlinear Receding Horizon Optimization

The results shown in the preceding section hint at a large available potential, but of course the implicit assumption of perfect knowledge of the future speed profile of the predecessor is not realistic. Also on-line solution of the problem with a very high number of manipulated variables (500) seems out of question for some time. However, Table 10.1 shows that the potential fuel benefits approaches a saturation for increasing floating margin, so that we might expect something similar as far as the optimization time is concerned.

So we consider addressing the problem by a moving horizon strategy, as done in model predictive control (MPC), see e.g. [18, 19]. Indeed, in MPC, only a relatively short horizon is considered, thus a short string of values of the manipulated variables is

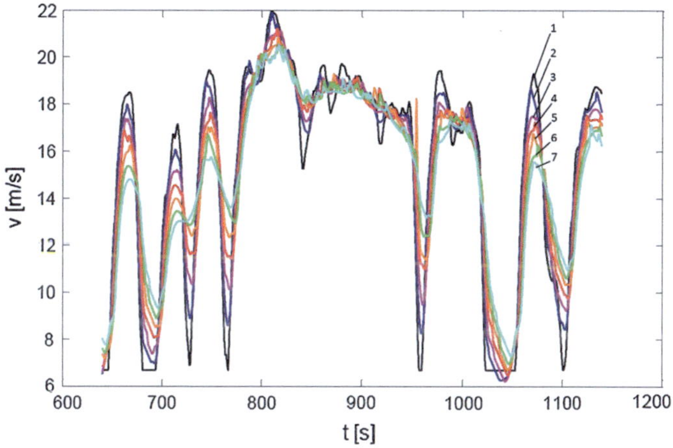

Fig. 10.4 Speed profiles of a string of vehicles (vehicle *1*: string leader, vehicles *2–7*: controlled vehicles)

Table 10.3 Comparison of potential fuel benefits (floating margin = 40 m)

Prediction horizon	5 s	10 s	15 s	20 s	30 s	40 s	unlimited
Potential fuel benefits (%)	2.8	12.3	17.1	18.6	19.2	19.7	20.2

computed, but only the first value is really used. At the next time step, the optimization is repeated. Although it is well known that the strategy does not necessarily lead to the optimal solution it is usually accepted as a sensible approximation. Figure 10.5 presents the dependency of the achieved fuel consumption benefit on the length of the finite prediction horizon for different margins using a nonlinear solver of the receding time version of the original problem. As expected, the same saturation occurs not only in terms of floating margin, but also in terms of prediction time, albeit with some cross dependency. Table 10.3 compares some of the obtained results to the optimal solution depicted in Table 10.1. As the prediction horizon becomes longer, the potential benefits converge to the optimal values.

The impact of the floating margin on traffic capacity is shown in Fig. 10.6 for an average average vehicle length of 4 m. As expected, the traffic capacity decreases with increasing floating margin, see Fig. 10.7.

10.5 Approximate Control Law Within the Linear MPC Framework

While the solution of the preceding section is already easier to implement than first one a further computational simplification can be obtained by using the linear MPC framework, using local linearizations as proposed in [20]. Figure 10.8 shows

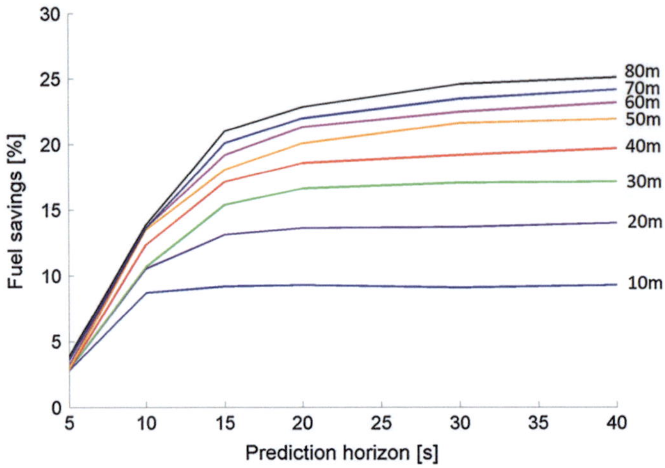

Fig. 10.5 Dependency of the achieved fuel consumption benefit on the length of the finite prediction horizon for different margins

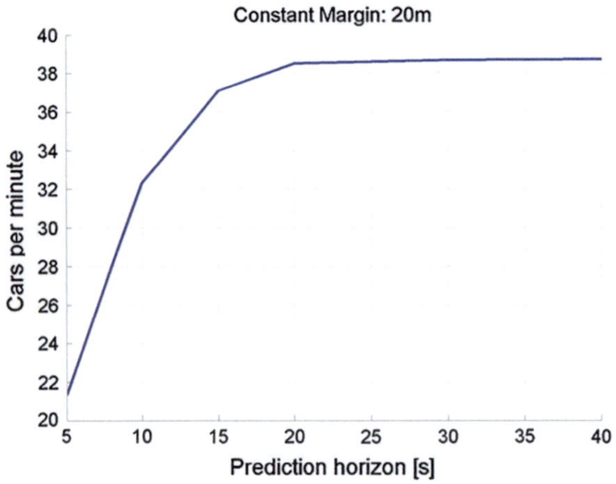

Fig. 10.6 Traffic capacity with respect to the prediction horizon

the behavior of the MPC-CACC with respect to the preceding vehicle, comparing speed and acceleration with a prediction horizon of 20 s and a control horizon 10 s. All together the behavior of the MPC-CACC resembles the behavior and findings presented by Sect. 10.3. The corresponding values of fuel benefit and the influence of different prediction horizon, are presented in Table 10.4. As expected, fuel benefits increase in a string, but converge also to a limit after a few vehicles as shown in Table 10.5. The effect of floating margin is quite similar to those of Fig. 10.5.

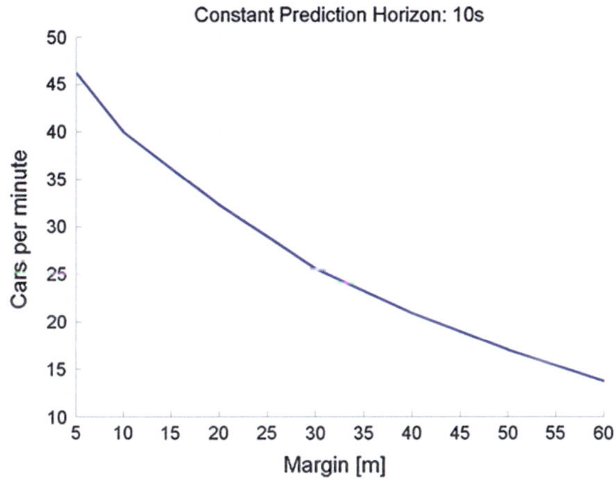

Fig. 10.7 Traffic capacity with respect to the provided margin

Table 10.4 Comparison of average fuel consumption and benefit of MPC-CACC with different prediction horizons

	Preceding vehicle	MPC-CACC			
		PH = 5	PH = 10	PH = 15	PH = 20
l/100 km	4.71	4.19	3.95	3.85	3.85
Relative fuel benefit in (%)		11.04	16.14	18.28	18.27
Average capacity in vehicles/min		20.16	20.8	25.62	26.16
Average capacity in (%)		92.7	105.0	117.9	120.3

Table 10.5 Comparison of average fuel consumption and benefit of MPC-CACC within a string of vehicles

Vehicle	Preceding vehicle	Following vehicle #				
		1	2	3	4	5
l/100 km	4.71	3.85	3.61	3.49	3.45	3.42
In-line benefit in (%)		18.27	6.18	3.28	1.16	0.95

10.6 Approximate Control Law Utilizing an Identified Hammerstein–Wiener Model

There is growing interest in applying MPC on more automotive issues, and progresses in algorithms and computational power make this choice the longer the more realistic [21]. However, for practical applications a further approximation level may be interesting based on the design of a fixed network whose output mimicries the behavior of the optimal control without the corresponding computational burden.

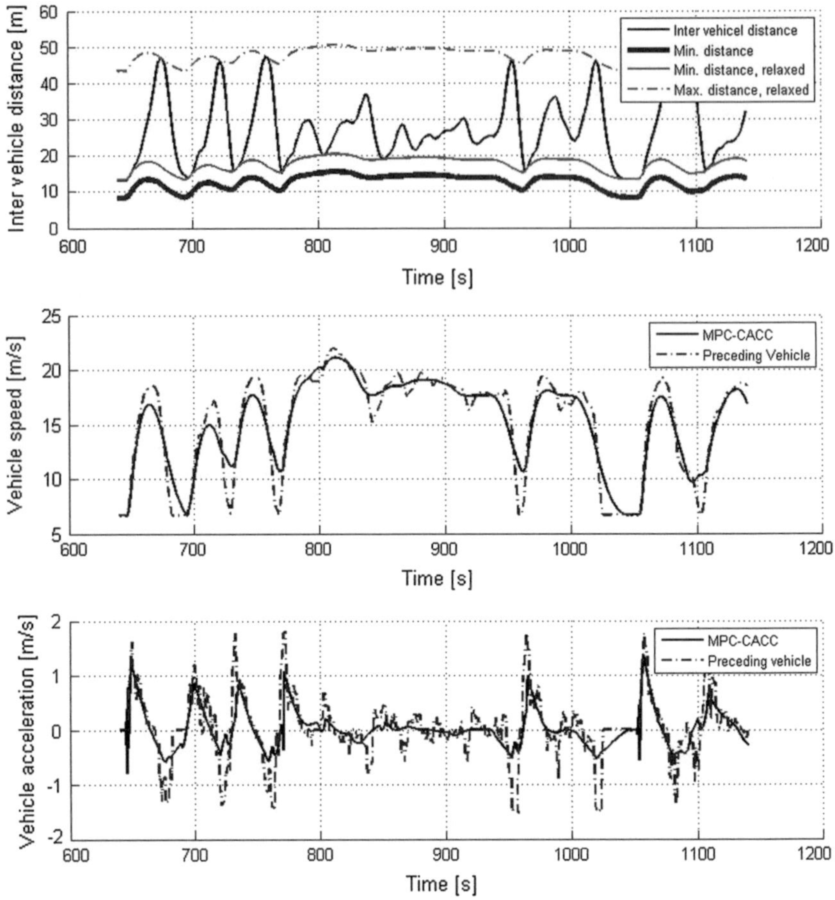

Fig. 10.8 Inter-vehicle distance including constraints and vehicle speed

Indeed, the optimal controller uses the velocity of the preceding vehicle v_P as input and delivers the optimal speed as output. If we use this quantity to compute the optimal distance to the preceding vehicle $\Delta x(t)$, and use v_P as input and $\Delta x(t)$ as output, it turns out that a nonlinear Hammerstein–Wiener (nlHW, see Fig. 10.9) model can be tuned to yield a good, cycle-independent approximation.

The nonlinearities of the identified model can be interpreted as an approximation of the distance constraints imposed on the corresponding optimization problem [22]. Of course, such a fixed controller needs additional elements to take into account the predictive action. This is recovered by adding a non causal smoothing filter, requiring a prediction of approximately 10 s. The smoothing filter additionally ensures the drivability of the approximated inter-vehicle distance. The schematic of this approximation, including an external branch to prevent an offset is portrayed in Fig. 10.10.

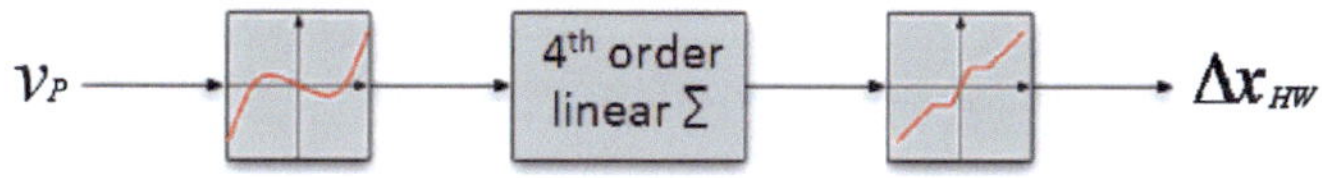

Fig. 10.9 Structure of nlHW

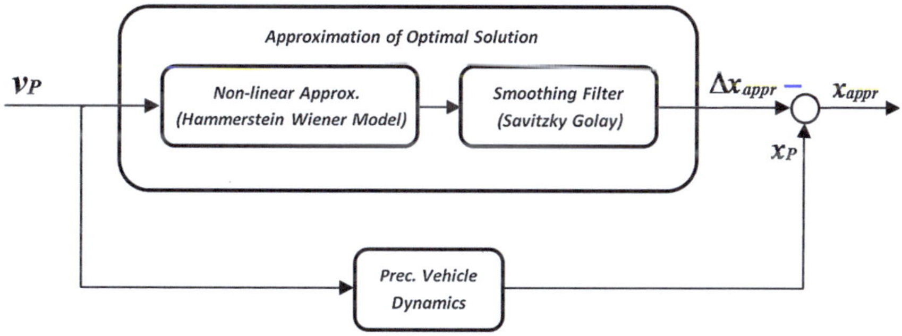

Fig. 10.10 Basic structure of approximate control

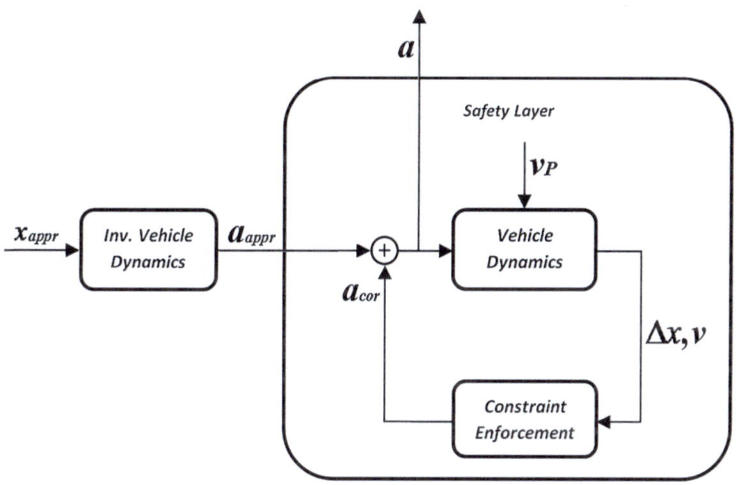

Fig. 10.11 Basic structure of safety layer with constraint enforcement controllers

Naturally, an additional safety layer is required to ensure the strict fulfillment of the range bounds by modifying the approximated vehicle acceleration, but this would be needed in any real application anyway. The basic structure of the proposed safety layer is shown in Fig. 10.11.

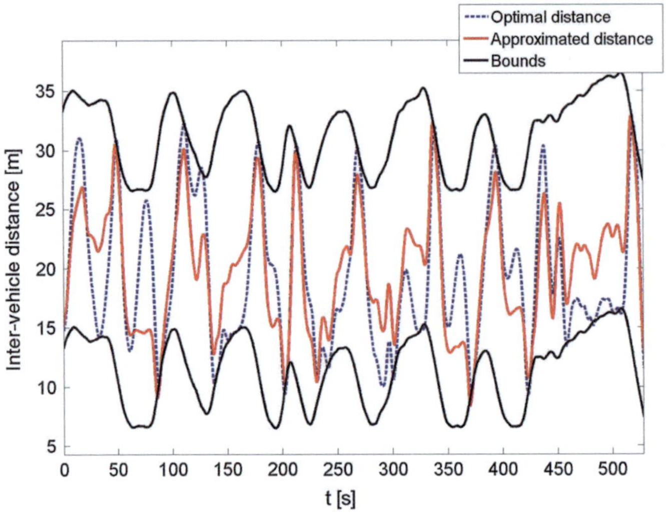

Fig. 10.12 Optimal inter-vehicle distance and its approximation

Table 10.6 Comparison of potential consumption reductions

Prediction horizon (s)	Benefits at different margins			
	10 m (%)	20 m (%)	30 m (%)	40 m (%)
10	9.9	11.4	11.9	12.3
15	10.8	14.4	16.3	17.1
20	11.2	15.1	17.5	18.6
30	11.3	15.3	18.4	19.2
40	11.3	15.3	18.5	20.2
Approximation	10.3	13.4	15.7	17

An example of an approximated inter-vehicle distance is displayed in Fig. 10.12. The resultant potential fuel benefits are shown in Table 10.6. The approximated solution yields benefits close to those achievable with a receding horizon of 15 s. A more detailed discussion and further results can be found in [22].

10.7 Traffic Prediction Model From Data

All methods presented until now assumed some knowledge on the future behavior of the preceding driver. This information, however, is usually not available. This section presents some results on possible prediction methods using a set of nonlinear, autoregressive (NARX) models, both in the case of interconnected and nonconnected vehicles using data gathered in suburban road traffic near to Linz, Austria. These

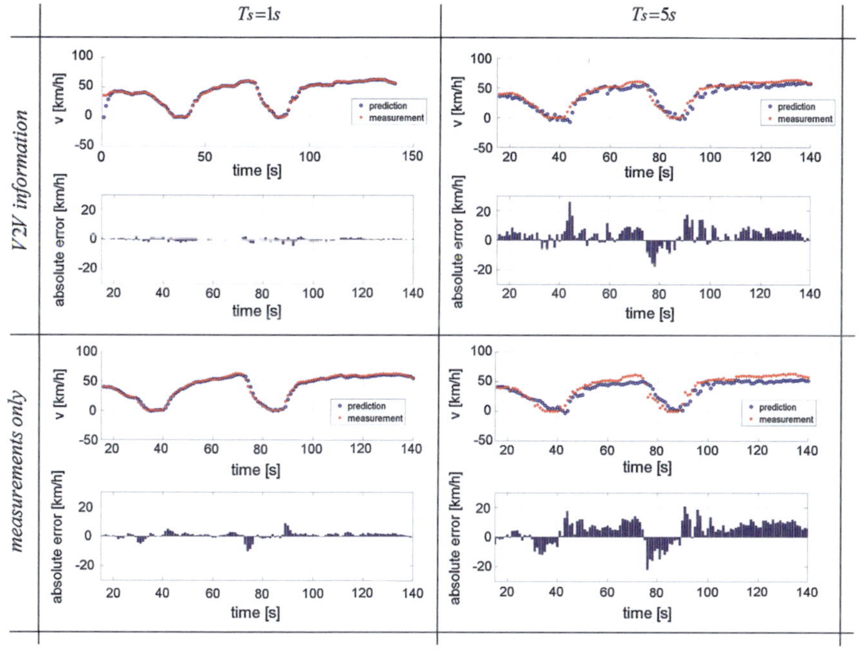

Fig. 10.13 1 and 5 s ahead prediction, employing 'V2V assisted approach' and the 'measurements only approach'

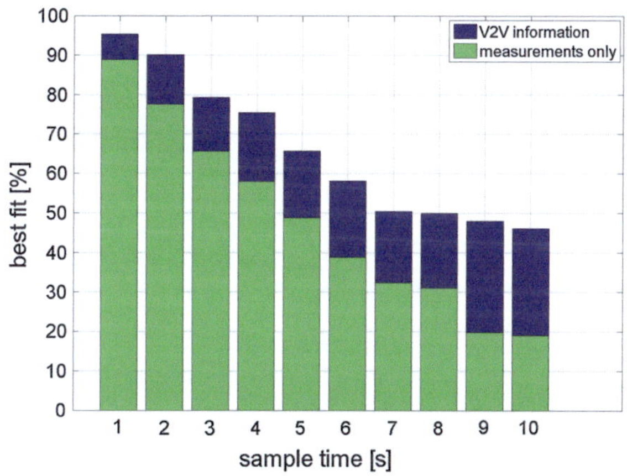

Fig. 10.14 Comparison of of the prediction quality between the two approaches

models are given by

$$y(t+i) = f_i(u(t), u(t-i)) \tag{10.7}$$

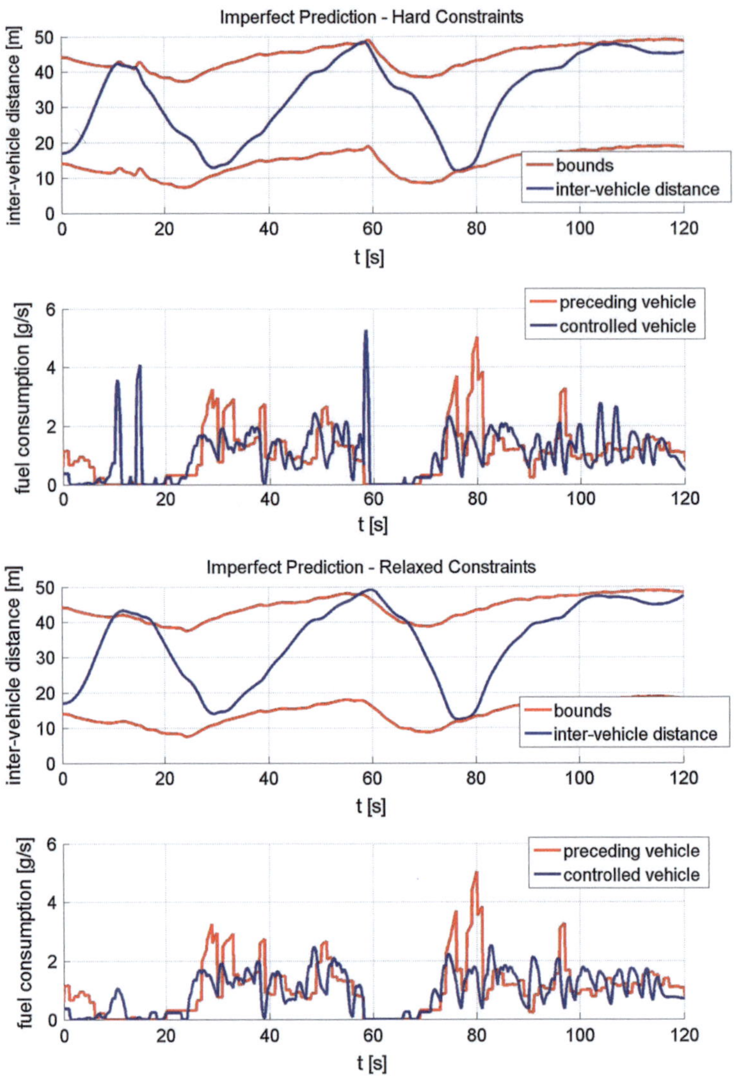

Fig. 10.15 Impact of imperfect prediction on fuel benefit

while $y(t) = v(t)$ is the vehicle speed. As the traffic can include both connected and independent vehicles, we need to consider both the case in which the preceding vehicle transmits in real time a set of data (its position, speed as well as the acceleration, gas, brake and clutch pedal positions) and the case in which only information on position, speed and acceleration of the preceding vehicle can be used by measurement or reconstructed by sensors on the controlled vehicle alone. Examples of the performance of the predictors for 1–5 s ahead for both approaches are given in

Fig. 10.13. The results correspond to the intuition that the estimate becomes less precise for an increasing time distance. In addition to this, it is clearly visible that the V2V approach yields more accurate results as shown in Fig. 10.14.

The top subplot in Fig. 10.15 shows results for a prediction horizon of 10 s using hard constraints and the traffic estimation with V2V based estimation with a reduced fuel benefit of 2.3 % due to the imperfect prediction. However, the strong reduction is also due to the hard constraints as—together with the imperfect prediction—they lead to sudden corrections as in the bottom subplot in Fig. 10.15. If the hard constraints on the upper bound are relaxed, then more moderate acceleration (bottom subplot in Fig. 10.15) are required, and a large part of the fuel savings is recovered (9.4 %).

10.8 Conclusions and Outlook

Coordinated control is receiving much attention, as the availability of cost effective communication and computational systems allows new possibilities. There are different ways to exploit this new opportunity, and much work is being done at the safety or at the infrastructure level. However, as this exposition meant to explain, there are many opportunities also at the local level, centered around the single vehicle, which should be used as well.

The authors gratefully acknowledge the sponsoring of this work by the COMET K2 Center "Austrian Center of Competence in Mechatronics (ACCM)". The COMET Program is funded by the Austrian federal government, the Federal State Upper Austria and the Scientific Partners of ACCM.

References

1. Vagg C, Brace CJ, Hari D, Akehurst S, Poxon J, Ash L (2013) Development and field trial of a driver assistance system to encourage eco-driving in light commercial vehicle fleets. IEEE Trans Intel Transport Syst 14(2):796–805
2. Rupp J, King A (2010) Autonomous driving: a practical roadmap. SAE technical paper 2010-01-2335
3. Li S, Li K, Rajamani R, Wang J (2011) Model predictive multi-objective vehicular adaptive cruise control. IEEE Trans Control Syst Technol 19(3):556–566
4. Lou L, Liu H, Li P, Wang H (2010) Model predictive control for adaptive cruise control with multi-objectives: comfort, fuel-economy, safety and car-following. J Zhejiang Univ Sci A 11(3):191–201
5. Naus G, Ploeg J, van de Molengraft R, Steinbuch M (2008) Explicit MPC design and performance-based tuning of an adaptive cruise control stop-&-go. In: IEEE on intelligent vehicles symposium 2008, pp 434–439
6. Naus G, van den Bleek R, Ploeg J, Scheepers B, van de Molengraft R, Steinbuch M (2008) Explicit MPC design and performance evaluation of an ACC stop-&-go. In: American control conference 2008, pp 224–229
7. Bu F, Tan HS, Jihua H (2010) Design and field testing of a cooperative adaptive cruise control system. In: American control conference (ACC 2010), pp 4616–4621

8. Naus G, Vugts R, Ploeg J, van de Molengraft R, Steinbuch M (2010) Cooperative adaptive cruise control, design and experiments. In: American control conference (ACC 2010), pp 6145–6150
9. Naus GJL, Vugts RPA, Ploeg J, van de Molengraft MJG, Steinbuch M (2010) String-stable CACC design and experimental validation: a frequency-domain approach. IEEE Trans Veh Technol 59(9):4268–4279
10. Ploeg J, Scheepers BTM, van Nunen E, van de Wouw N, Nijmeijer H (2011) Design and experimental evaluation of cooperative adaptive cruise control. In: 14th international IEEE conference on intelligent transportation systems (ITSC 2011), pp 260–265
11. McDonough K, Kolmanovsky I, Filev D, Yanakiev D, Szwabowski S, Michelini J, Abou-Nasr M (2011) Modeling of vehicle driving conditions using transition probability models. In: IEEE international conference on control applications (CCA 2011), pp 544–549
12. McDonough K, Kolmanovsky I, Filev D, Yanakiev D, Szwabowski S, Michelini J (2012) Stochastic dynamic programming control policies for fuel efficient in-traffic driving. In: American control conference 2012, pp 3986–3991
13. McDonough K, Kolmanovsky I, Filev D, Yanakiev D, Szwabowski S, Michelini J (2013) Stochastic dynamic programming control policies for fuel efficient vehicle following. In: American control conference (ACC 2013)
14. Ericson C, Westerberg B, Egnell R (2005) Transient emission predictions with quasi stationary models. SAE technical paper 2005-01-3852
15. Brackstone M, McDonald M (1999) Car-following: a historical review. Transp Res Part F: Traffic Psychol Behav 2(4):181–196
16. Marzbanrad J, Karimian N (2011) Space control law design in adaptive cruise control vehicles using model predictive control. J Automobile Eng 225:870–884
17. Lang D, Stanger T, del Re L (2013) Opportunities on fuel economy utilizing v2v based drive systems. SAE technical paper 2013-01-0985
18. Camacho EF, Bordons C (2004) Model predictive control. Springer, London
19. Maciejowski JM (2002) Predictive control: with constraints. Prentice Hall, Upper Saddle River
20. Stanger T, del Re L (2013) A model predictive cooperative adaptive cruise control approach. In: American control conference (ACC 2013)
21. del Re L, Allgöwer F, Glielmo L, Guardiola C, Kolmanovsky I (eds) (2010) Automotive model predictive control. Springer, London
22. Lang D, Stanger T, del Re L (2013) Fuel efficient quasi optimal adaptive cruise control by control identification. In: Control Applications (CCA), 2013 IEEE international conference, pp 229, 234. doi:10.1109/CCA.2013.6662763

Part III
Powertrain Optimization

Chapter 11
Topology Optimization of Hybrid Power Trains

Theo Hofman and Maarten Steinbuch

Abstract Topology optimization methods for continuum systems (structural topology, shape, material) are well-established. However, these methods do not apply to non-continuum or dynamic systems with discrete components with unique characteristics as with hybrid vehicles. This chapter examines the power train topology and control design optimization problem at vehicle system level. The design space related to power train and control system optimization level is rapidly increasing with new developments in power train, auxiliary technologies, system architectures (topologies) and cyber-physical systems. The multi-objective, mixed or hybrid (continuous/discrete time) character on both *coupled* levels of the problem requires relative long computation time. Therefore, it requires a bi-level (nested) or simultaneous system design approach. Since, sequential or iterative design procedures fail to prove system-level optimality. In this chapter, some illustrative examples are discussed related to nested control and design optimization problems related to continuous/stepped-gear transmission shifting, power split control and/or in combination with topology optimization.

11.1 Introduction

The design of a hybrid vehicle propulsion system is a complex task. The problem is often multi-objective (e.g., minimizing emissions, maximizing acceleration performing while minimizing system and component costs) and requires solving control and design problems that have continuous and discrete dynamics. In literature often case studies are selected, where the technology or topology choice have been fixed

T. Hofman (✉) · M. Steinbuch
Eindhoven University of Technology, P.O. Box 513, 5600 MB Eindhoven, The Netherlands
e-mail: t.hofman@tue.nl

M. Steinbuch
e-mail: m.steinbuch@tue.nl

H. Waschl et al. (eds.), *Optimization and Optimal Control in Automotive Systems*, 181
Lecture Notes in Control and Information Sciences 455,
DOI: 10.1007/978-3-319-05371-4_11, © Springer International Publishing Switzerland 2014

based on engineering knowledge (or heuristics) and the size of components (kW, kWh) is chosen as the main design variable. This simplifies the problem at hand, yet may still impose difficulties and challenges related to the control design.

Analytical methods based on optimal control theory (Pontryagin's Minimum Principle, [1]) are very efficient in solving the control problem for a class of hybrid vehicle systems [2–6]. However, difficulties arise when discrete-state variable, e.g. engine on/off, or gear position with a stepped-gear box with comfort constraints, are included in the design problem [7, 8]. Practically, these problems are circumvented by calculating, e.g., separate Hamiltonian functions (or values) related to each discrete-controlled state individually and accordingly are solved for the admissible control input, which minimizes this set of Hamiltonian functions [9]. Inherently, a singular solution for the optimal control may occur with this method [10], and heuristics is required in choosing a particular mode resulting in a suboptimal solution.

The control problem is nonconvex, and often, e.g., the engine on/off and gear-position state are optimized *a priori*, or by using simplified (heuristic) rules. The remaining control problem and constraints are sometimes convexified to speed up the design process [11].

11.1.1 Co-design Methods

In literature, some authors address the combined control and propulsion system design [11–13]. Often the design space related to technology choice or topology variation is limited to one or two main possibilities reducing the design space significantly. However, in order to create fundamentally new configurations or topologies new and efficient methods are required. Particular if dynamics are important it becomes very challenging to make proper design decisions [14, 15]. This process of coupled system and control optimization can be done utilizing a sequential, iterative, bi-level or simultaneous strategy [16]. In [17] the vehicle system design and control are optimized in an iterative manner: first, the independent system design parameters are selected using a Latin-Hypercube Sampling (LHS) method, and given a (heuristic) controller the fuel economy subject to performance constraints are optimized. Accordingly, at the optimal set of design parameters for the propulsion system, the control rules for the system are optimized using deterministic Dynamic Programming (DP). In a second iteration, again the optimal set of design parameters is found given the updated controller. The process converges till a stopping criteria related to the relative improvement is satisfied. From theory it follows that the control and system design problems are coupled and that sequential or iterative strategies fail to guarantee system-level optimality [16]. So far in literature, bi-level or simultaneous strategies applied to the hybrid vehicle system design is limited. In [18], the multiobjective design problem is nested where for every feasible system design (topology and technology constant) the control design is optimized separately using DP. Previously, in [19] also a multiobjective co-design (control and propulsion system) for a parallel hybrid problem is addressed, yet using a rule-based control strategy (ADVISOR, [20]) limiting the achievable control performance. The nonlinear Pareto frontier is computed and the sensitivity of the main design parameters to the targets is systematically investigated. Particle swarm optimization, generic algorithms, and deterministic search methods, such as Nelder-Mead simplex algorithm, was also suggested as efficient search methods to solve optimal sizing problem. Yet, the authors in [18] claim computing the separate objective values for different sizes to be more efficient, since the sizing problem can be solved repeatedly with minimal computational effort for any combination of weights.

11.1.2 Problem Definition: System Design Optimization

In this section, we would like to introduce the general system design problem, denoted as P. In the following section of the article, this problem is used to derive the different types of control and system design problems individually based on the analyzed case under consideration, or assumptions. Let

$$F_i(d_c) := \phi(x(t_f), t_f) + \int_{t_0}^{t_f} L(x(t), u(t), t)dt \text{ for } i \in \{n+1, \dots, N\} \quad (11.1)$$

and

$$d_c := [u^T(t)\; x^T(t)\; t_0\; t_f]^T \quad (11.2)$$

be the controller design objectives and the vector of the controller design variables. The control inputs and state variables are denoted as $u(t)$ and $x(t)$. The initial and final times are denoted as t_0 and t_f respectively. The global system design problem P is formulated as, minimization of the vehicle system $F_i(d_p)$ and controller design objectives $F_i(d_c)$ to the system and controller design variables (d_p, d_c), which are elements of all the feasible system/controller designs $\mathscr{D}$. The system and controller designs are subjected to (in-)equally (input and state) constraints denoted as $h_i(.)$ and $g_i(.)$ respectively for $i = \{c, p\}$.

$$P := \left\{ \begin{array}{r} \min\limits_{\{d_p, d_c\} \in \mathscr{D}} : J = \{F_1(d_p), \dots, F_n(d_p), F_{n+1}(d_c), \dots, F_N(d_c)\} \\ \text{subject to} \\ \mathscr{D} := \left\{ \begin{array}{r} d_p, d_c : h_p(d_p) = 0,\ g_p(d_p) \le 0,\ \dot{x}(t) = f(x(t), u(t), t, d_p), \\ h_c(x(t), u(t), t, d_p) = 0,\ g_c(u(t), t, d_p) \le 0,\ g_c(x(t), t, d_p) \le 0 \\ \phi(x(t_f), t_f) = 0,\ x(t_0) = x_0 \end{array} \right\} \end{array} \right. \quad (11.3)$$

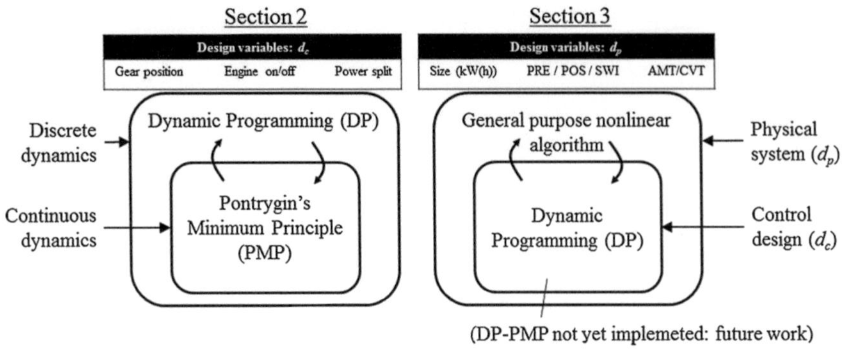

Fig. 11.1 Control and propulsion system design problems: a bi-level approach

The influence of physical system design on the control is accounted for by $h_c(.)$, $g_c(.)$ and $\dot{x}(t)$ to be a function of d_p as well. Further, the formulation assumes that all the state variables can be directly measured and that the physical system and controller optimization objectives are *separable* [16].

11.1.3 Outline and Contribution of the Chapter

A bi-level optimization framework has been proven to identify a system optimal solution. Traditionally, the outer loop optimizes the overall system performance by varying the plant design. For every feasible plant design tested by the outer loop, the inner loop calculates the optimal control for the given plant design [16]. In this work (see also Fig. 11.1), a novel nested solution approach is discussed for

(i) a control design problem alone with discrete (outer loop using dynamic programming) and continuous control inputs (inner loop using Pontryagin's Minimum Principle) reducing computation time significantly; and,
(ii) a novel co-design problem related to the control (using dynamic programming) and the vehicle propulsion design (transmission technology choice, sizing of power converters *and* topology, using a general purpose solver).

In the later case (ii), the general optimization problem P can also be formulated as,

$$P := \begin{cases} \min\limits_{d_p \in \Theta} : J = \{F_1(d_p), \ldots, F_n(d_p), F_{n+1}(d_c), \ldots, F_N(d_c)\} \\ \qquad\qquad\qquad\qquad\qquad\qquad\qquad\qquad \text{subject to} \\ \Theta := \left\{ \begin{array}{r} d_p, d_c^* : h_p(d_p) = 0,\ g_p(d_p) \leq 0, \\ d_c^* = \arg \min\limits_{d_c:(d_p,d_c)\in\mathscr{D}} \{F_{n+1}(d_c), \ldots, F_N(d_c)\} \end{array} \right\} \end{cases} \tag{11.4}$$

Under the assumption of the existence of a feasible combined physical system and control design for every feasible physical system design. The backward coupling from the control design to the physical system design (coupling parameters as output from the control design to the general physical design problem) is ignored. The solutions to the Eqs. (11.3) and (11.4) are mathematical equivalent [16].

The control design optimization problem, as described by case (i), including the gear shift position (denoted as state x_1), engine on/off (x_2) and power split (x_3) has been visited by others [13], yet no comfort constraints on the gear shift command has been taken into account (see Sect. 11.2).

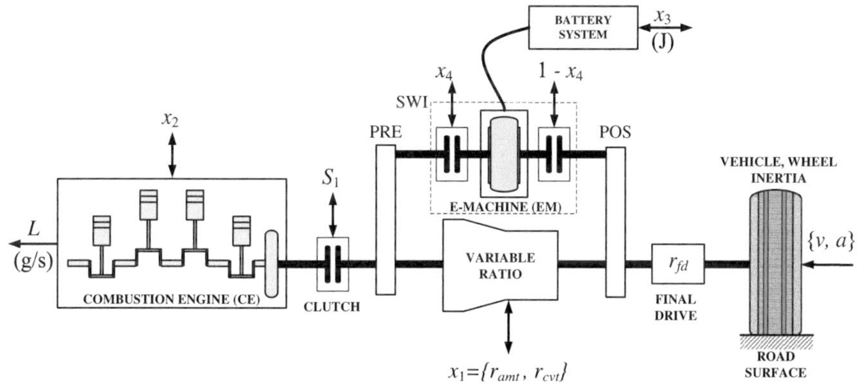

Fig. 11.2 Controlled switching topology: electric machine coupling is selectable (input and output signals are included; the controlled-state variables x_i for $i \in \{1, 2, 3, 4\}$ are indicated by *doubled-headed arrows*) [8]

In both discussed case studies, a single objective related to the control design, i.e., fuel consumption, or CO_2 emissions is used respectively. In case study (ii) (see Sect. 11.3), also a performance constraint (acceleration performance) is imposed onto the problem by an inequality constraint. Moreover, case study (ii), is an alternative hybrid propulsion system model where not only the topology choice (pre-(PRE) or post-coupled electric machine (POS) to the transmission) is as a propulsion design variable separately analyzed, but also considered as a controlled-state variable (x_4). This is referred to as a *switching topology* (SWI) (see Fig. 11.2). In literature, so far known to the authors, individually analyzed hybrid design models of *static* topology cases can only be found. The results are based on previously published articles (by the authors) and the corresponding optimizing problem is shown in Table 11.1. Without loss of generality, the work proceeds to show with the nested approach that a system-level optimality is guaranteed and proposes the integration of both problems as future work (see Fig. 11.1). Finally, we state our conclusions and recommendations in Sect. 11.4.

Table 11.1 Control and propulsion design cases under consideration

	d_c: control			d_p: propulsion system			
Section	Gear ratio	Engine on/off	Power split	Topology	Technology	EM size (kW)	Ref.
11.2	x_1	x_2	x_3	$x_4 = 1$	AMT	6	[7]
11.3	x_1	x_2	x_3	x_4	{AMT, CVT}	$[0, 50]$	[8]

AMT Automated Manual Transmission; *CVT* Continuously Variable Transmission; *EM* Electric Machine

11.2 Control Design Optimization: Gear Shift Strategies with Comfort Constraints for Hybrid Vehicles

Similar to the problem presented by the authors in [13], we formulate our control design problem. Yet, an additional comfort constraint is added to the (discrete) gear shift command limiting the admissible gear position [7]. Moreover, the main contribution of this section is on an alternative computational efficient solution procedure, which will be discussed in the following parts of this section. Taking into account physical constraints from the transmission, battery, engine and electric motor, the optimal control problem under study consists in minimizing the fuel consumption of the vehicle along a prescribed vehicle cycle. The gear position is defined as a discrete state, denoted as x_1. Further, let $x_2 \in \{0, 1\}$ be a discrete state representing the state of the engine: $x_2 = 0$ means the engine is off, while $x_2 = 1$ means the engine is on. Let x_3 be a state corresponding to the state of charge of the battery. We denote by $x = \{x_1, x_2, x_3\} \in \mathscr{X}$ the state vector and define the control vector $u = \{u_1, u_2, u_3\} \in \mathscr{U}(x_1, x_2)$ by

$$u = \begin{cases} u_1 & \text{the gear shift command,} \\ u_2 & \text{engine on/off command,} \\ u_3 & \text{battery power flow,} \end{cases} \tag{11.5}$$

with

$$\mathscr{U}(x_1, x_2) = \{(u_1, u_2, u_3) : u_1 \in \mathscr{U}_1(x_1), u_2 \in \mathscr{U}_2(x_2), u_3 \in \mathscr{U}_3(x_2)\} \tag{11.6}$$

where $\mathscr{U}_1(x_1)$ and $\mathscr{U}_2(x_2)$ are discrete values $\{-1, 0, 1\}$ depending on x_1 or x_2 value. A full hybrid vehicle is considered, hence the engine can be stopped when the torque is provided only the electric machine. Introducing the engine state allows to take into account an extra fuel consumption corresponding to the necessary energy to crank the engine [13]. If these losses are not engine state dependent, then the optimal engine state (and corresponding engine on/off command) may follow from the optimal power split minimizing the cost function. This similar assumption holds for associated state dependent losses affecting the fuel consumption for the main drive clutch (S_1) or gear shift losses. Although, gear shift losses are neglected here, the gear shift state is considered, since the following gear position $x_1(t^+)$ depends on the current gear position $x_1(t)$ and the gear shift command $u_1(t)$ that limits the admissible gear position state. The parallel hybrid topology is depicted in Fig. 11.2. It is assumed that the vehicle wheel speed and acceleration as a function of time, denoted as the disturbance (or external state) vector,

$$w(t) = [v(t), a(t)]^T, \tag{11.7}$$

are known. Let $L(x(t), u(t), w(t))$ be the instantaneous fuel consumption in (g/s), given by

$$L(x(t), u(t), w(t)) = \begin{cases} L(u_1(t), u_3(t), w(t)) & \text{if} \quad x_2(t) = 1 \text{ (engine running)}, \\ 0 & \text{if} \quad x_2(t) = 0 \text{ (engine stopped)}. \end{cases} \tag{11.8}$$

The continuous state $x_3(t)$ is a governed by

$$\dot{x}_3(t) = f_3(u_3(t)). \tag{11.9}$$

Accordingly, the dynamic optimization problem can be formulated as follows:

$$P_1 := \begin{cases} \min\limits_{u(t)\in\mathscr{U},\, x\in\mathscr{X}} : J = \int_{t_0}^{t_f} L(x(t), u(t), w(t))\, dt \\ \text{subject to} \begin{cases} g_1 := u_1(t) \in \mathscr{U}_1 = \{-1, 0, 1\} \subseteq \mathbb{Z} \\ g_2 := x_1(t) \in \mathscr{X}_1 = \{1, \dots, 5\} \subseteq \mathbb{N}^+ \\ g_3 := u_2(t) \in \mathscr{U}_2(x_2) = \begin{cases} \{-1, 0\} \text{ if } x_2 = 1 \\ \{0, 1\} \text{ if } x_2 = 0 \end{cases} \subseteq \mathbb{Z} \\ g_4 := x_2(t) \in \mathscr{X}_2 = \{0, 1\} \subseteq \mathbb{Z}_n \\ g_5 := u_3(t) \in \mathscr{U}_3(x_2) = \{\underline{u}_3(x_2), \overline{u}_3(x_2)\} \subseteq \mathbb{R} \\ g_6 := x_3(t) \in \mathscr{X}_3 = \{\underline{x}_3, \overline{x}_3\} \subseteq \mathbb{R} \\ g_7 := x_1(t^+) - x_1(t) - u_1(t) = 0 \\ g_8 := x_2(t^+) - x_2(t) - u_2(t) = 0 \\ g_9 := \dot{x}_3(t) - f_3(u_3(t)) = \dot{x}_3(t) - u_3(t) = 0 \\ g_{10} := x_1(t_0) - x_{10} = 0; \quad g_{11} := x_1(t_f) - x_{1f} = 0 \\ g_{12} := x_2(t_0) - x_{20} = 0; \quad g_{13} := x_2(t_f) - x_{2f} = 0 \\ g_{14} := x_3(t_0) - x_{30} = 0; \quad g_{15} := x_3(t_f) - x_{3f} = 0 \\ g_{16} := x_{1f} - x_{10} = 0; \quad g_{17} := x_{2f} - x_{20} = 0 \\ g_{18} := x_{3f} - x_{30} = 0; \end{cases} \end{cases} \tag{11.10}$$

The length of the optimization horizon is t_f. The final-state penalty term is $\phi(x(t_f), t_f) = 0$ (see Eq. (11.1)).

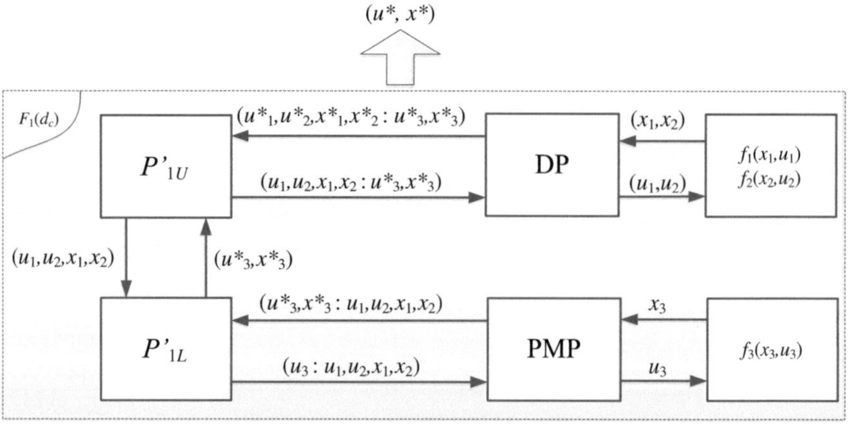

Fig. 11.3 Bi-level control design optimization: interactions between optimization levels and algorithms. *DP* Deterministic Dynamic Programming; *PMP* Pontryagin's Minimum Principle [7]

11.2.1 Bi-level Optimization: Control Problem

The above described problem (11.10) contains continuous and discrete dynamical properties. In order to speed up the computation time, a bi-level optimization procedure is proposed. The original problem is cascaded into two hierarchical problem levels, where solutions to the upper level problem P'_{1U} requires that the lower level problem P'_{1L} is optimized. The assumption is that both levels have a single objective. The optimization problem has a two-way coupling, which is presented in Fig. 11.3. The lower level problem is optimized for feasible discrete control variables and states from the upper level problem. On the other hand, the objective function at the upper level is affected by solutions at the lower level. The optimal variable values are denoted by *. The above described problem can be given as follows [21, 22],

$$P'_1 := \left[\begin{array}{c} P'_{1U} := \\ \min\limits_{(u_1,u_2,x_1,x_2)_j} \end{array} \left[\begin{array}{r} P'_{1L} := \min\limits_{(u_3,x_3)} : J((u_3,x_3) : (u_1,u_2,x_1,x_2)_j) \\ \text{s.t.} \left\{ \begin{array}{r} g_i((u_3,x_3) : (u_1,u_2,x_1,x_2)_j) \leq 0 \\ (u_3,x_3) \in \mathcal{U}_3 \times \mathcal{X}_3 \\ (.)_j := (u_1,u_2,x_1,x_2) \in \{\mathcal{U}_1 \times \mathcal{U}_2 \times \mathcal{X}_1 \times \mathcal{X}_2\} \cap \mathcal{V}_1 \end{array} \right. \end{array} \right] \right]. \tag{11.11}$$

The subscript $_j$ is used to index the feasible NLP subproblems. Let us also define the set $\mathcal{V}_1$ of all integer assignments (u_1, u_2, x_1, x_2)'s for which there exist feasible solution in the (u_3, x_3) variables as

$$\mathcal{V}_1 = \left\{ \begin{array}{r} (u_1,u_2,x_1,x_2) : \exists (u_3,x_3) \in \mathcal{U}_3 \times \mathcal{X}_3 \text{ with } g_i(u,x) \leq 0, \\ i \in \{1,\ldots,18\} \subseteq \mathbb{N}^+ \end{array} \right\}. \tag{11.12}$$

11.2.1.1 Solution Procedure

In the following the solution procedure for the bi-level problem is given. As described it has similarities with the bi-level procedure presented by [23]. Yet, the procedure is adapted accordingly to the problem under investigation.

Algorithm 1. Bi-level optimization

Let the bi-level optimization problem be defined as in (11.11). A suitable method to handle mixed-integer non-linear problems is deterministic Dynamic Programming (DP). Hence, the upper level problem is solved using DP [24]. The lower level problems contains continuous control variables and Pontryagin's Minimum Principle (PMP) is used. Then the solution procedure for such a problem can be defined as follows.

1. *Initialize the optimization method (DP) selected for the upper-level optimization*: since DP is a numerical algorithm used here to solve a continuous and discrete control problem, the continuous-time model (11.9) must also be discretized. Let the total discrete-time model describing the system dynamics be given by

$$x_i(k+1) = x_i(k) + f_i(x_i(k), u_i(k), w(k), k) \cdot \Delta t, \quad k = \{0, 1, \ldots, n-1\} \tag{11.13}$$

for $i \in \{1, 2, 3\}$. The criterion to be minimized in discrete-time format and expressed in a non-recursive form becomes,

$$\min_{u(k) \in \mathscr{U}, x(k) \in \mathscr{X}} : J \approx \left\{ \phi(x(0), 0) + \sum_{k=1}^{n-1} L'(x(k), u(k), w(k), k) \cdot \Delta t \right\} \tag{11.14}$$

with the small time step, denoted as Δt, defined as the constant difference: $\Delta t \equiv t(k+1) - t(k)$. Final state penalty terms $\phi(.)$ at $t_f \equiv t(k = N)$ related to the constrained final states has been omitted in this analysis. Since, it is assumed that the initial states equal the final states ($g_{16} - g_{18}$). The initial fuel cost at $t_0 \equiv t(k = 0)$ is defined as $\phi(x(0), 0)$. The second term relates to the cumulated *equivalent* fuel-mass flow cost, denoted as L', as a function of the control inputs and states.

2. *Compute optimal cost-to-arrive and trajectory-to-arrive matrices (J and H respectively)*: after defining the objective function, a vector of optimization variables and constraints for the upper-level optimization problem (including the lower-level optimization problem parameterized by the upper-level optimization variables) then perform optimization as follows. Do for each calculation step for time-indices $k = 0$ to $n - 1$, as follows.

   a. Let vector $(u_1(k), u_2(k), x_1(k), x_2(k))_j$ contain the current (feasible) values for the optimization variables on the upper level and solve the dynamic process model, $f_{1,2}(u_{1,2}(k), x_{1,2}(k), k)$. Define the corresponding lower-level optimization problem P'_{1L} at $(u_1(k), u_2(k), x_1(k), x_2(k))_j$.

b. Find the optimal input and state for the lower level $(u_3^*(k), x_3^*(k))$ with PMP, i.e., perform the optimization based on *Algorithm 2* (see below).

c. Based on $u_3^*(k)$ and $x_3^*(k)$, evaluate the objective function on the upper level and provide the objective function value to the optimization method.

Based on the principle of optimality [24], DP is the algorithm, which evaluates the optimal *cost-to-arrive* function $J(x(k), k)$ at every node in the discretized state-time space by proceeding *forward*-in-time:

i. with the initial cost calculation step

$$J(x(0), 0) = \phi(x(0), 0) \tag{11.15}$$

ii. and, proceeds in time for each intermediate calculation step for time-indices $k = 1$ to $n - 1$,

$$J(x_{1,2}(k+1), k+1) = \min_{(u_1(k), u_2(k)) \in \mathcal{U}_1 \times \mathcal{U}_2} \left\{ \begin{array}{c} J(x_{1,2}(k), k) + \\ L'(x_{1,2}(k), u_{1,2}(k), w(k), k : x_3^*(k), u_3^*(k)) \cdot \Delta t \end{array} \right\} \tag{11.16}$$

The optimal control $(u_1^*(k), u_2^*(k))$ is given by the argument that minimizes the righthand side of Eq. (11.16) for each $x_1(k)$ and $x_2(k)$ of the discretized state-time space (to simplify notation denoted as $x_{1,2}(k)$). The time direction may also be reversed[1] and infeasible nodes are assigned with a very high (or, infinite) cost. As an output of the algorithm (11.15)–(11.16) the associated optimal *trajectory-to-arrive* at every node can be effectively stored in a matrix

$$H(k+1, x_{1,2}^*(k+1)) = x_{1,2}^*(k), \tag{11.17}$$

which contains the optimal state of the preceding time step (optimal control signal map). This matrix is used to find the optimal control signals during a backward simulation, starting from the given final states $x_{1,2}^*(n) = x_{1,2}(0)$ from g_{16} and g_{17} of Eq. (11.10), to generate the optimal state trajectory with

$$x_{1,2}^*(k-1) = H(k, x_{1,2}^*(k)) \tag{11.18}$$

for $k = \{n, n-1, \ldots, 1\}$.
End Do

[1] The *backward*-in-time algorithm starts at a given final state; either algorithm is convenient to used when both (initial and final) states are fixed. The forward algorithm is more convenient as it will optimize the final state.

3. The optimal solutions for the bi-level optimization problem are J at $(u_1^*(k), u_2^*(k), x_1^*(k), x_2^*(k))$ with the corresponding optimal lower-level optimization solution L' at $(u_3^*(k), x_3^*(k))$.

Algorithm 2. Lower-level optimization

In the following the function $\mathscr{H}$, called the Hamiltonian [1], defined as

$$\begin{aligned} L'(x_3(k), u_3(k), p(k), k : (u_1(k), u_2(k), x_1(k), x_2(k))_j) &\equiv \mathscr{H}(x_3(k), u_3(k), p(k), k) \\ &= L((x_3(k), u_3(k), p(k), k)) \cdot x_2(k) + p(k) \cdot [f_3(x_3(k), u_3(k), k)] \end{aligned} \quad (11.19)$$

is used for convenience. The solution procedure is as follows.

1. Initialize the optimization method (PMP) at the given values for the costate $p(k)$ and control inputs and states $(u_1(k), u_2(k), x_1(k), x_2(k))_j$ following from Algorithm 1, step 2a.
2. Define the objective functions, a vector of the optimization variables and constraints in P'_{1L} for the given parameters $(u_1(k), u_2(k), x_1(k), x_2(k))_j$. Start the optimization procedure as follows. Do for the given calculation step k, as follows.
   a. Solve the dynamic process model, $f_3(u_3(k), x_3(k), k)$ with the current optimization variables $u_3(k)$.
   b. Evaluate objective L' at $u_3(k)$ based on the state variables $x_3(k)$ for the given parameters $(u_1(k), u_2(k), x_1(k), x_2(k))_j$.
   End Do
3. Save the optimal values of the optimization variables $u_3(k)$ and the corresponding state variable values $x_3(k)$, as well as L', which is needed in Algorithm 1.

The above described solution procedure is shown in Fig. 11.4. To simplify the notation the control input variables in the diagram are omitted. A constant costate value satisfying the end constraint on the battery energy can not be found, since, e.g., engine on and off solutions can be found minimizing both the Hamiltonian $(\mathscr{H}_{x_2=1} = \mathscr{H}_{x_2=0})$). To circumvent this problem, a mildly-tuned integrator I is used to update the costate value p over time. The found solution is computed very efficiently using the nonlinear *nonconvex* models and is close to global optimal (relative error $< 0.4\,\%$, see next section).

In literature, other approximated methods may use *convex* modeling to solve the problem. For example in [25] a heuristic strategy is used in a sequential manner to solve for the discrete control inputs (gear position, x_1, engine on/off, x_2) and accordingly a convex solver for the power split, x_3; or, improved (in terms of accuracy vs. computation speed) in [26] by iteratively updating the costate p using the output of PMP (for engine on/off decision with the assumption switching the engine on if $\mathscr{H}_{x_2=1} \le \mathscr{H}_{x_2=0}$)) as the high-level problem and again a convex solver (for the power split, x_3) as the low-level problem.

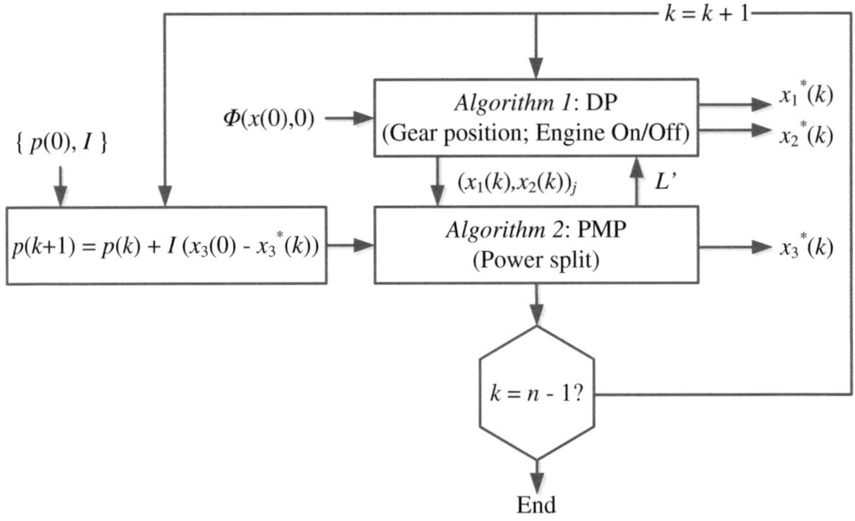

Fig. 11.4 Solution procedure

Table 11.2 Comparison of simulation results

Fuel (g)			Time (s)			
DP	DP-PMP	Rel. diff. [%]	DP	DP-PMP	Reduction factor	Cycle
300	301	0.4	3964	9	440	NEDC
473	474	0.4	4799	12	400	FTP75

11.2.2 Simulation Result: Bi-level Control Design

Based on the vehicle model and (physical) constraints given in [7], the above describe problem P_1' has been solved and the optimal control signals were computed. In Fig. 11.5, the controlled states in inputs are shown as a function of time for the NEDC. The difference between DP and DP-PMP method is found to be negligible small (less than 0.4 %). Differences are also caused by numerical errors caused by griding the state and input space. However, the computation time is significantly reduced from approximately 4,000 to 9 s (factor 440).[2] These results and for the FTP75 cycle are listed in Table 11.2. In particular for large-scale complex propulsion system optimization problems, this method can speed up the design procedure significantly.

[2] HP EliteBook 8530w, Core 2 Due, CPU T9600 @ 2.8 GHz, RAM 4.0 GB

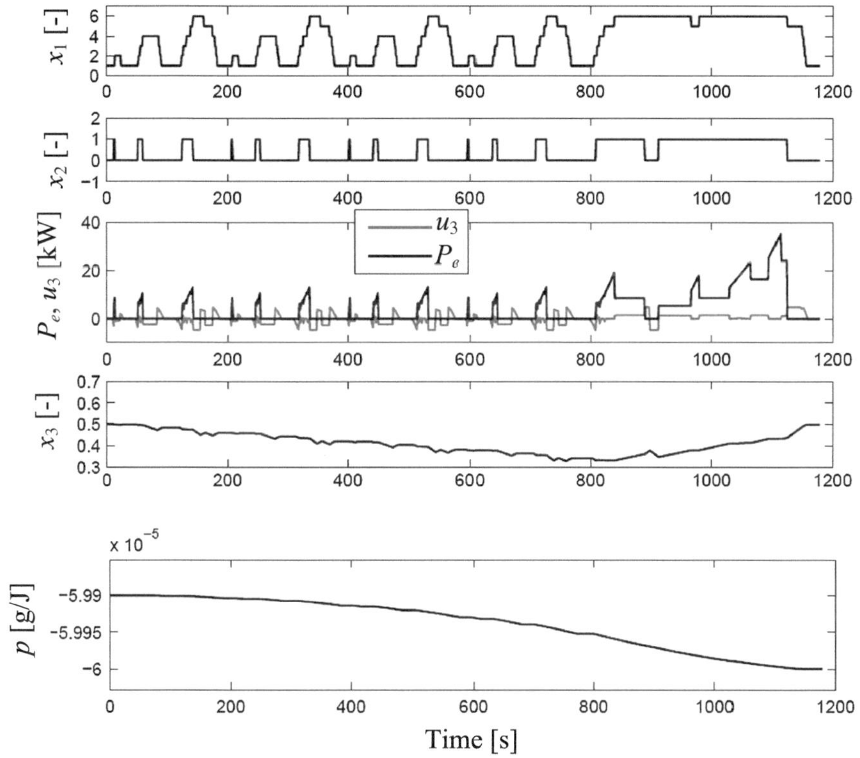

Fig. 11.5 DP-PMP simulation result for a parallel hybrid vehicle with discrete gear box and engine start-stop [7]. P_e = engine power; u_3 = battery power

11.3 Control and Propulsion System Design Optimization: Topology, Transmission, Size and Control Optimization for Hybrid Vehicles

In this section, the combined control and propulsions system optimization for a hybrid vehicle is discussed [8]. The control problem is different from P_1 of the previous section: the topology selection (pre or post coupling electric machine to the transmission) is an additional control input. Moreover, the transmission technology, pushbelt Continuously Variable Transmission (CVT) and Automated Manual Transmission (AMT), and the sizing in terms of maximum output power for the electric machine and engine are optimized. Again a bi-level method is used, where in this case at the upper level the topology and sizing of the components is optimized, P'_{2U}, whereas at the lower level the control problem P'_{2L} is optimized.

From literature [16] it follows that if the bi-level system design (plant)/controller optimization strategy is convergent, then it is guaranteed to converge to an optimal combined plant/controller design. Next, a scalar substitute objective function is defined as the weighted sum of the propulsion system objective (hybrid system cost, which is proportional assumed with electric machine and battery size), denoted as F_1, and controller objective (fuel consumption costs), denoted as F_2; and, the points on the Pareto set will be generated by varying the weights w_i. Given these conditions the combined propulsion/controller optimization problem becomes:

$$P_2 := \left\{ \begin{array}{r} \min\limits_{\{d_p, d_c\} \in \mathscr{D}} : J = w_p \cdot F_1(d_p) + w_c \cdot F_2(d_c) \\ \text{subject to} \\ \mathscr{D} := \left\{ \begin{array}{r} d_p, d_c : g_p(d_p) \leq 0,\ \dot{x}(t) = f(x(t), u(t), t, d_p), \\ g_c(d_c, t, d_p) \leq 0 \\ \phi(x(t_f), t_f) = 0,\ x(t_0) = x_0 \end{array} \right\} \end{array} \right. \tag{11.20}$$

and is similar to,

$$P_2' := \left[\begin{array}{c} P_{2U}' := \\ \min\limits_{(d_p)_j} \end{array} \left[\begin{array}{r} P_{2L}' := \min\limits_{d_c} : J = w_p \cdot F_1((d_p)_j) + w_c \cdot F_2(d_c) \\ \text{subject to} \left\{ \begin{array}{r} g_p((d_p)_j) \leq 0 \\ g_c(d_c : (d_p)_j) \leq 0 \\ h_c(d_c : (d_p)_j) = 0 \\ d_c \in \mathscr{U}_c \times \mathscr{X}_c \\ (d_p)_j := d_p \in \mathscr{D} \cap \mathscr{V}_2 \end{array} \right. \end{array} \right] \right]. \tag{11.21}$$

Let us also define the set $\mathscr{V}_2$ of all integer assignments d_p's (engine and electric machine power specification) for which there exist feasible solution in the d_c variables as

$$\mathscr{V}_2 = \left\{ \begin{array}{r} d_p : \exists d_c \in \mathscr{U}_c \times \mathscr{X}_c \text{ with } g_i(d_c, d_p) \leq 0, \\ i \in \{1, \ldots, 24\} \subseteq \mathbb{N}^+ \end{array} \right\}. \tag{11.22}$$

Complementary to the previous defined control problem, we introduce a *new* discrete state related to the topology choice, denoted as x_4 (see Fig. 11.2). If $x_4 = 1$, then the electric machine pre-coupled to the transmission and vice versa (post-coupled) if $x_4 = 0$. Moreover, the following inequality constraint related to the power specification of the electric machine is added. Finally, initial and final state constraints on x_4 are imposed.

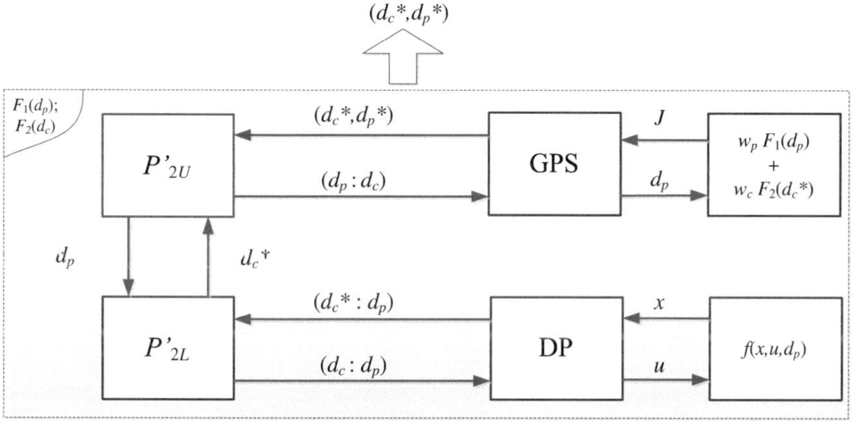

Fig. 11.6 Bi-level control and propulsion system design optimization: interactions between optimization levels and algorithms. *DP* Deterministic Dynamic Programming; *GPS* General Purpose Solver [8]

$$g_c \begin{cases} g_5 := u_3(t) \in \mathscr{U}_3(x_2, d_p) = \{\underline{u}_3(x_2, d_p), \overline{u}_3(x_2, d_p)\} \subseteq \mathbb{R}; \\ g_{19} := u_4(t) \in \mathscr{U}_4(x_4) = \begin{cases} \{-1, 0\} & \text{if } x_4 = 1, \\ \{0, 1\} & \text{if } x_4 = 0 \end{cases} \subseteq \mathbb{Z}; \\ g_{20} := x_4(t^+) - x_4(t) - u_4(t) = 0; \\ g_{21} := x_4(t_0) - x_{40} = 0; \\ g_{22} := x_{4f} - x_{40} = 0; \end{cases}$$
$$g_p \begin{cases} g_{23} := \Delta t_{0-100\ km/h}(d_p) - 11 = 0, \text{ acceleration time (s)}; \\ g_{24} := d_p \in \mathscr{D} \subseteq \{\mathbb{R} :\ d_p \geq 0 \wedge d_p - \overline{u}_3 = 0\} \end{cases} \quad (11.23)$$

Under the assumption that the maximum battery power equals the absolute minimum battery power, i.e., $\overline{u}_3 = -\underline{u}_3$.

11.3.1 Simulation Result: Bi-level Propulsion System and Control Design

A General Purpose Solver (GPS) (e.g., sequential quadratic programming) can be used in finding the optimal plant design in combination with DP for the control design (see Fig. 11.6). Here, the responses of the objective functions $F_1(d_p)$ (hybrid system costs) and $F_2(d_c)$ (fuel consumption costs) has been computed for three topologies (pre-, post-coupled and switching topology) on two different drive cycles (NEDC, CADC urban). The results are shown in Fig. 11.7 and the cost function values are scaled with their maximum values. From observation it followed that for

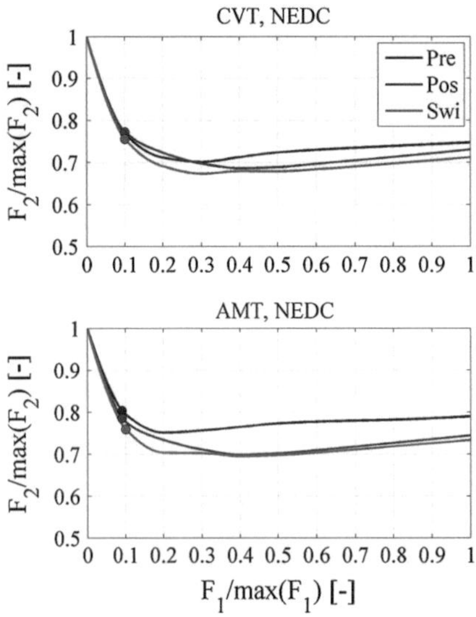

Fig. 11.7 Combined propulsion system and control design responses for different topologies and two drive cycles. A Pareto optimal point for $w_p = w_c = 0.5$ is indicated with a *circle*. Pre = Pre-coupled; Pos = Post-coupled; Swi = Switching topology. F_1 = hybrid system cost; F_2 = fuel cost

any combination of the weighting factors w_i with $\sum w_i = 1$ only a single solution for the different optimal electric machine and engine sizes (sizes are constraint by performance requirements, g_p) can be found. The circle denotes the particular choice of $w_p = w_c = 0.5$. Moreover, it can be observed that the switching (dynamic) topology outperforms the static topologies.

For the automated manual transmission (AMT) the fuel benefit from a fixed pre- or post coupled electric machine as a function of a constant transmission efficiency is computed (on two drive cycles). The results are shown in Fig. 11.8. Moreover, it can be observed that the optimal topology strongly depends on the transmission characteristics and driving loads (cycle).

11.4 Conclusions

This work discussed two design frameworks based on bi-level optimization for the control design with discrete and continuous control variables; and, for the co-design of the vehicle propulsion system (topology, size, transmission technology) and control design for a parallel hybrid vehicle. In the last decade, significant control performance improvement is realized by the transition from heuristic control design methods (fuzzy, rule based) to methods based on optimal control theory (e.g., equivalent consumption minimization strategy—ECMS, PMP, DP). Traditionally, the vehicle propulsion design (technology and topology selection, and sizing of power converter

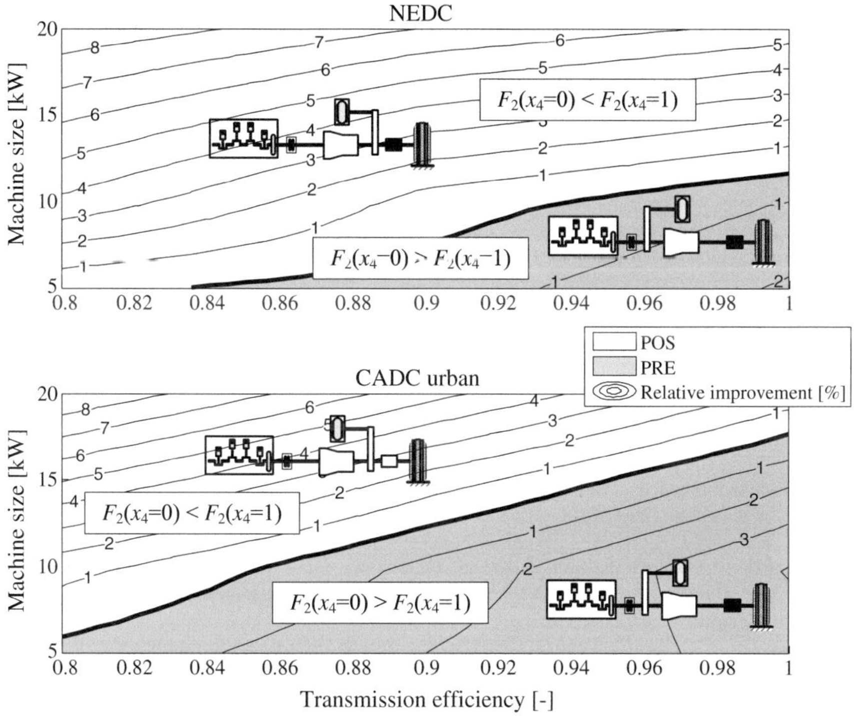

Fig. 11.8 The optimal topology depending on transmission efficiency and electric machine size (kW)

and energy storage systems) is based on engineering intuition and creativity (heuristics). To handle complexity and reducing the design space effectively, fixed topologies (with technologies) are fitted on the solution rather than generating complete new propulsion systems. In future, challenges are seen to develop new efficient design methods for *co-design* (by integration of both discussed design problems) including topology optimization, which creates the required transition from the (classical) heuristic approaches to more optimal design based methods improving the overall system design performance.

References

1. Kirk D (1970) Optimal control theory: an introduction. Prentice-Hall, Englewood Cliffs
2. Delprat S, Lauber J, Guerra T, Rimaux J (2004) Control of a parallel hybrid powertrain: optimal control. IEEE Trans Veh Technol 53(3):872–881
3. Serrao L, Sciarretta A, Grondin O, Chasse A, Creff Y, di Domenico D, Pognant-Gros P, Querel C, Thibault L (2011) Open issues in supervisory control of hybrid electric vehicles: a unified

approach using optimal control methods. In: Les rencontres scientifiques IFP energies nouvelles internat scient conf hybrid and electric vehicles, Rueil-Malmaison, France
4. Sciarretta A, Back M, Guzzella L (2004) Optimal control of parallel hybrid electric vehicles. IEEE Trans Control Sys Technol 12(3):352–363
5. Sciarretta A, Guzzella L (2007) Control of hybrid electric vehicles. IEEE Control Sys Mag 27(2):60–70
6. Koot M, Kessels J (2005) Energy management strategies for vehicular electric power systems. IEEE Trans Veh Technol 54:771–782
7. Ngo V, Hofman T, Steinbuch M, Serrarens A (2012) Optimal control of the gearshift command for hybrid electric vehicles. IEEE Trans Veh Technol 61(8):3531–3543
8. Hofman T, Ebbesen S, Guzzella L (2012) Topology optimization for hybrid electric vehicles with automated transmissions. IEEE Trans Veh Technol 61(6):2442–2451
9. Ambühl D, Sundström O, Sciarretta A, Guzzella L (2010) Explicit optimal control policy and its practical application for hybrid electric powertrains. Control Eng Pract 18(12):1429–1439
10. Wei X, Guzzella L, Utkin V, Rizzoni G (2007) Model-based fuel optimal control of hybrid electric vehicle using variable structure control systems. Jf Dyn Sys Meas Control 129:13–19
11. Murgovski N, Johannesson L, Hellgren J, Egardt B, Sjöberg J (2011) Convex optimization of charging infrastructure design and component sizing of a plug-in series HEV powertrain. In: Proceedings of the 18th world congress the international federation of automatic control, IFAC, Ed. Milano, Italy: IFAC, Aug 2011, pp 13052–13057
12. Patil RM (2012) Combined design and control optimization: application to optimal PHEV design and control for multiple objectives. Ph.D. dissertation, University of Michigan
13. Sinoquet D, Rousseau G, Milhau Y (2011) Design optimization and optimal control for hybrid vehicles. Optim Eng 12:199–213
14. Zhang X, Li C-T, Kum D, Peng H (2012) Prius$^+$ and volt$^-$: configuration analysis of power-split hybrid vehicles with a single planetary gear. IEEE Trans VehTechnol 61(8):3544–3552
15. Liu J, Peng H (2008) Automated modelling of power-split hybrid vehicles. In: Proceedings of the 17th world congress, IFAC, Seoul, Korea, Jul 2008, pp 4648–4653
16. Fathy H, Reye J, Papalambros P, Ulsog A (2001) On the coupling between the plant and controller optimization problems. In: Proceedings of American control conference, ACC, Ed., Arlington, VA, USA, Jun 2001, pp 1864–1869
17. Filipi Z, Louca L, Daran B, Lin C-C, Yildir U, Wu B, Kokkolaras M, Assanis D, Peng H, Papalambros P, Stein J, Szkubiel D, Chapp R (2004) Combined optimization of design and power management of the hydraulic hybrid propulsion system for the 6—6 medium truck. Int J Heavy Veh Sys 11(3/4):372–402
18. Ebbesen S, Elbert P, Guzzella L (2013) Engine downsizing and electric hybridization under consideration of cost and drivability. Oil Gas Sci Technol—RevIFP Energies nouvelles 68(1): 109–116
19. Assanis D, Delagrammatikas G, Fellini R, Filipi Z, Liedtke J, Michelena N, Papalambros P, Reyes D, Rosenbaum D, Sales A, Sasena M (1999) An optimization approach to hybrid electric propulsion system design. J Mech Struct Mach, Automot Res Cent Spec Ed 27:393–421
20. Wipke K, Cuddy M, Burch S (1999) ADVISOR 2.1: user-friendly advanced powertrain simulation using a combined backward/forward approach. IEEE Trans Veh Technol 48(6):1751–1761
21. Floudas CA (1995) Nonlinear and mixed-integer optimization: fundamentals and applications. Oxford University Press, New York
22. Fletcher R, Leyffer S (1994) Solving mixed integer non linear programs by outer approximation. Math Prog 66:327–349
23. Linnala M, Madetoja E, Ruotsalainen H, Hämäläinen J (2011) Bi-level optimization for a dynamic multiobjective problem. Eng Optim 44(2):195–207
24. Betsekas D (2000) Dynamic programming and optimal control. Athena Scientific, Belmont
25. Murgovskia N, Johannesson L, Sjöberg J, Egardt B (2012) Component sizing of a plug-in hybrid electric powertrain via convex optimization. Mechatronics 22(1):106–120
26. Elbert P (2013) Noncausal and causal optimization strategies for hybrid electric vehicles. Ph.D. dissertation, ETH, Zurich

Chapter 12
Model-Based Optimal Energy Management Strategies for Hybrid Electric Vehicles

Simona Onori

Abstract Methods from optimal control theory have been used since the past decade to design model-based energy management strategies for hybrid electric vehicles (HEVs). These strategies are usually designed as solutions to a finite-time horizon, constrained optimal control problem that guarantees optimality upon perfect knowledge of the driving cycle. Properly adapted these strategies can be used for real-time implementation (without knowledge of the future driving mission) at the cost of either high (sometime prohibitive) computational burden or high memory requirement to store high-dimensional off-line generated look-up tables. These issues have motivated the research reported in this chapter. We propose to address the optimal energy management problem over an infinite time horizon by formulating the problem as a nonlinear, nonquadratic optimization problem. An analytical supervisory controller is designed that ensures stability, optimality with respect to fuel consumption, ease of implementation in real-time application, fast execution and low control parameter sensitivity. The approach generates a drive cycle independent control law without requiring discounted cost or shortest path stochastic dynamic programming introduced in the prior literature.

12.1 Introduction

In response to the present and future environment and energy challenges worldwide the automotive industry has been focusing on improving vehicle fuel efficiency. Although there is no "silver-bullet" technology to replace the existing ones, at least in the near future, one possible answer to the challenges posed by the automotive and transportation sectors is found in electrification of both the mobility and transport systems. New concepts and new technologies are being developed to realize efficient

S. Onori (✉)
Automotive Engineering Department, Clemson University, Clemson, USA
e-mail: sonori@clemson.edu

H. Waschl et al. (eds.), *Optimization and Optimal Control in Automotive Systems*,
Lecture Notes in Control and Information Sciences 455,
DOI: 10.1007/978-3-319-05371-4_12, © Springer International Publishing Switzerland 2014

hybrid and electric vehicles suited for both individual and public mobility and for goods distribution in urban areas [7]. This chapter deals with the energy management in HEVs.

12.2 Optimization Problems in HEVs

From a design prospective, a hybrid powertrain is much more complicated than a conventional powertrain as selection hybrid architecture (e.g., series, paralllel, power-split [18]) and component sizing, is not always an easy task because of many design options and the rapidly developing technologies in the automotive industries. Design optimization tools, such as neural networks, genetic algorithms and particle swarm optimization, have been successfully used for powertrain optimization design to maximize fuel economy and minimize emission, weight and cost while guaranteeing vehicle performance (see, for instance, [17], and references therein).

Given a predefined optimized powertrain, a second problem in a HEV is the power split on-board of the vehicle. This is generally referred to as energy management problem or supervisory vehicle control.

Realistic figures of achievable improvement in fuel economy in HEVs range from 10 % for mild hybrids to more 30 % for full hybridized vehicles [10]. This potential can be realized only with a sophisticated control system that optimizes energy flow within the vehicle. The adoption of systematic model-optimization methods using meaningful objective functions has been the pathway to go in order to achieve near-optimal results in designing the vehicle energy management system. In this chapter we focus on *model-based energy management strategy design* techniques. The chapter is organized as follows. In Sect. 12.3 we present a heavy-duty pre-transmission hybrid truck model, which is used as a case study. Section 12.4 presents the standard optimal energy management problem formulation. In Sect.12.5, we review results from the literature to solve the optimal control problem. In the same section, we present the basics of Pontryagin's Minimum Principle (PMP), Equivalent Consumption Minimization Strategy (ECMS) and Adaptive-PMP (A-PMP). Issues related to the real-time implementation of A-PMP are analyzed that motivate the design of a new energy management control framework presented in Sect. 12.6. Section 12.7 reports on some mathematical background used later in Sect. 12.8 where an analytical control law, referred to as nonlinear optimal control strategy (NL-OCS), is presented. Section 12.9 presents a comparison in simulation of the NL-OCS against PMP and A-PMP and the effectiveness of the new control design is shown both from a calibration and implementation standpoint.

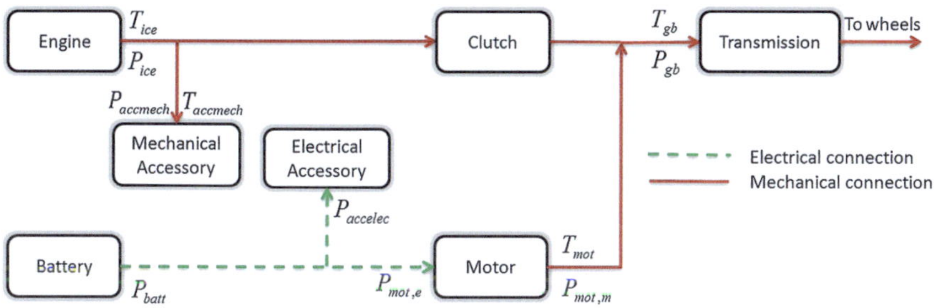

Fig. 12.1 Power flow diagram of pre-transmission parallel HEV

12.3 Case Study: Pre-transmission Parallel Hybrid

A heavy-duty pre-transmission parallel HEV is used as a case study along the chapter. The vehicle architecture and the power flow among the different components (the arrowheads denote the positive power sign convention) are illustrated in Fig. 12.1. The main specifications of the powertrain components are reported in Table 12.1. With the clutch closed, the parallel mode of operation uses both the devices, engine (ice) and motor (mot), to propel the vehicle and their speed is directly determined by the vehicle velocity. The additional degree of freedom available in this mode is used to optimize the vehicle energy usage. The torque/power balance equations are:

$$\begin{cases} T_{mot}(t) + T_{ice}(t) = T_{gb}(t) + T_{accmech}(t), \\ P_{batt}(t) = P_{mot,e}(t) + P_{accelec}(t), \\ \omega_{mot}(t) = \omega_{ice}(t) = \omega_{gb}(t). \end{cases} \tag{12.1}$$

where T_{gb}, ω_{gb} are the instantaneous gearbox torque and speed; T_{mot}, ω_{mot} are the instantaneous electric motor torque and speed; $P_{accelec}$ is the instantaneous electrical accessory power and $P_{mot,e}$ is the instantaneous electrical power at input/output terminals of the electric motor. The battery power (P_{batt}) can be represented as a function of engine power (P_{ice}) and the requested power (P_{req}) as:

$$\begin{cases} P_{batt}(t) = -\frac{1}{\eta_{mot}} P_{ice}(t) + \frac{1}{\eta_{mot}} P_{req}(t), \\ P_{req}(t) = P_{gb}(t) + \frac{1}{\eta_{mot}} P_{accelec}(t) + P_{accmech}(t). \end{cases} \tag{12.2}$$

The vehicle model has been implemented in PSAT (Powertrain Simulation Analysis Toolkit) environment [1].

An analytical model of the engine fuel consumption, based on Willans line approximation is used [10], which expresses the engine chemical power (P_{chem}) as an affine function of the engine power (P_{ice}) and speed (ω_{ice}):

$$P_{chem}(t) = e_0(\omega_{ice}(t)) + e_1(\omega_{ice}(t)) \cdot P_{ice}(t) \tag{12.3}$$

Table 12.1 Vehicle characteristics

Component	Size
Vehicle mass	19,878 kg
Engine capacity	6.7 L Diesel
Engine power	194 kW
Motor power	200 kW
Battery energy capacity	7.5 kWh (27 MJ)
Electrical accessory	7 kW
Mechanical accessory	4 kW

where $P_{chem} = \dot{m}_f \cdot Q_{LHV}$ [Q_{LHV} is the lower heating calorific value of diesel in (kJ/kg)] is the chemical power input to the engine and $P_{ice} = T_{ice}\omega_{ice}$ is the engine power output. The coefficient $e_0(\omega_{ice})$ represents the engine friction losses and $e_1(\omega_{ice})$ the conversion efficiency of the machine. A good approximation of the friction losses and conversion efficiency coefficients is given by expressing e_0 and e_1 as a quadratic fitting with respect to engine speed [25], as:

$$\begin{cases} e_0(\omega_{ice}(t)) = e_{00} + e_{01} \cdot \omega_{ice}(t) + e_{02} \cdot \omega_{ice}^2(t) \\ e_1(\omega_{ice}(t)) = e_{10} + e_{11} \cdot \omega_{ice}(t) + e_{12} \cdot \omega_{ice}^2(t) \end{cases} \tag{12.4}$$

where $e_{ij} > 0$, $i, j = 0, 1, 2$ are the constant Willans line coefficients. Hence, the fuel consumption rate can be written as:

$$\dot{m}_f(t) = \frac{1}{Q_{LHV}}[e_0(\omega_{ice}(t)) + e_1(\omega_{ice}(t)) \cdot P_{ice}(t)] \tag{12.5}$$

or:

$$\dot{m}_f(t) = p_0(\omega_{ice}(t)) + p_1(\omega_{ice}(t))P_{ice}(t) \tag{12.6}$$

with $p_0(\omega_{ice}(t)) = \dfrac{e_0(\omega_{ice}(t))}{Q_{LHV}}$, and $p_1(\omega_{ice}(t)) = \dfrac{e_1(\omega_{ice}(t))}{Q_{LHV}}$.
Note: The Willans line fuel consumption rate model, together with a suitable description of the battery model, is used to reformulate the energy management control problem as an infinite-time horizon optimal problem including stability in Sect. 12.8.

12.4 Problem Formulation

One important characteristic of the energy management problem is that the control objectives are mostly integral in nature (for instance, fuel consumption, emissions per mile of travel, battery life or a combination of the above, [9, 12, 16, 30]), while the control actions are local in time. In addition to that, the control objectives are subject to constraints which are both integral or global, such as maintaining battery

SOC within a prescribed range, and local constraints, such as physical limitation of the actuators. The very nature of this problem has made the task of finding a near-optimal implementable solution a challenging goal motivating a wealth of research over the past decade [23].

12.4.1 Optimal Energy Management Problem in HEVs

In this chapter, we consider the problem of minimizing the total mass of fuel, m_f (g), during a driving mission. This is equivalent to minimizing the following cost J_T:

$$J_T = \int_0^T \dot{m}_f(u(t))dt \tag{12.7}$$

where $\dot{m}_f$ (g/s) is the instantaneous fuel consumption rate, $u(t)$ is the control action, and T is the optimization horizon. The objective function (12.7) is minimized under a set of both local and global constraints, as outlined in the following.
System Dynamics. The system dynamics is given in terms of SOC variation with respect to time according to:

$$S\dot{O}C(t) = -\alpha\frac{I(t)}{Q_{nom}} \tag{12.8}$$

where α represents the Coulombic efficiency [10]; $I(t)$ (A) is the current flowing in (positive) and out (negative) of the battery and Q_{nom} (Ah) is the nominal battery charge capacity. The battery is modeled through the zero-*th* order equivalent circuit model [28], whose parameters are: the equivalent resistance, R_{eq} and the open circuit voltage, V_{oc}. For the application at hand, i.e., charge sustaining HEVs, the battery is used over a range of SOC (typically between 0.5 and 0.8 SOC), where the parameters are not dependent on SOC [28]. Following the discussion in [28] we can express the current $I(t)$ as a function of $P_{batt}(t)$ and write the system dynamics as:

$$S\dot{O}C(t) = -\alpha\frac{V_{oc} - \sqrt{(V_{oc})^2 - 4R_{eq}P_{batt}(t)}}{2R_{eq}Q_{nom}}. \tag{12.9}$$

Global Constraints. In a charge sustaining HEV, the net energy from the battery should be zero over a given driving mission, meaning that the SOC at the end of the driving cycle, $SOC(T)$, should be the same as the SOC at the beginning of the driving cycle, $SOC(0)$, and equal to a reference SOC value, i.e., SOC_{ref}:

$$SOC(T) = SOC(0) = SOC_{ref}. \tag{12.10}$$

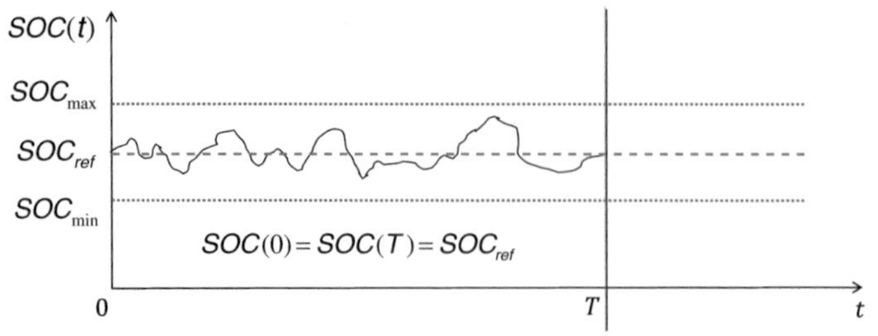

Fig. 12.2 Typical optimal SOC behaviour obtained solving Problem 12.1

Condition (12.10) is justified mainly as a way to compare the results of different solutions by guaranteeing that they start and reach the same level of battery energy. In real vehicles, it is sufficient to keep the SOC between two boundary values.

Local Constraints. Local constraints are imposed on the state and control variables. These constraints mostly concern physical operation limits, such as the maximum engine torque and speed, the motor power, or the battery SOC. For the pre-transmission parallel HEV powertrain local constraints are expressed as:

$$\begin{aligned} P_{batt,min} &\leq P_{batt}(t) \leq P_{batt,max}, \\ SOC_{min} &\leq SOC(t) \leq SOC_{max}, \\ T_{x,min} &\leq \;\; T_x(t) \;\; \leq T_{x,max}, \\ \omega_{x,min} &\leq \;\; \omega_x(t) \;\; \leq \omega_{x,max}, \;\; x = ice, \; mot. \\ T_{mot,min} &\leq T_{mot}(t) \leq T_{mot,max} \end{aligned} \tag{12.11}$$

where the last two inequalities in (12.11) represent limitations on the instantaneous engine and motor torque and speed, respectively; $(\cdot)_{min}$, $(\cdot)_{max}$ are the minimum and maximum value of power/SOC/torque/speed at each instant. Moreover, at each instant the supevisory controller ensures that the total power request at the wheels is satisfied.

Problem 12.1 The **energy management problem** in a charge sustaining HEV consists in finding the optimal control sequence u^* that minimizes the cost function (12.7) while meeting the dynamic state constraint (12.9), the global state constraint (12.10) and local state and control constraints (12.11).

Problem 12.1 by its very nature is a finite-time horizon (the cost function (12.7) is being minimized over a finite time horizon $[0,\;T]$), constrained (constraints on the state and control are being enforced at each instant of time), nonlinear [the system dynamics (12.9) are nonlinear], nonquadratic (the cost function is the fuel consumption map of the engine), optimal control problem. We refer to Problem 12.1 as the standard HEV energy management problem. A typical SOC behavior resulting from solving Problem 12.1 is shown in Fig. 12.2.

12.5 Finite-Time Horizon Energy Management Strategies

Several approaches have been proposed over the years to solve Problem 12.1. Those can be grouped into [24]:

- non-causal or non-realizable strategies. They require a priori knowledge of the driving cycle and are not applicable in real conditions [e.g., Dynamic Programming (DP), PMP];
- causal or realizable strategies. They do not require a priori knowledge of the driving cycle and are developed with the primary objective of realizability and do not guarantee optimality [e.g., Adaptive-PMP, Stochastic DP, rule-based, equivalent consumption minimization strategy (ECMS)].

Although, the primary objective is to design and implement causal strategies that can be eventually tested on real vehicles, the importance of finding non-causal optimal solutions resides in that: (1) they provide a benchmark solution (global optimum) any causal strategy can be compared against, and (2) properly modified they can be used to develop on-line strategies [27, 28]. In [16], for the first time, and in [4, 5] later, results from DP were analyzed with the aim of gaining insights to generate reproducible rules to design a rule-based strategy capable to mimic the DP behaviour. Although rule-based energy management strategies are relatively easy to develop and implement in a real vehicle, a significant amount of calibration effort is required to guarantee performances within a satisfactory range for any driving cycle. Moreover, rules are not necessarily scalable to different powertrain architectures and different component sizes. In addition to the DP [6, 31, 32], that finds the global solution recursively going backwards in time using Bellman's principle of optimality [3], local optimization methods have also been extensively used to find the global optimum. These methods can be used to find the optimum, by performing an offline optimization when the drive cycle is known, and they are also employed to design adaptive optimal strategies to achieve near optimal performances when the driving cycle is unknown. Much of the literature on local optimization methods pertain to PMP and/or ECMS [8, 27, 29].

The PMP [22] formulates and minimizes the Hamiltonian function (a function of the instantaneous cost and the state constraint) at each instant to obtain the optimal solution. PMP conditions, which in principles, are only necessary conditions of optimality in the case of Problem 12.1 become also sufficient.[1] This makes PMP a design tool to find the global optimal solution. Given Problem 12.1, PMP states that the optimal control solution $u^*(t)$ must satisfy the following conditions:

- $u^*(t)$ minimizes at each instant of time the Hamiltonian associated to the system:

$$H(u(t), SOC(t), \lambda(t)) = \lambda(t) \cdot S\dot{O}C(t) + \dot{m}_f(u(t))) \tag{12.12}$$

i.e.:

[1] Results from [13] and [14] prove the uniqueness of the solution of the optimal control problem under the satisfied assumption of constant battery efficiency over the SOC range of operation.

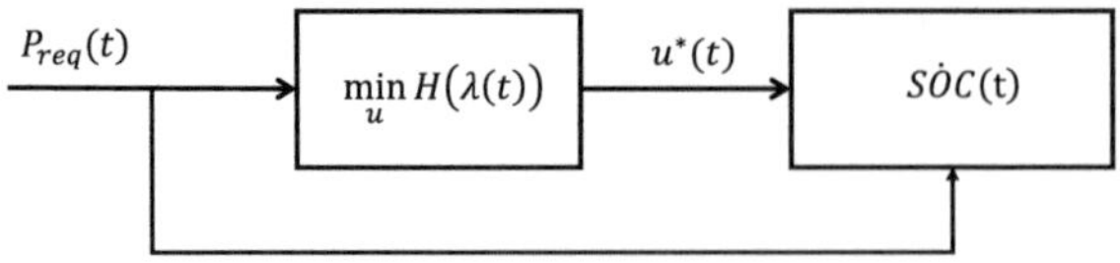

Fig. 12.3 Open-loop PMP-based energy management control scheme

$$u^*(t) = \min_{u \in \mathscr{U}} H(u(t), SOC(t), \lambda(t)) \tag{12.13}$$

$\mathscr{U}$ is the set of admissible solutions;

- the optimization variable $\lambda(t)$, also known as adjoint state or co-state must satisfy the dynamic equation along the optimal solution:

$$\dot{\lambda}(t) = -\left.\frac{\partial H}{\partial SOC}\right|_{u^*, SOC^*} \tag{12.14}$$

The optimal control sequence generated by (12.13) operates in open-loop as shown in Fig. 12.3.

Hence, the optimal solution u^* can only be obtained in simulation where the power request is known a-priori. In particular, the optimality of PMP resides in the perfect knowledge of the optimal co-state λ^* whose value varies from cycle to cycle. In [29] it is mathematically shown that the minimization of the Hamiltonian H is equivalent to the minimization of an equivalent fuel consumption function, used in the ECMS. ECMS, initially proposed by Paganelli et al. [21], is based on accounting for the use of stored electrical energy, in units of chemical fuel use (g/s), such that one can define an equivalent cost function taking into account the cost of electricity:

$$\dot{m}_{f,eq}(t) = s(t)\frac{E_{batt}}{Q_{lhv}} \cdot S\dot{O}C(t) + \dot{m}_f(t) \tag{12.15}$$

where E_{batt} is the battery energy and $s(t)$ is the *equivalent factor* that assigns a cost to the use of electricity, and the equivalent cost function $\dot{m}_{f,eq}(t)$ is equivalent to the Hamiltonian in PMP. If, on one hand, PMP/ECMS are practical tools to find the optimal solution to Problem 12.1 using a forward looking simulator, they can also been employed for real-time implementation.

In fact, the only control parameter in the PMP (or ECMS) is the co-state (or equivalent factor), which is cycle-dependent. The key idea to use the PMP (or ECMS) as a causal strategy resides in adapting the co-state as a function of driving conditions. From the PMP solution one can observe that the variation of the co-state as driving conditions change is correlated to the divergence of the actual SOC from its charge-sustaining reference value [20]. This observation has led to the development of an adaptation scheme based on feedback from SOC to be used in combination to the minimization of H [20]. The role of adaptation is to update the value of the co-state without using past driving information or prediction of future driving behavior, but

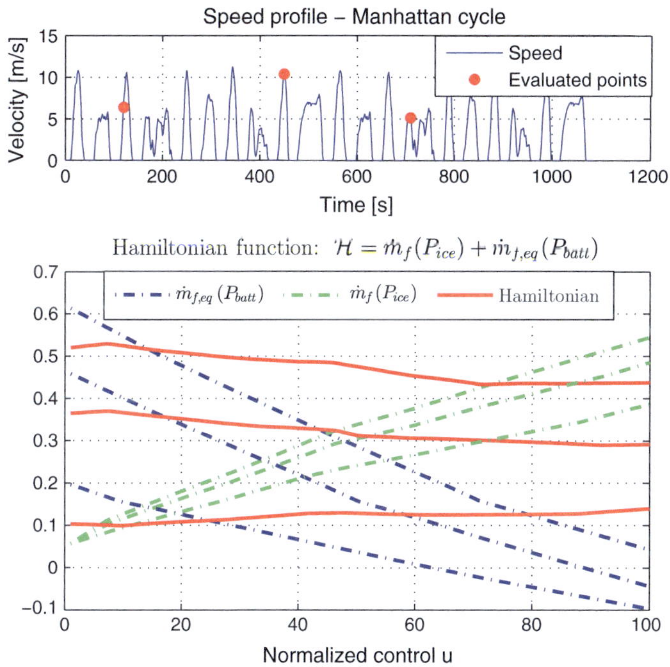

Fig. 12.4 Hamiltonian function H (*bottom*) evaluated for different instances of a Manhattan driving cycle (*top*) [19]

just using information of current SOC. For example, the adaptation can be performed via PI-like controller [15], or via an autoregressive moving average (ARMA) mechanism of the type [20]:

$$\lambda(k) = \frac{\lambda(k-1) - \lambda(k-2)}{2} + K \cdot (SOC_{ref} - SOC(k)) \tag{12.16}$$

which allows the adaptation to take place at regular intervals of duration T_s, ($t = kT_s, k = 1, 2, \ldots$), rather than at each time instant as in the case of PI-like based correction.

12.6 Motivation for Infinite-Time Horizon Optimization

The real-time controller based on A-PMP (12.16) requires the Hamiltonian function to be minimized instantaneously. This operation, that needs to be executed on-board at each tick of the clock, despite being computationally expensive, can lead in some cases to unpredictable no-optimal results, due to the fact that the Hamiltonian is in many instances of the driving cycle not a convex function of the control variable, as one can see from Fig. 12.4. Different control values could be in principle equally suitable in the minimization process, leading to a not unique solution of the optimal control problem, thus causing undesirable chattering in the control outputs [26]. These issues have suggested to move towards a new research direction to find optimal solutions that would not have such a detrimental behaviour when used in a real-time setting. Inspired by Bernstein and Haddad's work [2, 11] on theoretical results on optimal nonlinear regulation problem involving non quadratic cost functionals, a first attempt to propose a new control framework for the energy management problem was done in [25]. The authors cast the energy management problem into a nonlinear optimal regulation problem where the battery SOC is optimally regulated to its reference target in the case of zero disturbance ($P_{req} = 0$). Preliminary results showed the feasibility of the closed-form control law in the simple case of vehicle at standstill and series hybrid architecture. Reduction in computational complexity and decreased sensitivity of the control parameter with respect to driving conditions were also showed. Nonetheless, two issues were not properly addressed in [25]: the stability definition and the extension of finite-time cost function into an infinite-time functional (needed to formally use the results from [2, 11]). In [19], a rigorous framework is developed where stability of the energy management state trajectory is finally defined while guaranteeing optimality by means of an analytical, cycle-independent control law. The novel framework is summarized in the next section and new simulation results comparing the performances of the new analitycal supervisory controller against PMP (used as a benchmark) and A-PMP (for on-line strategies comparison) are presented in Sect. 12.9.

12.7 From Finite-Time to Infinite-Time Horizon Optimal Control Problem

The energy management problem is reformulated as a nonlinear-nonquadratic infinite-time optimization problem. The new control framework consists in re-thinking the standard finite-time optimal control problem in HEV (Problem 12.1) as an infinite time horizon problem. To ensure optimality of vehicle operation when $t > T$, the $[0,\ T]$ optimization horizon is extended into the infinite horizon $[0,\ \infty]$, leading to a new cost function, J_∞ [19]:

$$J_\infty = \int_0^\infty \dot{m}_f(u(t)) \cdot g(t)dt \tag{12.17}$$

by means of the scalar positive function, $g(t)$:

$$g(t) = \frac{1 + \alpha \left(\frac{t}{T} \right)^q}{1 + \left(\frac{t}{T} \right)^q} \qquad 0 < \alpha < 1, \quad q > 0 \tag{12.18}$$

The role of the function $g(t)$ is to penalize the action of the control $u(t)$ for $t > T$ in order to approximate the finite-time cost J_T defined in (12.7) to the infinite-time functional (12.17). The system dynamics is reformulated in order to fit the problem in the form used in [11] (as discussed in [19]), where the nonlinear system is required to be dissipative with respect to a supply rate function.

12.7.1 System Dynamics Reformulation

In the new control framework, a Lyapunov-based approach is used to obtain a state-feedback control law to find the optimal torque/power split where the power requested (P_{req}) is regarded as a $\mathscr{L}_2$ disturbance. The battery state of energy (SOE), defined as the amount of battery energy stored at the present time ($E(t)$) to the maximum battery energy capacity (E_{max}), is used as state variable in this discussion. SOE is related to SOC by the following relationship [29]:

$$SOE(t) = SOC(t) \frac{V_L(t)}{V_{oc}^{max}} = \frac{E(t)}{E_{max}} \tag{12.19}$$

where V_L is the battery terminal voltage and V_{oc}^{max} the maximum open circuit voltage. Hence, the SOE dynamics:

$$\begin{cases} \dot{SOE} = -\eta_{batt} \dfrac{P_{batt}}{E_{max}} \\ E_{max} = Q_{max} \cdot V_{oc}^{max} \end{cases} \tag{12.20}$$

Defining $k = \frac{\eta_{batt}}{E_{max}\eta_{mot}}$, the battery SOE error $\zeta = SOE_{ref} - SOE$ is introduced, whose dynamics is described as a function of the control input (P_{ice}) and the disturbance (P_{req}) by virtue of Eq. (12.2):

$$\dot{\zeta} = -kP_{ice} + kP_{req} \tag{12.21}$$

Note that in parallel mode the power requested is the sum of accessory powers ($P_{accelec} + P_{accmecc}$) and the gearbox power (P_{gb}). When the vehicle is not moving ($v = 0$), instead, the power requested P_{req} only accounts for the accessory loads power. Thus, the disturbance power P_{req} is:

$$P_{req} = \begin{cases} P_{gb} + \eta_{mot} P_{accelec} + P_{accmecc} & v > 0 \quad \forall t \in [0,\ T] \\ \eta_{mot} P_{accelec} + P_{accmecc} & v = 0 \quad \forall t \in [T,\ \infty] \end{cases} \tag{12.22}$$

Consider an open set $\mathscr{X} \subset \mathbb{R}$ such that $\zeta \in \mathscr{X}$, a set $\mathscr{U} \subset \mathbb{R}$ such that $P_{ice} \in \mathscr{U}$, and a set $\mathscr{W} \subset \mathbb{R}$ such that $P_{req} \in \mathscr{W}$ and P_{req} in $\mathscr{L}_2$. The compact sets for the control, state and disturbance are:

$$\begin{cases} \mathscr{X} = \left[SOE_{ref} - SOE_{max}, SOE_{ref} - SOE_{min}\right] \\ \mathscr{U} = \left[0,\ P_{ice}^{max}\right] \\ \mathscr{W} = \{P_{req} : P_{req} \in \mathscr{L}_2\} \end{cases} \tag{12.23}$$

Consider the following control system:

$$\begin{cases} \dot{\zeta} = -kP_{ice} + kP_{req}, \quad \zeta(0) = \zeta_0 \\ z = \zeta \end{cases} \tag{12.24}$$

where $\zeta = 0$ is an equilibrium point of the autonomous system and z is the performance output variable. Also consider the following functional cost [in virtue of (12.6)]:

$$J_\infty = \int_0^\infty \dot{m}_f(P_{ice}(t))dt = \int_0^\infty \frac{p_0(\omega_{ice}) + p_1(\omega_{ice}) \cdot P_{ice}(t)}{Q_{LHV}} dt \tag{12.25}$$

Problem 12.2 The **infinite-time optimal energy management problem** consists in minimizing the cost function (12.25) under system dynamics (12.24), with state and control variables lying in the compact sets $\mathscr{X}$ and $\mathscr{U}$, and $P_{req} \in \mathscr{W}$.

Definition 12.1 Consider Problem 12.2 with $P_{req} \equiv 0$ and let $\phi(\zeta(t))$ be its optimal solution. Then the origin $\zeta(t) = 0$ of the closed-loop system under $\phi(\zeta(t))$ is asymptotically stable if $\zeta(t) \to 0$ for $t \to \infty$.

A typical SOC behavior obtained as a solution of Problem 12.2, is shown in Fig. 12.5. It can be noticed that the global constraint used in Problem 12.1 requiring $SOC(T)$ to be equal to the reference value SOC_{ref} is not met in this case as the convergence of SOC to SOC_{ref} is guaranteed only as $t \to \infty$.

12.8 Infinite-Time Nonlinear Optimal Control Strategy (NL-OCS)

With respect to the system (12.24) and the infinite cost function (12.25) [11] defines the Hamiltonian function H as following:

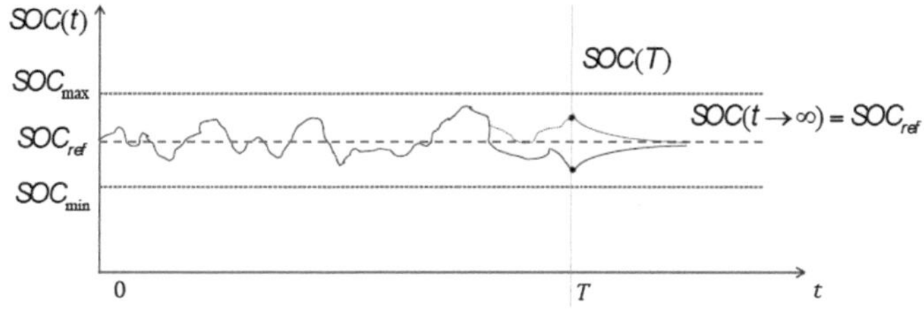

Fig. 12.5 Typical SOC profile as a solution of a infinite-time optimization problem including stability

$$H(\zeta, P_{ice}, \lambda) = \dot{m}_f(P_{ice}) + \Gamma(\zeta, P_{ice}) + \lambda \cdot (kP_{ice}) \tag{12.26}$$

where $\Gamma(\zeta, P_{ice})$ is a positive scalar function (to be selected), and λ is the co-state variable. In order to have the Hamiltonian function zero at the minimum value, as requested in [11], a shifting of the H is operated as follows:

$$\bar{H}(\zeta, P_{ice}, \lambda) = H(\zeta, P_{ice}, \lambda) - p_0(\omega_{ice}) \tag{12.27}$$

Theorem 12.1 *Consider the system (12.24) with functional cost (12.25). Then, the feedback control law* $P^*_{ice}(\zeta)$ *defined as:*

$$P^*_{ice} = \phi(\zeta) = \begin{cases} \frac{2k^2(\mu^4\zeta^3)^2}{(k\mu^4\zeta^3 - p_1(\omega_{ice})\, g(t))\gamma^2} & \zeta > \bar{\zeta} \vee \zeta \le 0 \\ \zeta^2 & 0 < \zeta \le \bar{\zeta} \end{cases} \tag{12.28}$$

with $\bar{\zeta} = \left(\frac{p_1(\omega_{ice})}{k\mu^4}\right)^{\frac{1}{3}}$*, is such that:*

1. *the solution* $\zeta(t) = 0, t \ge 0$ *of the closed-loop system is locally asymptotically stable in accordance to Definition 12.1.*
2. *the adjoint performance functional* $\mathcal{J}(\zeta, P_{ice}(\zeta))$

$$\mathcal{J}(\zeta, P_{ice}) = \int_0^\infty \left[\dot{m}_f(P_{ice}) + \Gamma(\zeta, P_{ice})\right] dt \tag{12.29}$$

is minimized.

Proof Consider the candidate Lyapunov function

$$V(\zeta) = \frac{1}{4}\mu^4\zeta^4, \quad \mu > 0 \in R \tag{12.30}$$

then we can define the storage function $\Gamma(\zeta, P_{ice})$ and the supply rate function $r(\zeta, P_{req})$, associated to the system (12.24) and the Lyapunov function (12.30), as:

$$\begin{cases} \Gamma(\zeta, P_{ice}) = \dfrac{1}{\gamma^2}\left(\dfrac{\partial V}{\partial \zeta}\right)^2 k^2 \cdot \left(1 + \log(P_{ice}^2)\right) \\ r(\zeta, P_{req}) = \gamma^2 P_{req}^2 - \zeta^2 \end{cases} \tag{12.31}$$

The proof of Theorem 12.1, following the same reasoning provided in [11], is based on a series of sufficient conditions that ensure optimality and stability that are shown to hold true when the optimal feedback control $\phi(\zeta) = P_{ice}^*(\zeta)$ is used.

1. The Lyapunov function $V(\zeta)$ assumes its minimum value of 0 at the origin.

$$V(0) = 0 \tag{12.32}$$

2. $V(\zeta)$ is a positive definite function because it is a quadratic scalar function with the minimum at the origin.
3. The optimal feedback control law is zero at the origin, i.e., from (12.28):

$$P_{ice}^*(0) = 0 \tag{12.33}$$

4. The optimal control law (12.28) makes the origin $\zeta(t) = 0$ asymptotically stable when $P_{req} = 0$, equivalently:

$$\frac{\partial V}{\partial \zeta} \cdot k P_{ice}^*(\zeta) < 0, \quad \zeta \neq 0 \tag{12.34}$$

In order to show (12.34), without loss of generality we consider P_{batt} as new control variable with $P_{req} = 0$. Thus:

$$P_{batt}^* = \begin{cases} -\frac{2k^2(\mu^4\zeta^3)^2}{(k\mu^4\zeta^3 - n_1(\omega_{mot})\ g(t))\gamma^2} & \zeta > \bar{\zeta}^* \vee \zeta \leq 0 \\ -\zeta^2 & 0 < \zeta \leq \bar{\zeta}^* \end{cases} \tag{12.35}$$

where $\bar{\zeta}^* = \left(-\dfrac{n_1(\omega_{mot})}{k\mu^4}\right)^{\frac{1}{3}}$, and this makes (12.34) become:

$$\mu^4 \cdot \zeta^3 \cdot k \cdot k P_{batt}^*(\zeta) < 0, \quad \zeta \neq 0 \tag{12.36}$$

In the domain $0 < \zeta \leq \bar{\zeta}^*$, it is immediate to see that

$$-\mu^4\, \zeta^3\, k\, \zeta^2 < 0 \tag{12.37}$$

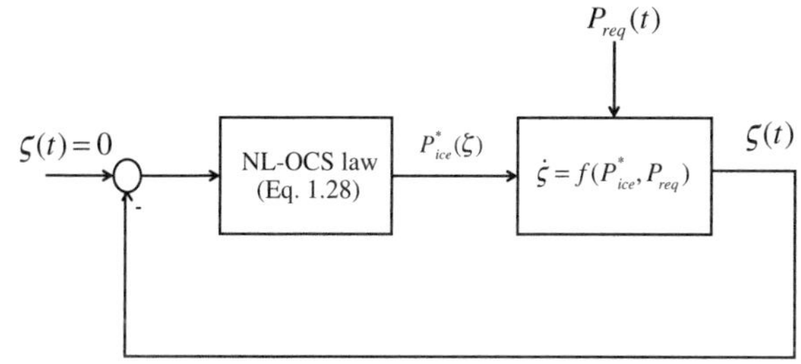

Fig. 12.6 Closed-loop energy management control scheme based on the analytical NL-OCL solution

In the domain $\zeta > \bar{\zeta}^* \vee \zeta \leq 0$ denominator of (12.36) is positive when ζ is positive and negative otherwise, thus leading to:

$$\begin{cases} -\mu^4\zeta^3 k \cdot 2k^2(\mu^4\zeta^3)^2 < 0 \; \zeta > \bar{\zeta}^* \\ \mu^4\zeta^3 k \cdot 2k^2(\mu^4\zeta^3)^2 < 0 \quad \zeta \leq 0 \end{cases} \tag{12.38}$$

5. The Hamiltonian function (12.27) takes on the minimum value when the optimal control law (12.28) is applied. The shifted hamiltonian $\bar{H}$,

$$\bar{H}\left(\zeta, P^*_{ice}, \frac{\partial V}{\partial \zeta}\right) = \dot{m}_f + \Gamma(\zeta, P^*_{ice}) + \frac{\partial V}{\partial \zeta} k P^*_{ice}(\zeta) \tag{12.39}$$

becomes

$$\bar{H} = p_1\, P_{ice} + \frac{1}{\gamma^2}k^2\left(\mu^4\zeta^3\right)^2\left(1 + \log(P^2_{ice})\right)^2 + \mu^4\zeta^3 k P_{ice} \tag{12.40}$$

for the system (12.24) and cost function (12.25). It can be easily shown that the closed-loop controller (12.28) is a minimum of the $\bar{H}$ (the stationary first order conditions and the second order convexity conditions are verified).

6. The passivity condition with respect to the disturbance input P_{req} requires that the following inequality is satisfied:

$$\left(\frac{\partial V}{\partial \zeta}\right) \cdot k \cdot P_{req} \leq r(\zeta, P_{req}) + \dot{m}_f + \Gamma(\zeta, P^*_{ice}) \tag{12.41}$$

A second order algebraic inequality in P_{req} is obtained which is verified when $\gamma \leq \bar{\gamma} = 2.369$. *Q.D.E.*

In virtue of Theorem 12.1, the origin $\zeta = 0$ of the closed-loop system is optimally locally asymptotically stable when $P_{req} = 0$. Moreover, P^*_{ice} is optimal with respect to the adjoint functional $\mathscr{J}(\zeta, P_{ice}(\cdot))$, which is an upper bound for J_∞.

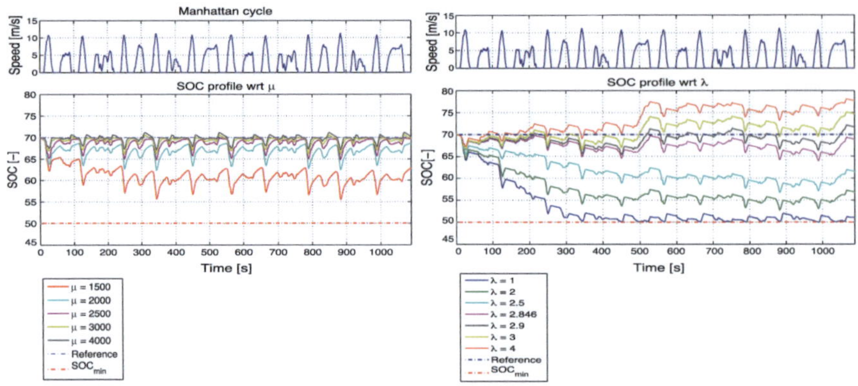

Fig. 12.7 SOC trajectories from: (1) NL-OCS as μ varies (*left*) and (2) PMP as λ varies (*right*) for the Manhattan driving cycle

The optimal control law obtained from Theorem 12.1 is referred to as Nonlinear Optimal Control Strategy (NL-OCS) and it is implemented according to the closed-loop system scheme shown in Fig. 12.6. To the best knowledge of the author of this article, this is the first time that an analytical supervisory controller is proposed to solve the energy management problem in HEVs. In the optimal control law (12.28) that operates from SOC feedback, the values of k, $p_1(\omega_{ice})$ are known from the vehicle models, γ is a constant whose upper bound was obtained from the theorem's proof, and μ is the only calibration parameter that needs to be selected for on-board implementation.

12.9 Strategies Comparison: Simulation Results

In this section, we first evaluate the novel closed-loop supervisory controller against the benchmark solution from PMP and then we compare the NL-OCS against the real-time implementable A-PMP to show the effectiveness of the proposed control-law for on-board implementation. Offline simulations are performed to test the sensitivity of the new model-based strategy against the calibration parameter μ. Results are shown on the left plot of Fig. 12.7 where different SOC profiles from NL-OCS are shown for different value of μ. On the left plot of Fig. 12.8 the fuel consumption (FC) is plotted together with $\Delta SOC = SOC(T) - SOC(0)$ (for different driving cycles) to measure the ability of the control law to guarantee charge-sustainability. On the right hand side of Fig. 12.7 and Fig. 12.8 we show: (1) the solution obtained from PMP for different values of the co-state λ and, (2) the high sensitivity of charge-sustainability to the co-state λ.

It is well known, in fact, that performance of PMP is highly dependent on the co-state λ, both in terms of charge-sustainability and fuel consumption (see, for instance

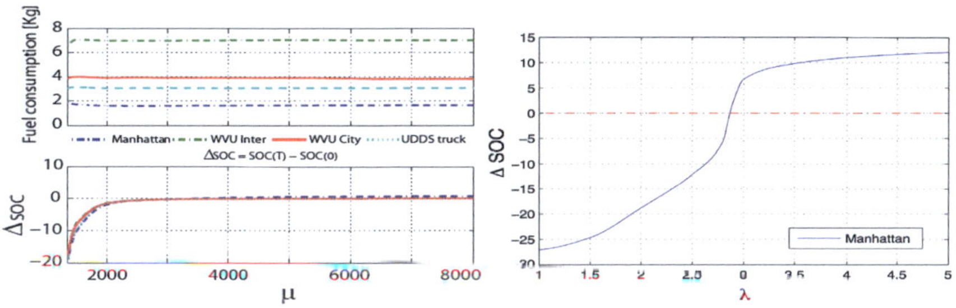

Fig. 12.8 NL-OCS: fuel consumption and $\Delta SOC = SOC(T) - SOC(0)$ as a function of μ for four different driving cycles (*left*). PMP: ΔSOC as a function of λ for Manhattan driving cycle (*right*)

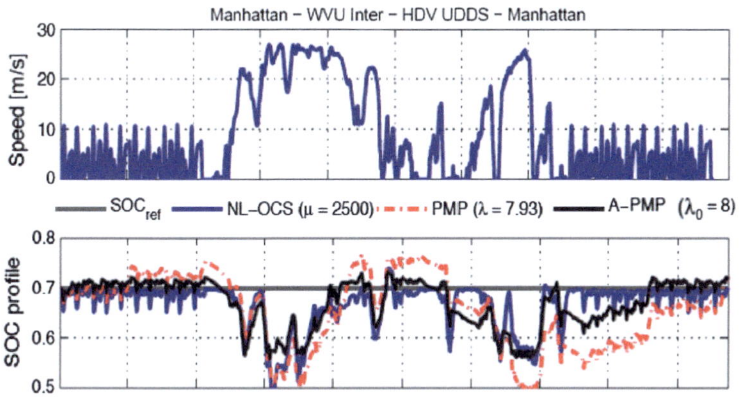

Fig. 12.9 SOC profiles from PMP, NL-OCS and A-PMP

[28]). The results of this analysis are used to calibrate the NL-OCS for on-board implementation. A combined driving cycle obtained by concatenating a Manhattan, West Virginia Urban (WVU) Interstate, Heavy-Duty UDDS, and Manhattan driving cycles is used to validate and compare the NL-OCS against the PMP solution and the real-time controller A-PMP. The three SOC profiles are shown in Fig. 12.9 and a quantitative analysis in terms of fuel economy and engine efficiency of the three control strategies is reported in Table 12.2. Not only does the analytical control law guarantee optimality (with values within 1 % from the PMP benchmark solution) for a wide range of values of the control parameter μ (see, Fig. 12.8), but also it guarantees low sensitivity against driving characteristics, making the performance of the new strategy driving cycle independent. In addition, the calibrated NL-OCS also shows better performance in terms of fuel consumption than the real-time A-PMP. Above all, the main advantage of having an analytical solution is in the fast execution of the control action as opposed to the computational burden required by the instantaneous minimization operation of A-PMP. In [19], it is reported that the NL-OCS solution is up to 5 times faster than the A-PMP. The NL-OCS can be implemented in the

Table 12.2 Fuel consumption and engine efficiency comparison between the PMP, A-PMP and NL-OCS solutions

Controller	FC m_f (kg)	Norm. fuel cons. %	ICE eff.
PMP	13.11	100	0.319
A-PMP	13.36 (< 2 %)	98.13	0.309
NL-OCS	13.24 (< 1 %)	99.02	0.310

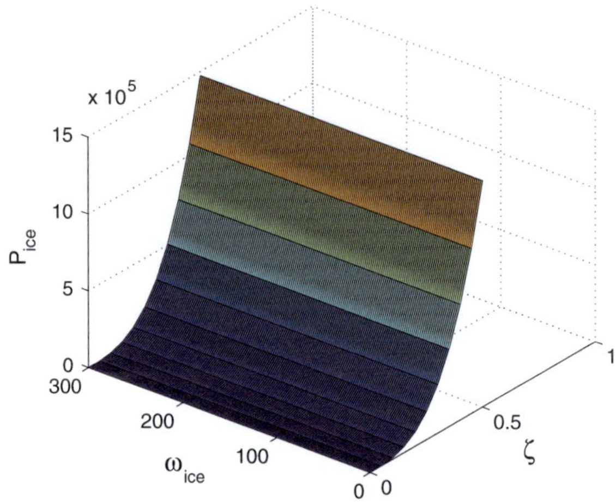

Fig. 12.10 Engine power map: $P_{ice} = f(\omega_{ice}, \zeta)$

form of a look-up table, by mapping the power issued by the control law (12.28) as a function of ζ and the engine speed ω_{ice}, as shown in Fig. 12.10.

12.10 Conclusions

In this chapter, we have first presented the standard formulation of the energy management problem in HEVs and reviewed the PMP and A-PMP methods. As a real-time implementable strategy, if on one hand the A-PMP is very promising as it performs near to the global optimum, on the other hand, the high computational burden due to the instantaneous minimization can make the use of this strategy prohibitive for in vehicle operation. A new framework centered around the theory of nonlinear, nonquadratic optimal control has been developed and presented in this chapter. An analytical, cycle-independent, state-feedback supervisory controller has been proposed that achieves optimality with respect to an infinite time horizon performance functional while guaranteeing asymptotic stability. The proposed control law was implemented in a pre-transmission parallel hybrid heavy-duty vehicle and the performances of the closed-loop system were compared to the benchmark solution provided by the PMP and the real-time solution provided by A-PMP. The advantages offered by the newly designed solutions are: (1) low calibration effort (only one parameter needs to be calibrated); (2) low sensitivity to the control parameter; (3) fast execution for on-board applications; (4) close-to-the-optimum performance despite the driving mission.

Acknowledgments The author would like to deeply thank Roberto Mura for taking this research a step forward, Lorenzo Serrao for the enjoyable and productive discussions, Yann Guezennec, Giorgio Rizzoni, and Stephen Yuorkovich for the productive iterations.

References

1. Argonne National Laboratory: Powertrain system analysis toolkit (PSAT) documentation. DuPage County, IL, http://web.anl.gov/techtransfer/pdf/PSAT.pdf
2. Bernstein DS (1993) Non quadratic cost and nonlinear feedback control. Int J Robust Nonlinear Control 3:211–229
3. Bertsekas DP (1995) Dynamic programming and optimal control. Athena Scientific, Belmont
4. Bianchi D, Rolando L, Serrao L, Onori S, Rizzoni G, Al-Khayat N, Hsieh TM, Kang P (2011) Layered control strategies for hybrid electric vehicles based on optimal control. Int J Electric Hybrid Veh 3:191–217
5. Biasini R, Onori S, Rizzoni G (2013) A rule-based energy management strategy for hybrid medium duty truck. Int J Powertrains 2(2/3):232–261
6. Brahma A, Guezennec Y, Rizzoni G (2000) Optimal energy management in series hybrid electric vehicles. In: Proceedings of the 2000 American control conference, vol 1, issue 6, pp 60–64.
7. Chan CC (2002) The state of the art of electric and hybrid vehicles. Proc IEEE 90(2):247–275
8. Cipollone R, Sciarretta A (2006) Analysis of the potential performance of a combined hybrid vehicle with optimal supervisory control. In: Proceedings of the 2006 IEEE international conference on control applications, pp 2802–2807.
9. Ebbesen S, Elbert P, Guzzella L (2012) Battery state-of-health perceptive energy management for hybrid electric vehicles. IEEE Trans Veh Technol 61(7):2893–2900
10. Guzzella L, Sciarretta A (2007) Vehicle propulsion systems: introduction to modeling and optimization, 2nd edn. Springer, Berlin
11. Haddad WM, Chellaboina V (2008) Nonlinear dynamical systems and control: a lyapunov-based approach. Princeton University Press, NJ
12. Johnson VH, Wipke KB, Rausen DJ (2000) HEV control strategy for real-time optimization of fuel economy and emissions. SAE Paper number No. 2000–01-1543.
13. Kim N, Cha S, Peng H (2011) Optimal control of hybrid electric vehicles based on Pontryagin's minimum principle. IEEE Trans Control Systems Technol 19(5):1279–1287
14. Kim N, Rousseau A (2012) Sufficient conditions of optimal control based on Pontryagin's minimum principle for use in hybrid electric vehicles. Proc Inst Mech Eng, Part D: J Automobile Eng 226:1160–1170
15. Koot M, Kessels J, de Jager B, Heemels W, van den Bosh PPJ, Steinbuch M (2005) Energy management strategies for vehicular electric systems. IEEE Trans Veh Technol 54(3):771–782
16. Lin CC, Peng H, Grizzle JW, Kang JM (2003) Power management strategy for a parallel hybrid electric truck. IEEE Trans Control Syst Technol 11(6):839–849
17. Mi C, Masrur MA, Gao DW (2011) Hybrid electric vehicles: principles and applications with practical perspectives. Wiley, New York
18. Miller JM (2003) Propulsion systems for hybrid vehicles. The Institution of Electrical Engineers, London

19. Mura R, Utkin V, Onori S (2013) Ecasting the HEV energy management problem into an infinite-time optimization problem including stability. In: 52nd IEEE CDC.
20. Onori S, Serrao L, Rizzoni G (2010) Adaptive equivalent consumption minimization strategy for hybrid electric vehicles. In: Proceedings of the 2010 ASME DSCC, pp 499–505.
21. Paganelli G, Ercole G, Brahma A, Guezennec Y, Rizzoni G (2001) General supervisory control policy for the energy optimization of charge-sustaining hybrid electric vehicles. JSAE Rev 22:511–518
22. Pontryagin L, Boltyanskii VG, Gamkrelidze RV, Mishchenko EF (1962) The mathematical theory of optimal processes. Wiley, NJ
23. Rizzoni G, Peng H (2013) Hybrid and electric vehicles: the role of dynamics and control. In: ASME dynamic systems and control magazine, pp 10–17.
24. Sampathnarayanan B (2013) Analysis and design of stable and optimal energy management strategies for hybrid electric vehicles. Ph.D. Dissertation The Ohio State University.
25. Sampathnarayanan B, Onori S, Yurkovich S (2012) An optimal regulation strategy for energy management of hybrid electric vehicles. In: 51st IEEE CDC.
26. Sciarretta A, Back M, Guzzella L (2004) Optimal control of parallel hybrid electric vehicles. IEEE Trans Control Syst Technol 12:352–363
27. Sciarretta A, Guzzella L (2007) Control of hybrid electric vehicles. IEEE Control Syst Mag 27:60–70
28. Serrao L, Onori S, Rizzoni G (2011) A comparative analysis of energy management strategies for hybrid electric vehicles. ASME JDSMC 133(3):1–9
29. Serrao L, Onori S, Rizzoni G (2009) ECMS as a realization of Pontryagin's minimum principle for HEV control. In: Proceedings of the 2009 American control conference, pp 3964–3969.
30. Serrao L, Onori S, Sciarretta A, Guezennec Y, Rizzoni G (2011) Optimal energy management of hybrid electric vehicles including battery aging. In: Proceedings of the 2011 American control conference, pp 2125–2130.
31. Sundstrom O, Guzzella L (2009) TA generic dynamic programming Matlab function. In: Control applications (CCA) intelligent control (ISIC), 2009 IEEE, pp 1625–1630.
32. Sundstrom O, Guzzella L, Soltic P (2008) Optimal hybridization in two parallel hybrid electric vehicles using dynamic programming. In: Proceedings of the 17th IFAC world congress.

Chapter 13
Optimal Energy Management of Automotive Battery Systems Including Thermal Dynamics and Aging

Antonio Sciarretta, Domenico di Domenico, Philippe Pognant-Gros and Gianluca Zito

Abstract Hybrid-electric vehicles (HEV) has been the subject of intensive research as a field of application of optimal control in the past decade. In particular, researchers have proven that energy management (or supervisory control) can be effectively designed using optimal control-based techniques (Guzzella and Sciarretta, Vehicle Propulsion Systems. Introduction to Modeling and Optimization. Springer, Berlin, 2013. Such methods have been applied to charge-sustaining hybrids implementing various architecture, as well as, more recently, to plug-in hybrids (Stockar et al. IEEE Trans Vehr Technol, 60(7):2949–2962, 2011; Sivertsson 2012). Plug-in hybrids (PHEV) are characterized by much higher battery capacities and energies than charge-sustaining hybrids, thus the proper description of battery behavior plays an even more fundamental role in energy management design.

13.1 Introduction

Hybrid-electric vehicles (HEV) has been the subject of intensive research as a field of application of optimal control in the past decade. In particular, researchers have proven that energy management (or supervisory control) can be effectively designed using optimal control-based techniques [1]. Such methods have been applied to charge-sustaining hybrids implementing various architecture, as well as,

A. Sciarretta (✉) · D. di Domenico · P. Pognant-Gros · G. Zito
IFP Energies nouvelles, Rueil-Malmaison Cedex, France
e-mail: Antonio.Sciarretta@ifpen.fr

D. di Domenico
e-mail: domenico.didomenico@ifpen.fr

P. Pognant-Gros
e-mail: Philippe.Pognant-Gros@ifpen.fr

G. Zito
e-mail: Gianluca.Zito@ifpen.fr

H. Waschl et al. (eds.), *Optimization and Optimal Control in Automotive Systems*, Lecture Notes in Control and Information Sciences 455,
DOI: 10.1007/978-3-319-05371-4_13,

more recently, to plug-in hybrids [2, 3]. Plug-in hybrids (PHEV) are characterized by much higher battery capacities and energies than charge-sustaining hybrids, thus the proper description of battery behavior plays an even more fundamental role in energy management design. Usually, energy management aims at minimizing a cost function that is fuel consumption (or a weighted sum of consumption and regulated emissions), under a global constraint on the battery state of charge (SOC). Therefore, battery dynamics is usually described only in terms of macroscopic SOC variation as a function of current. However, the battery is a crucial component of HEV and PHEV that affects significantly the cost and the performance of the whole system. While installation cost is a consequence of the battery design, battery end of life induces replacement costs if not matching vehicle lifetime. Moreover, as battery power and energy capabilities deteriorate with aging, its buffer role tends to fade, decreasing the energy efficiency of the powertrain with time. Battery aging is thus a key factor, even more for PHEV, which depends on the design choices but also on the use of the battery, i.e., on the energy management strategy adopted. Including aging in the cost function to be minimized needs a proper modeling of the aging factors, particularly capacity loss and internal resistance increase. A first attempt of integrating an aging model in the cost function of an optimal-control based energy management was reported in [4]. The aging model there used is sensitive to the current only, similarly to a few other contributions [5]. However, a more accurate and experimentally validated model is needed to draw general conclusions. This chapter considers mainly the capacity loss factor, which has been the subject of several experimental investigations [6–8]. Such investigations pointed at the importance of both current and temperature as the main capacity loss factors. The relevance of temperature implies an additional dynamics, other than that of SOC, to be taken into account in the optimal control solution. A few researches have dealt with battery temperature variations [9, 10], without explicitly considering aging. Additional temperature states imply additional costates in the optimal control solution, which are particularly difficult to treat both in an online controller and in offline optimization, as it has been shown for other thermal states, such as engine or catalyst temperatures [11]. For example, costate own dynamics are unstable for stable state dynamics and this fact makes the optimal solution very sensitive to the prediction of the costate values (online) or to the guess of the initial costate values (offline, e.g., with shooting-like methods). As an additional complexity, the established capacity loss models exhibit an internal state that is the total ampere-hour (Ah) throughput, which is an equally undesired feature from the viewpoint of the optimal control solution.

The contribution of this chapter is a simulation study, albeit based on experimental measures, on the minimization of a mixed fuel consumption–battery aging cost function during PHEV operation. A validated capacity loss model is modified in order to eliminate the Ah-throughput state. However, thermal dynamics are explicitly considered in the optimal control problem and for this purpose a validated lumped-parameter battery thermal model is used. Dynamic programming (DP) and Pontryagin's minimum principle (PMP) are compared to numerically compute the optimal solutions, for various scenarii in terms of ambient temperature and optimization criterion. A specific analysis is conducted on the costate variations, in view of an online implementation of the derived strategies.

13.2 Case Study and Motivation

The case study presented in this chapter is a plug-in parallel hybrid–electric demo car designed by IFPEN. The baseline vehicle, a gasoline-engine utility vehicle (Renault Kangoo), has been transformed to a full hybrid vehicle by integrating off-the-shelf electric components (battery pack, DC–DC converter, and electric motor) and replacing the original manual gearbox with an entry-level automated manual transmission. The electric motor has been coupled to the primary shaft of the transmission through a gear reducer, thus ensuring a wide vehicle speed range for the pure electric mode. A large Li-ion battery allows to obtain a range of about 33 km in zero-emission mode when running on a NEDC cycle limited to 55 km/h. The characteristics of the main vehicle components are resumed in Table 13.1.

The on-board vehicle energy management is based on an Equivalent Consumption Minimization Strategy (ECMS) [1], a technique derived from optimal control theory and PMP, see Sect. 13.4.2, using the engine fuel consumption as the minimization criterion. The optimal torque split between the engine and the electric motor is computed at a low frequency of 5–10 Hz. According to the ECMS, at each time step the optimal split of the required wheel torque is chosen among a fixed (relatively small) number of hybrid split values, and one purely electric mode. The optimal torque split is computed making the assumption that the corresponding driveline configuration is instantaneously available. This is not generally true, since transient phases are required to change the driveline configuration as, for example, starting and clutching the engine or performing a gear change.

Embedded implementation of the ECMS in a vehicle control unit (VCU) has proved that is possible to approach the minimal fuel consumption whilst integrating powertrain component limits as well as drivability constraints. Plausibility of ECMS can be assessed by comparing the fuel consumption measured on an urban mission profile with an a-posteriori optimization computed with DP, see Sect. 13.4.1) by imposing the measured wheel torque demanded by the driver, vehicle speed profile, as well as initial and final battery state of charge (SOC). Figure 13.1 shows such a comparison in terms of fuel energy consumption and SOC. Fuel consumption is measured from the injection signals in the engine controller. In the DP, engine on/off state and gear are enforced to match the measured quantities.

It should be noted that the engine torque and motor torque setpoints issued by the VCU are not always coincident with the raw output of the ECMS, since dynamic corrections are made downstream of the ECMS to manage driveline transients. Such corrections are not performed in the DP. However, the result presented shows that, at least for this particular application, experimental fuel energy consumption and battery discharge are close to the optimal trajectories found with the DP. That is an inherent effect of the PMP/DP assumptions (quasistatic modeling, absence of transient

Table 13.1 Flex hybrid components main characteristics

Component	Speed range (rpm)	Voltage range (V)	Power (kW)	Capacity (Ah)
ICE	0–6,000	–	63	–
Electric motor	0–20,000	400–550	37	–
Battery	–	145–216	–	39
DCDC	–	400–600	40	–
Gearbox		AMT (5 gears)		
Vehicle		1,700 kg curb weight		

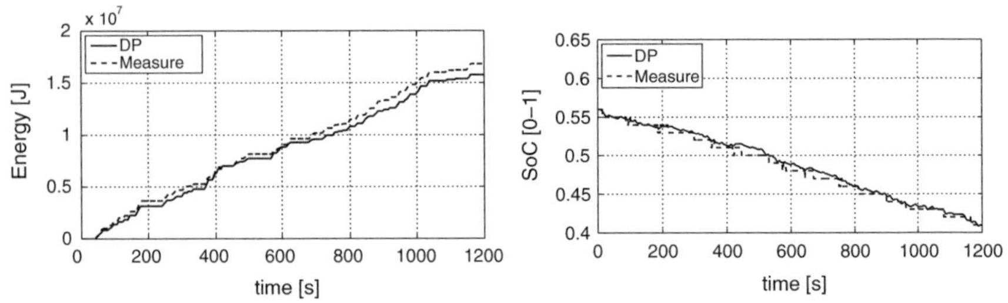

Fig. 13.1 Comparison between measured (online ECMS) and DP traces: fuel energy (*left*); battery SOC (*right*)

maneuvers). On the one hand, during transient maneuvers (engine stop/start, gear change), PMP/DP overestimate fuel consumption because they assume that the powertrain is instantaneously in its setpoint configuration. On the other hand, PMP/DP quasistatic modeling underestimates the fuel consumption during transients. These contrasting effects are partially compensating, which explains the good agreement of Fig. 13.1.

Standard optimal-based energy management considers only the fuel consumption in the cost function to be minimized. However, resulting battery current can exhibit a rather aggressive behavior, as shown in Fig. 13.4. The Root Mean Square (RMS) of current is of about 40 A in this test, or higher than C1 in terms of C-rate (peak values are limited to C3). This fact enlightens the relevance of considering the aging of the battery, of which current RMS is one important factor, in the global powertrain energy management.

13.3 Optimal Control Problem Formulation

In [4], the fraction of battery life depleted was given as

$$\frac{1}{\Gamma}\sigma(I(u(t), w(t)), \xi(t), \theta(t)) \cdot |I(u(t))|, \tag{13.1}$$

where I is the battery current, u is the set of controlled variables, $w(t)$ is the set of uncontrolled quantities related to the drive cycle, i.e., the required torque at the wheels, the vehicle speed, etc., ξ is the battery SOC, θ is the battery temperature, the function $\sigma(\cdot)$ is the severity factor, i.e., the relative aging effect with respect to a nominal cycle, and Γ is the total Ah-throughput corresponding to the nominal cycle. The severity factor was calculated using a postulated map.

In this chapter, battery life depletion is particularized as capacity loss and described by the aging factor Y, yielded by a semi-physical model described in Sect. 13.3.3. Consequently, the severity factor concept is replaced by the rate of capacity loss, $\dot{Y}$. Using such a model, the optimal control problem consists in minimizing the combined criterion

$$J = (1-\alpha)\int_0^T \dot{m}_f(u(t)) \cdot LHV\, dt + \alpha \int_0^T \beta \cdot \dot{Y}(I(u(t), w(t)), \xi(t), \theta(t))\, dt, \tag{13.2}$$

where β is a transformation coefficient to make the capacity loss rate dimensionally compatible with the fuel consumption, and α is a weighting factor to adjust the relative importance of the two cost contributions (α is arbitrary, while β has the meaning of a physical parameter). For the pre-transmission parallel architecture considered in this chapter, the variable $u(t)$ is the engine torque.

Neglecting electrochemical and mechanical dynamics, the current $I(t)$ is assumed to be an algebraic function of $u(t)$, $w(t)$, $\xi(t)$ and $\theta(t)$ (the latter two dependencies are through the battery inner parameters). However, the two variables $\theta(t)$ and $\xi(t)$ have too relevant dynamics to be neglected, and therefore are treated as state variables. Generally, the state equations are

$$\dot{\xi} = f_\xi(I(t)), \tag{13.3}$$

$$\dot{\theta} = f_\theta(\xi(t), \theta(t), I(t)). \tag{13.4}$$

The next three subsections will describe the models used to calculated the functions $\dot{m}_f$, I, $\dot{Y}$, f_ξ, and f_θ.

Global constraints to the optimal control problem presented are on the initial and final states, i.e., $\xi(0) = \xi_0$, $\xi(T) = \xi_f$ (SOC target value), $\theta(0) = \theta_0$ (ambient temperature), while $\theta(T)$ is free. Of course, the state variables and the control variable u are submitted to physical and drivability-based local constraints.

13.3.1 Powertrain Modeling

The calculations reported in this chapter are performed in an offline approach, where a driving cycle is prescribed and assumed to be perfectly followed. The vector $w(t)$ contains values of wheel torque $T_w(t)$ and wheel speed $\omega_w(t)$. Gear $n(t)$ is chosen according to gear shift maps as a function of T_w and ω_w. Consequently, the engine

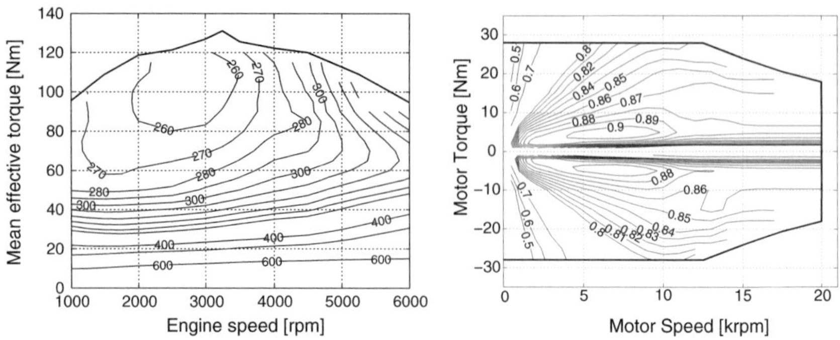

Fig. 13.2 Engine fuel map (*right*, [g/kWh]) and motor efficiency map (*left*) used in this study

speed is $\omega_e(t) = R_t(n(t))\cdot\omega_w(t)$ and the motor speed is $\omega_m(t) = R_m\cdot R_t(n(t))\cdot\omega_w(t)$, where R_m is the motor reduction gear ratio and R_t the variable transmission ratio including gearbox and final gear.

Engine torque $T_e(t)$ coincides with the controlled variable $u(t)$ for this architecture. Fuel consumption rate is evaluated from a quasistatic engine fuel map,

$$\dot{m}_f(T_e, \omega_e) = \dot{m}_f(u, w), \tag{13.5}$$

whose tabulated data are shown in Fig. 13.2. Engine torque and wheel torque demand define motor torque,

$$T_m(t) = \frac{1}{R_m}\left(\frac{T_w(t)}{R_t(n(t))\eta_t^{sign(T_w(t))}} - T_e(t)\right), \tag{13.6}$$

where η_t is the transmission efficiency. Motor electric power is evaluated from from an efficiency map $P_m(T_m, \omega_m) = P_m(u, w)$, whose tabulated data are shown in Fig. 13.2.

Battery SOC and temperature, as well as aging factor rate, are evaluated from the battery power $P_b(t) = P_m(t)$, according to the battery models described in the next sections. Note that, although the original democar presented in Sect. 13.2 is equipped with a SAFT module of 54 series-arranged cells, the model considered in this chapter refers to an A123System Li-ion cell, for which a sufficient amount of experimental data was available. In order to match the original pack's energy and power, the number of cells arranged in series and parallel has been set to $N_s = 58$, $N_p = 17$.

13.3.2 Battery Modeling

The state dynamics are described by modeling the cell with an equivalent circuit approach. The state of charge ξ is computed based on the equivalence between the electrochemical and electric charge, by means of an ampere-hour counting

$$\dot{\xi} = f_\xi(I(t)) = -\frac{I(t)}{C_{nom}}, \tag{13.7}$$

where C_{nom} is the cell nominal capacity (in C).

The cell current $I(t)$ is evaluated from the definition of battery power $P_b(t) = I(t) \cdot V(t)$ and the following model for the battery voltage:

$$V(t) = U_0(\xi(t)) + R(\xi(t), \theta(t)) \cdot I(t), \tag{13.8}$$

where the open circuit voltage U_0 is a function of the state of charge and the internal resistance R, capturing the ohmic, charge transfert and diffusion effects, is a function of state of charge and temperature.

Thermal dynamics are described with a lumped-parameter thermal balance that reads

$$\dot{\theta} = f_\theta(\xi(t), \theta(t), I(t)) = \frac{q_{gen} - q_{tra} - q_{cool}}{MC}, \tag{13.9}$$

where

$$q_{gen}(\xi(t),\ \theta(t), I(t)) = R \cdot I(t)^2 + \theta(t) \cdot I(t) \cdot \frac{dU}{d\theta}(\xi(t)), \tag{13.10}$$

$$q_{tra}(\theta(t)) = h \cdot A \cdot (\theta(t) - \theta_0), \tag{13.11}$$

are the heat generated by the cell and the heat exchanged with the ambient, θ_0 the ambient absolute temperature, q_{cool} is the heat provided by the cooling system, M is bulk mass, C is calorific thermal capacity, h the heat transfer coefficient, A the thermal exchange surface, $\frac{dU}{d\theta}$ the entropic heat. The time constant of such dynamics is thus of the order of MC/hA which, for the cell considered, amounts to about 800 s. This value makes the thermal dynamics comparable with that of SOC (that is almost a pure integrator).

The cooling system is modeled as it is realized in the original democar battery. A thermostatic controller activates liquid coolant circulation when the measured surface battery temperature reaches an upper value. However, the electricity consumption of the water pump is neglected in this study.

Battery pack characteristics are scaled from the corresponding cell values that were identified experimentally on a commercial 2.3 Ah A123 System $LiFePO_4$ Li-ion cell. In particular, $U_0 = N_s \cdot U_{0,cell}$, $R = R_{cell} \cdot N_s/N_p$, $M = M_{cell} \cdot N_s \cdot N_p$, $A = A_{cell} \cdot N_s \cdot N_p$, $dU/d\theta = (dU/d\theta)_{cell} \cdot N_s$. Figure 13.3 shows the dependence

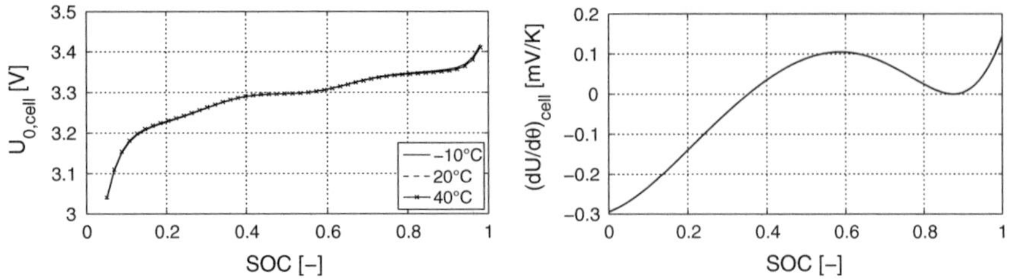

Fig. 13.3 Open-circuit voltage (*left*) and entropic heat (*right*) of the A123 cell as a function of SOC

of $U_{0,cell}$ and $\left(\frac{dU}{d\theta}\right)_{cell}$, respectively, on the SOC [12]. More details on model and parameter identification can be found in [12, 13].

13.3.3 Battery Aging Modeling

Due to several microscopic aging phenomena, Li-ion cells suffer a progressive performance degradation, that usually implies cell capacity loss and impedance increasing.The mechanisms leading to the capacity fade include the contact loss of active material particle, metallic lithium plating, cracking formation in particles and Solid Electrolyte Interphase (SEI) formation [13]. The relative impact of each mechanism varies with the cell technology and usage [14]. In [6], a large experimental test campaign on the A123 System $LiFePO_4$ Li-ion cell is described, consisting in a large range of constant C-rates and temperatures charges and discharges. It shows that the loss of active lithium associated with the anode degradation is a primary cause of capacity fade for this technology of Li-ion of cell. An empirical cycle-life model is also identified and validated on the experimental data. It assumes that the percent of capacity loss is a function of temperature and C-rate (or, equivalently, current), and can be expressed as

$$Q_{loss} = B_0 \cdot \exp\left(-\frac{E_a}{R\theta}\right) \cdot \mathrm{Ah}^z \tag{13.12}$$

where B_0 and E_a are functions of C-rate, $z = 0.55$ is a constant parameter, and Ah is the ampere-hour throughput, representing the amount of charge delivered by the cell during cycling, that is

$$\mathrm{Ah} = \int \frac{|I|}{3{,}600 \cdot C_{nom}}\, dt. \tag{13.13}$$

The range of applicability of the proposed model correspond to experimental test conditions, i.e. constant C-rate and constant temperature. For the purpose of this chapter, a generalization of life model to time-varying temperature and current is necessary. The literature on this topic is still quite poor. RMS of current and temperature over the cycle are used in [15]. Similarly, in [7] an effective temperature and SOC are defined, taking the place of constant values.

Here a different approach is proposed. The Eq. (13.12) assumes that I and θ are constant. It can be derived with respect to the time, giving

$$\frac{dQ_{loss}}{dt} = B_0 \cdot \exp\left(-\frac{E_a}{R\theta}\right) \cdot \frac{d(\mathrm{Ah}^z)}{dt} = z \cdot B_0 \cdot \exp\left(-\frac{E_a}{R\theta}\right) \cdot \mathrm{Ah}^{(z-1)} \cdot \frac{d(\mathrm{Ah})}{dt}. \tag{13.14}$$

As

$$\frac{d(\mathrm{Ah})}{dt} = \frac{|I|}{3{,}600}, \tag{13.15}$$

it is

$$\frac{dQ_{loss}}{dt} = z \cdot B_0 \cdot \frac{|I|}{3{,}600} \cdot \exp\left(-\frac{E_a}{R\theta}\right) \cdot \mathrm{Ah}^{(z-1)}, \tag{13.16}$$

that is proposed here as a general differential life model. A similar differential approach has been proposed in [8], where the aging model has been experimentally validated on several vehicle usage scenarios. Unfortunately, considering this aging model for the optimization problem, introduces explicitly the Ah in the objective function, making the optimal control solution very computationally intensive. As a consequence, in order to simplify the problem, the Eq. (13.12) is modified by defining an auxiliary variable

$$Y = z \cdot Q_{loss}^{\frac{1}{z}} = z \cdot B_0^{\frac{1}{z}} \cdot \exp\left(-\frac{E_a}{zR\theta}\right) \cdot \mathrm{Ah}. \tag{13.17}$$

Applying the argumentations developed from Eqs. (13.14) to (13.16) to the new variable Y, gives

$$\dot{Y} = z \cdot B_0^{\frac{1}{z}} \cdot \frac{|I|}{3{,}600} \cdot \exp\left(-\frac{E_a}{zR\theta}\right), \tag{13.18}$$

which allows to define an approximated capacity loss as

$$\tilde{Q}_{loss} = Y^z = \left(\int z \cdot B_0^{\frac{1}{z}} \cdot \frac{|I|}{3{,}600} \cdot \exp\left(-\frac{E_a}{zR\theta}\right) dt\right)^z. \tag{13.19}$$

Figure 13.4 compares Q_{loss} as it predicted by (13.16) to $\tilde{Q}_{loss}$ predicted by (13.19). The results refer to the cell current profile measured during the test discussed in Sect. 13.2. For the sake of completeness, the RMS approach is also shown in the figure. The relationship between Q_{loss} and $\tilde{Q}_{loss}$ appears to be monotonic. Moreover, a sufficient quantitative agreement is observed between the reference and the approximated capacity loss (a relative divergence of about 6 % is observed after

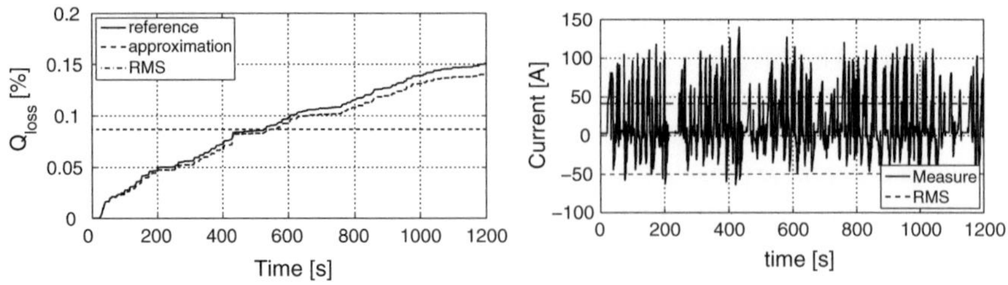

Fig. 13.4 Comparison between reference capacity loss Q_{loss} and approximated capacity loss $\tilde{Q}_{loss}$ (*left*) for a measured cell current profile (*right*)

1,200 s). These considerations allow the use of the aging factor Y as an optimization criterion in the rest of this study.

13.4 Optimal Control Problem Solution

The optimal control problem formulated in Sect. 13.3 is offline solved using two different numerical techniques: dynamic programming and Pontryagin's minimum principle. While DP generally approaches better the global optimum (at least for single-state problems), its computational burden increases dramatically with the number of states. On the other hand, PMP needs initial values of the costates, which however can be pre-tuned from the DP results, and its structure is closer to that of an online-implementable solution.

13.4.1 Dynamic Programming

Dynamic programming is implemented following the DPM algorithm [16]. Time step is 1 s. For the SOC state, 200 grid points equally spaced between 0 and 70 % are defined. A hard constraint on the final SOC ξ_f is implemented, with a tolerance of ± 1 %. For the temperature state, 100 grid points equally spaced between $\theta_0 - 5$ and $\theta_0 + 10$ are defined. Final temperature is free. The control variable u is engine torque and it is discretized in 23 values as follows. 20 values are equally spaced between the engine minimum and maximum torque corresponding to the engine speed at current time step. Additional three values are for ICE-only torque, zero torque, and ZEV mode (both latter terms are zero, but in the ZEV case the engine speed is also set to zero).

Discrete states, i.e., engine on/off state and gear engaged, are not considered in the optimization in order to avoid excessive computational effort. Therefore, too frequent engine start and stops cannot be regulated or avoided. This is a critical point that has to be verified a posteriori and possibly treated using PMP. The gear ratio is imposed with the drive cycle assuming the engine is on. If, however, the DP chooses the ZEV mode, the engine is disengaged (its speed is set to zero) and the gearbox is set to the second gear. The gear shift laws are pre-computed as a function of vehicle speed and acceleration pedal position. The latter is reconstructed from the drive cycle by computing the wheel torque demand and using information on maximum and minimum torque available from the powertrain at the current vehicle speed.

Dynamic programming is executed either with both states (DP2) or with SOC state only (DP1). In the latter case, first DP is executed using a temperature constant and equal to θ_0. The energy management law $u(t)$ calculated in this way is then run against the complete model with variable temperature, and results are accordingly recorded.

13.4.2 PMP

Pontryagin's Minimum Principle (PMP) is used in order to find a real-time implementable optimal solution to the energy management problem. The principle states that the optimal control $u(t)$ minimizes at each instant the Hamiltonian function, which is derived from the criterion (13.2) and the state functions (13.3) as

$$H(u, w, \xi, \theta, \lambda_\xi, \lambda_\theta) = (1-\alpha)\cdot \dot{m}_f(u, w)\cdot LHV + \alpha\cdot\beta\cdot\dot{Y}(u, w, \xi, \theta) + \lambda_\xi P_{ech}(u, w, \xi, \theta) + \lambda_\theta P_{th}(u, w, \xi, \theta). \quad (13.20)$$

In this definition of the Hamiltonian, the terms f_ξ and f_θ are converted into power units by multiplication by the factors $-C_{nom}U_0$ and $-MC$, respectively, resulting in the battery electrochemical power $P_{ech} = U_0 \cdot I$ and thermal power $P_{th} = -(q_{gen} - q_{tra} - q_{cool})$, see Sect. 13.3.2.

Consequently, the adjoint state dynamics are given by the following equations:

$$C_{nom}U_0\dot{\lambda_\xi} = \frac{\partial H}{\partial \xi} = \alpha\cdot\beta\cdot\frac{\partial \dot{Y}}{\partial \xi} + \lambda_\xi\frac{\partial P_{ech}}{\partial \xi} + \lambda_\theta\frac{\partial P_{th}}{\partial \xi}, \quad (13.21)$$

$$MC\dot{\lambda_\theta} = \frac{\partial H}{\partial \theta} = \alpha\cdot\beta\cdot\frac{\partial \dot{Y}}{\partial \theta} + \lambda_\xi\frac{\partial P_{ech}}{\partial \theta} + \lambda_\theta\frac{\partial P_{th}}{\partial \theta}. \quad (13.22)$$

PMP is executed either with both states (PMP2) or with SOC state only (PMP1). In the latter case, λ_θ is identically set to zero. Moreover, the residual variation of λ_ξ is neglected, in order to recover the usual implementation of 1-state PMP, where λ_ξ is constant [1]. This approximation corresponds to neglect the influence of SOC on the battery dynamics, see (13.21). As for the influence of SOC on the SOC dynamics itself (the term $\frac{\partial P_{ech}}{\partial \xi}$), it is usually neglected in the literature (e.g., in [5]) by virtue of the fact that charge-sustaining HEV only use a small amount of battery charge. In the case of this chapter, however, the application under study is a PHEV that, by its very nature, is designed to operate the battery over a wide range of SOC. Therefore, the validity of the aforementioned assumption will be checked a posteriori by observing the DP results, Sect. 13.5.1. Neglecting the term $\frac{\partial \dot{Y}}{\partial \xi}$ is justified here by the particular model used (13.18), where the dependency of the parameters on SOC was shown in

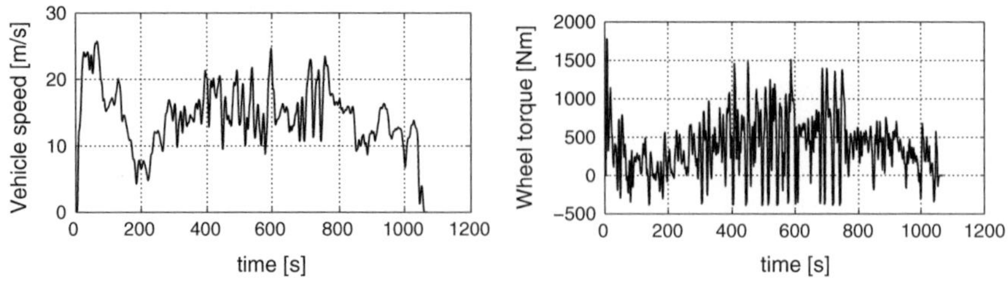

Fig. 13.5 Vehicle speed (*left*) and wheel torque (*right*) for the driving cycle considered in this study

[6] to be negligible. The same is not necessarily true for other battery chemistries. In those cases, (13.21) must be integrated as done in [4].

Similarly to DP, the energy management laws calculated with both PMP1 and PMP2 are later run against the complete model with variable temperature, and results are accordingly recorded.

13.5 Optimal Control Problem Results

Simulation results will be presented in the next two sections for several scenarii in terms of θ_0 and α. The driving cycle is the same for all simulations and it consists of a 15 km-long fraction of a driving profile recorded in the Italian Alps, with altitude variations [17]. Vehicle speed and wheel torque demand are shown in Fig. 13.5, together with gear setpoint. The latter quantity is calculated from gear shift laws and actually realized if the energy management strategy chooses the engine to be on (as explained in Sect. 13.4.1). The initial SOC is set to 60 %, while the target SOC is 20 %.

13.5.1 Dynamic Programming Results

Table 13.2 summarizes the results obtained with DP1 and DP2 for an ambient temperature $\theta_0 = 20\,^\circ\mathrm{C}$. Besides the values of α that define each single scenario, the table lists three metrics and four additional quantities. The three metrics are the total cost J and its two components, namely, the fuel energy E_f and the aging factor Y weighted by the transformation coefficient $\beta = 1 \times 10^9$. Since the final SOC is not exactly the same for all the tests (a tolerance of ± 1 % was enforced in the DP coding), raw results outputted by the DP corrected to compensate for SOC deviations from the target value of 20 %. The correction rules are

Table 13.2 Simulation results obtained with the DP for an ambient temperature of 20 °C

α	DP	J (MJ)	E_f (MJ)	βY (MJ)	ΔJ (%)	q_f (ℓ/hkm)	Q_{loss} (%/hkm)	$\lambda_1(0)$	$\lambda_2(0)$	max $\Delta\theta$ (°C)	$RMS(P_b)$ (kW)
0	1	35.7	35.7	19.9		7.2	0.75	2.09	–	6.4	14.9
0	2	35.7	35.7	20.4	0.2	7.2	0.76	2.13	0.45	6.7	15.1
1/3	1	30.0	35.9	18.3		7.3	0.72	0.92	–	5.8	14.5
1/3	2	30.0	36.2	17.7	0.1	7.3	0.70	0.61	−2.52	5.6	14.3
2/3	1	23.6	37.0	16.9		7.5	0.69	−0.19	–	5.4	14.0
2/3	2	23.5	37.5	16.5	−0.3	7.6	0.68	−0.77	−5.26	5.0	13.7
1	1	16.6	41.3	16.6		8.4	0.68	−1.34	–	5.3	13.9
1	2	16.2	41.4	16.2	−2.2	8.4	0.67	−2.17	−8.06	4.5	13.1

$$E_f = E_{f,raw} + \lambda_\xi^{\alpha=0}(0) \cdot (0.2 - \xi(T)) \cdot (C_{nom} U_0(\xi_0, \theta_0)), \tag{13.23}$$

$$\beta Y = (\beta Y)_{raw} + \lambda_\xi^{\alpha=1}(0) \cdot (0.2 - \xi(T)) \cdot (C_{nom} U_0(\xi_0, \theta_0)), \tag{13.24}$$

where the costate values $\lambda_\xi^{\alpha=0}(0)$ and $\lambda_\xi^{\alpha=1}(0)$ are the quantities shown in the table for the scenarii $\alpha = 0$ and, respectively, $\alpha = 1$. The costate traces are calculated from the value function outputted by the DP algorithm, according to their definition

$$\lambda_\xi(t) = -\left.\frac{\partial V(t,\xi,\theta)}{\partial \xi}\right|_{t,\xi^*(t),\theta^*(t)} \cdot \frac{1}{C_{nom} U_0(\xi_0, \theta_0)} \tag{13.25}$$

$$\lambda_\theta(t) = -\left.\frac{\partial V(t,\xi,\theta)}{\partial \theta}\right|_{t,\xi^*(t),\theta^*(t)} \cdot \frac{1}{MC}, \tag{13.26}$$

where the partial derivatives are calculated along the optimal state trace denoted by starred variables and the scaling factors that make the costates non-dimensional are added in the right-hand sides.

Together with the absolute value of the cost J, also shown are the relative variations when switching from DP1 to DP2. The last quantities listed in Table 13.2 are the maximum rise of the battery temperature during the test, and the RMS of the battery power, both being the main factors that increase capacity loss. Figure 13.6 shows time variations of the two states and the two costates obtained with DP2 and the four values of α considered.

Several considerations arise from the analysis of these results. Firstly, it is evident that the weighting factor α plays a fundamental role in modifying the nature of the solution. An increase of α clearly yields a decrease of βY, with a simultaneous increase of E_f. The transition from more fuel-efficient strategies ($\alpha = 0$) to more aging-reducing strategies ($\alpha = 1$) is accompanied by a reduction of temperature increase (max $\Delta\theta$), also visible in Fig. 13.6, and a reduction of the root mean square of the battery power ($RMS(P_b)$), as expected. Not shown in the table, the average engine efficiency generally decrease with an increase of α. Starting from a baseline strategy that is only sensitive to fuel consumption ($\alpha = 0$), a significant reduction

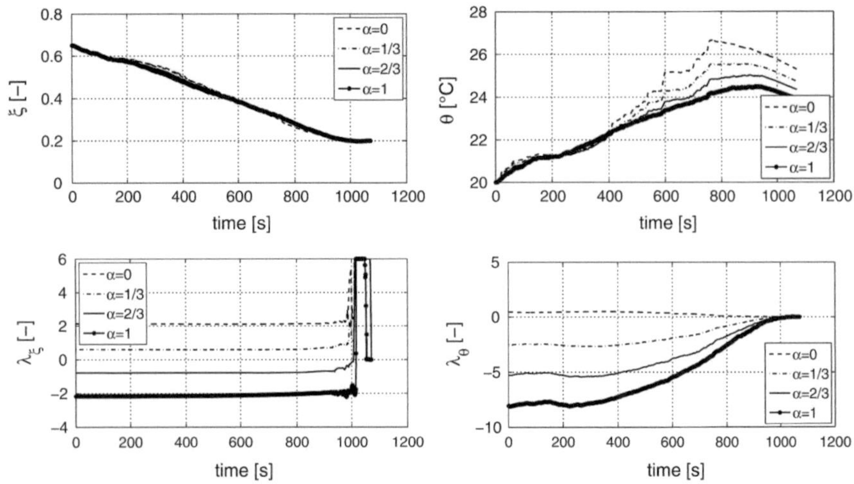

Fig. 13.6 Traces of SOC (*top-left*), temperature (*top-right*), SOC costate (*bottom-left*), and temperature costate (*bottom-right*) obtained with DP2

Fig. 13.7 Fuel energy versus aging factor as calculated with DP1 and DP2 ($\theta_0 = 20\,°C$)

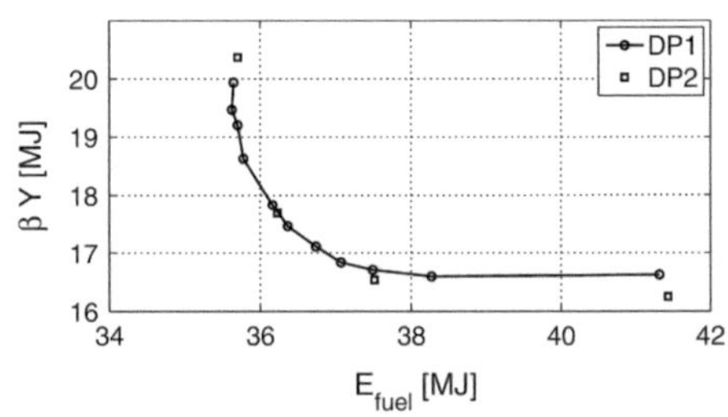

of the aging factor is possible with increasing α, only with a relatively small fuel-economy penalty. For example, switching to $\alpha = 1/3$, βY can be decreased of about 10 %, while E_f increases only of 1.4 %. The SOC costate λ_ξ is almost constant during the cycle, which justifies the assumption introduced in Sect. 13.4.2 for the PMP solution. Variations toward the cycle end are due to numerical reasons (unfeasible states) and the emergence of regenerative braking (SOC is no longer regulated by the control variable but by the disturbance). Its initial value decreases with an increase of α and becomes negative for aging-reducing strategies. In fact, while fuel consumption rate generally varies in the opposite direction as electric power consumption with respect to u, aging factor rate exhibits the same trend (i.e., it is lowered by a lesser use of the battery). The optimal compromise when the aging factor is the predominant criterion is then found for negative values of the costate. The temperature costate λ_θ has larger variations than the SOC costate and converges to zero at the end of the cycle as predicted by the fact that terminal temperature is not constrained.

The effect of α on the two optimization criteria is made more visible in Fig. 13.7 that shows the curves E_f–βY obtained with DP1 and DP2, respectively (note that

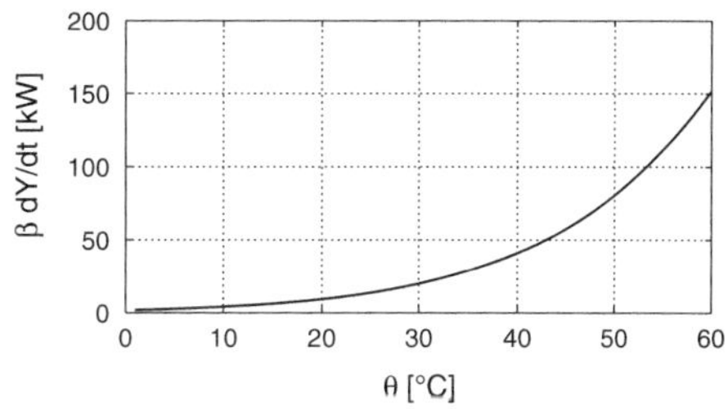

Fig. 13.8 Variation of the aging factor as a function of battery temperature

the data concerning DP1 are much numerous than those shown in Table 13.2). The figure clearly shows the asymptotic behavior of E_f and βY for α tending to 1 and 0, respectively. In contrast to charge-sustaining HEV [4, 5], the asymptotic value for βY does not approach zero because a positive net battery use is imposed by the charge-depleting PHEV operation. The DP2 curve should be deemed as the Pareto frontier of the multi-objective minimization of fuel energy and aging factor, in the sense that points below the curve cannot be found in principle. The DP1 curve is an approximation of the DP2 curve obtained with a simpler algorithm. It should be reminded that what differs between these two cases is only how DP evaluates the energy management strategy. Once the latter is evaluated, final metrics such as E_f and βY are calculated by the two-state version of the system model in both cases. Figure 13.7 and Table 13.2 show a very limited improvement with using DP2, facing a much higher computational effort (for reference, running DP1 over the cycle takes about 1 min on a 2.60 GHz personal computer, while DP2 takes approximately three hours). The largest discrepancy is obtained for $\alpha = 1$, where an improvement of 2–2.5 % is obtained in the cost J with DP2.

This result can be explained with the following considerations. Denote $J[U, \Theta]$ the functional dependency of the cost function on the energy management law and the temperature profile (influence of SOC profile is neglected in this analysis). Denote $U_k, \Theta_k, k = \{1, 2\}$ the profiles obtained with DP1 and DP2, while Θ_0 is the constant temperature profile. Consider the difference $J[U_k, \Theta_k] - J[U_k, \Theta_0]$. When $\alpha = 0$, it vanishes since $J = J[U]$. But such a difference increases for large α, consequently to the variation of $\dot{Y}$ with θ (illustrated in Fig. 13.8 for given values of C-rate and SOC) and can reach, e.g., 20 % for 20 °C. To explain why the difference $J[U_2, \Theta_2] - [U_1, \Theta_1]$ is much lower, it can be analyzed as the sum of two effects, $(J[U_2, \Theta_2] - J[U_2, \Theta_1]) + (J[U_2, \Theta_1] - J[U_1, \Theta_1])$. The first difference is typically negative for $\alpha > 0$ (controlled temperature profiles induce less aging). However, the second term is typically positive for $\alpha > 0$. In other terms, the benefit obtained with the former effect is partially compensated by the latter, which can explain the relatively low gains of using DP2 instead of DP1.

On the other hand, as DP2 is intrinsically less robust than DP1, and very sensitive to the gridding parameters, interpolation functions, etc. [18], the "true" two-state optimum may not be reached. Such a consideration motivates the use of a different technique, namely, PMP, to verify the two-state case.

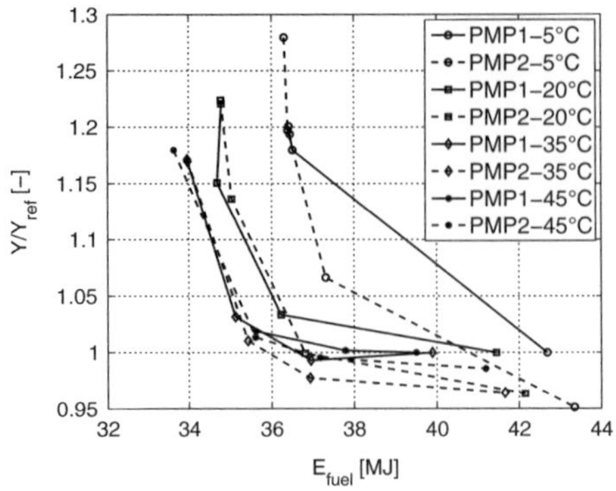

Fig. 13.9 Relative differences in aging factor versus fuel energy as calculated by PMP1 and PMP2

13.5.2 PMP Results

As described in Sect. 13.4.2, the difficulty in executing PMP resides in the evaluation of the values λ_ξ (for PMP1 and PMP2) and $\lambda_\theta(0)$ (for PMP2) for each scenario. In this study, the calibration of $\lambda_\theta(0)$ follows from the DP2 result, while λ_ξ is evaluated with a root-finding algorithm targeting the condition $\xi(T) = 0.2$.

Figure 13.9 summarizes the optimization results obtained with PMP. The figure shows the Pareto frontiers for four ambient temperatures, namely, 5, 20, 35, and 45 °C, with PMP1 and PMP2, respectively. Results are qualitatively and quantitatively close to those shown in Sect. 13.5.1, for instance, the addition of a second state is as more beneficial as the weighting factor α increases. Two additional conclusions can be drawn from these results. Firstly, larger relative improvements, with PMP2 with respect to PMP1, are obtained for colder conditions. For example, the reduction of βY with $\alpha = 1$ is of only 1.5 % at 45 °C, which is likely below the precision of the model forecast, but increases to about 5 % at 5 °C. The latter point is easily explained observing the dependency of the aging factor rate $\dot{Y}$ on the temperature θ, which is $\propto \exp(-1/\theta)$. Consequently, the relative variation $\frac{1}{\dot{Y}}\frac{\partial \dot{Y}}{\partial \theta} \propto \frac{1}{\theta^2}\exp(-1/\theta)$, which is a decreasing function of θ. A second conclusion is that also the slope of the Pareto frontier in the neighborhood of the baseline strategy $\alpha = 0$ decreases as the temperature increases. In other terms, the ratio of aging reduction on consumption increase is worse at larger temperatures.

Figure 13.10 shows the variation of the initial values of the costates $\lambda_\xi(0)$ and $\lambda_\theta(0)$ as a function of θ and α, both with PMP1 and PMP2. As already observed, an increase of α induces a less aggressive use of the battery and, consequently, the SOC target is obtained with a smaller λ_ξ (which, per se, favors battery discharge). An increase of θ_0 induces higher aging and thus is also accompanied by a decrease of λ_ξ, again to compensate battery underuse. Only slight differences can be observed between the values of λ_ξ calculated with PMP1 and PMP2. A similar trend is observed for $\lambda_\theta(0)$, whose values are generally larger (in absolute value).

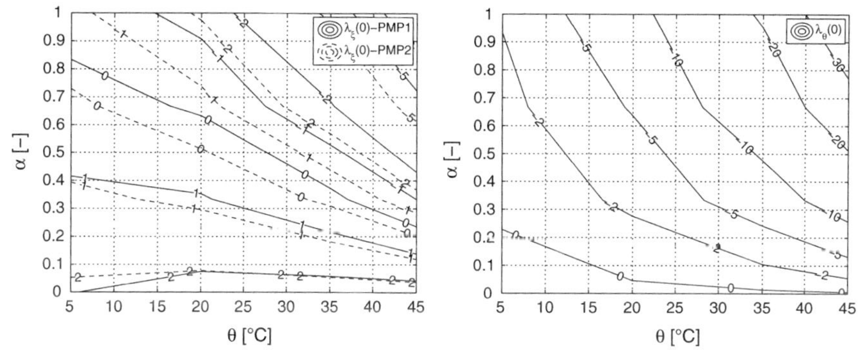

Fig. 13.10 Costates (initial values) as a function of temperature and α

13.6 Conclusions

The study presented in this chapters has shown that energy management in HEV and PHEV can be made sensitive to battery aging. Using a proper definition, an aging factor can be adjoint to the fuel consumption into a combined cost function to be minimized by the energy management law. By varying the relative weight of the two cost functions, a more fuel-efficient or a less battery-aggressive strategy can be obtained. For PHEV, a substantial improvement in battery aging is expected to be obtained at the expense of a few percent deterioration of fuel economy.

The chapter has also discussed the nature of the optimal control solutions. The comparison of single-state (SOC only) versus two-state (SOC and temperature) solutions has shown that the former is sufficient in most cases, while the computational-intensive addition of the temperature state has an added value only in the case of an extremely aging-biased cost function. An explanation for this somehow surprising evidence has been attempted. This result would suggest the use of single-state strategies even to treat scenarios when minimization of aging is relevant. The behavior of both costates has been explored as well. While the SOC costate is essentially constant for an optimal solution, the temperature costate varies substantially in time, tending to zero. Initial values of both costates have been shown to be monotonic functions of ambient temperature, a property that could be useful in view of their estimation in a real-time implementation of optimal control-based energy management.

The analyses and results presented have been computed for a given battery chemistry. Although not quantitatively generalizable to other battery systems, the main trends shown are expected to remain valid. In particular, some simplifying assumptions introduced are expected to need reconsideration when treating the case of chemistries with strongly SOC-dependent aging, combined with wide variations of the SOC itself (as in PHEV).

References

1. Guzzella L, Sciarretta A (2013) Vehicle propulsion systems. Introduction to modeling and optimization, 3rd edn. Springer, Berlin (chapter 7 and references there in)
2. Stockar S, Marano V, Canova M (2011) Energy-optimal control of plug-in hybrid electric vehicles for real-world driving cycles. IEEE Trans Veh Technol 60(7):2949–2962
3. Sivertsson, M (2012) Adaptive control using map-based ECMS for a PHEV. In: Proceedings of the 2012 IFAC workshop on engine and powertrain control, simulation, and modeling
4. Serrao L, Onori S, Sciarretta A, Guezennec Y, Rizzoni G (2011) Optimal energy management of hybrid electric vehicles including battery aging. In: Proceedings of the IEEE American control conference (ACC). San Francisco, CA, Jun 29–Jul 1 2011
5. Ebbesen S, Elbert P, Guzzella L (2012) Battery state-of-health perceptive energy management for hybrid electric vehicles. IEEE Trans Veh Technol 61(7):2893–2900
6. Wang J, Liu P, Hicks-Garner J et al (2011) Cycle-life model for graphite-$LiFePO_4$ cells. J Power Sour 196:3942–3948
7. Smith K, Earleywine M, Wood E, Neubauer J, Pesaran A (2012) Comparison of plug-in hybrid electric vehicle battery life across geographies and drive cycles. SAE Paper 2012-01-0666. doi:10.4271/2012-01-0666
8. Prada E, Petit M (2013) Aging modeling for advanced Li-ion battery pack sizing and management for HEV/EV through AMESim simulation platform. To be presented at the LMS European vehicle conference, 29–30 Oct 2013
9. Pham HT, van den Bosch PPJ, Kessels JTBA, Huisman RGM (2012) Integrated energy and thermal management for hybrid electric heavy duty trucks. In: Proceedings of the IEEE vehicle power and propulsion conference. Seoul, Korea, pp 9–12, Oct 2012
10. Zheng CH, Kim NW, Park YI et al (2013) The effect of battery temperature on total fuel consumption of fuel cell hybrid vehicles. Int J Hydrogen Energy 38(13):5192–5200
11. Merz F, Sciarretta A, Dabadie JC, Serrao L (2012) On the optimal thermal management of hybrid-electric vehicles with heat recovery systems. Oil Gas Sci Technol 67(4):610–612. doi:10.2516/ogst/2012017
12. Prada E, Bernard J, Mingant R, Sauvant-Moynot V (2010) Li-ion thermal issues and modeling in nominal and extreme operating conditions for HEV / PHEV's. In: Proceedings of the IEEE vehicle power and propulsion conference (VPPC), 1–3 Sept 2010
13. Di Domenico D, Prada E, Creff Y (2011) An adaptive strategy for Li-ion battery SoC estimation. In: Proceedings of the IFAC world congress, 28 Aug–2 Sept 2011
14. Prada E, Di Domenico D, Creff Y, Bernard J, Sauvant-Moynot V, Huet F (2012) Simplified electrochemical and thermal model of $LiFePO_4$-graphite Li-ion batteries for fast charge applications. J Electrochem Soc 159(9):A1508–A1519
15. Bishop J, Axon C, Bonilla D, Tran M, Banister D, McCulloch M (2013) Evaluating the impact of V2G services on the degradation of batteries in PHEV and EV. Appl Energy 111:206–218
16. Sundström O, Guzzella L (2009) A generic dynamic programming Matlab function. In: Proceedings of the IEEE conferences on control applications (CCA) and intelligent control (ISIC), 8–10 Jul 2009
17. Sciarretta A, Serrao L, Dewangan PC, Tona P et al (2014) A control benchmark on the energy management of a plug-in hybrid electric vehicle. Accepted for publication in control engineering practice
18. Elbert P, Ebbesen S, Guzzella L (2013) Implementation of dynamic programming for n-dimensional optimal control problems with final state constraints. IEEE Trans Control Syst Technol 21(3):924–931

Chapter 14
Optimal Control of Diesel Engines with Waste Heat Recovery System

Frank Willems, M. C. F. Donkers and Frank Kupper

Abstract This study presents an integrated energy and emission management strategy for a Euro-VI diesel engine with Waste Heat Recovery (WHR) system. This Integrated Powertrain Control (IPC) strategy optimizes the CO_2-NO_x trade-off by minimizing the operational costs associated with fuel and AdBlue consumption, while satisfying tailpipe emission constraints. The main contribution of this work is that the optimal solution is determined numerically for a given cycle and is compared with a real-time implementable strategy. Also, the WHR dynamics are explicitly included in the control design. In a simulation study, the potential of this IPC strategy is demonstrated over the World Harmonized Transient Cycle. It is shown that the real-time strategy can be applied with negligible loss of optimality. Using IPC, an additional 3.5 % CO_2 reduction is achieved, while complying with the NO_x emission limit, when compared to a baseline strategy.

F. Willems (✉)
Department of Mechanical Engineering, Eindhoven University of Technology and TNO Automotive, P.O. Box 513, 5600 MB Eindhoven, The Netherlands
e-mail: f.p.t.willems@tue.nl

M. C. F. Donkers
Department of Electrical Engineering, Eindhoven University of Technology, P.O. Box 513, 5600 MB Eindhoven, The Netherlands
e-mail: m.c.f.donkers@tue.nl

F. Kupper
Powertrains, TNO Automotive, 5708 HN HELMOND, The Netherlands
e-mail: frank.kupper@tno.nl

H. Waschl et al. (eds.), *Optimization and Optimal Control in Automotive Systems*, Lecture Notes in Control and Information Sciences 455,
DOI: 10.1007/978-3-319-05371-4_14,

14.1 Introduction

With the introduction of Euro-VI emission targets, current heavy-duty diesel engines meet ultra-low emission levels. Although vehicles have become extremely clean, fuel consumption levels have remained stable for the last two decades. To reduce fuel consumption and to meet upcoming CO_2 targets, heavy-duty diesel engines equipped with Waste Heat Recovery (WHR) systems seem very promising, especially for long-haul truck applications. This WHR system allows energy to be recovered from heat flows and to be converted into useful mechanical energy for propulsion. Up to 6 % fuel consumption reduction has been demonstrated in [1, 2].

Most literature on WHR control focusses on low-level control, see, e.g., [3–7]. Only very few studies concentrate on energy management strategies for the complete engine [8, 9]. For heavy-duty diesel engines with WHR system, studies that optimize overall engine-aftertreatment-WHR system performance by minimizing total operational costs and that explicitly deal with tailpipe emission constraints set by legislation are lacking. We, therefore, present a cost-based optimization strategy, which integrates energy and emission management: so-called Integrated Powertrain Control (IPC). Contrary to earlier work on IPC in [10, 11], the optimal solution of Pontryagin's Minimum Principle is computed numerically and used as a benchmark for a real-time implementable IPC strategy. Furthermore, WHR dynamics are explicitly incorporated in the control model.

This work is organized as follows. First, the studied powertrain and applied models are presented in Sect. 14.2. Sections 14.3 and 14.4 discuss the developed IPC strategies and their calibration, respectively. The results of these IPC strategies are compared with the results of a baseline control strategy over a World Harmonized Transient Cycle in Sect. 14.5. Finally, conclusions are drawn and directions for future research are sketched.

14.2 System Description

Figure 14.1 shows a scheme of the examined engine platform. It is based on a state-of-the-art 6 cylinder, 13 l, 340 kW Euro-VI diesel engine. This engine is equipped with a turbocharger with Variable Turbine Geometry (VTG) and a high pressure Exhaust Gas Recirculation (EGR) system with an EGR valve and EGR cooler. Furthermore, an exhaust gas aftertreatment system is installed. This system consists of a Diesel Oxidation Catalyst (DOC), a Diesel Particulate Filter (DPF) and an urea-based Selective Catalytic Reduction (SCR) system. The DPF system filters the particulates out of the exhaust flow. To avoid clogging of the filter, the trapped particulates are burned by periodically injecting fuel upstream of the DOC (*DPF regeneration*). The remaining NO_x emissions downstream of the DPF system are converted into harmless products over the Cu-Zeolite SCR catalyst. For this catalytic process, ammonia (NH_3) is required. This is formed upstream of the catalyst by decomposition of the

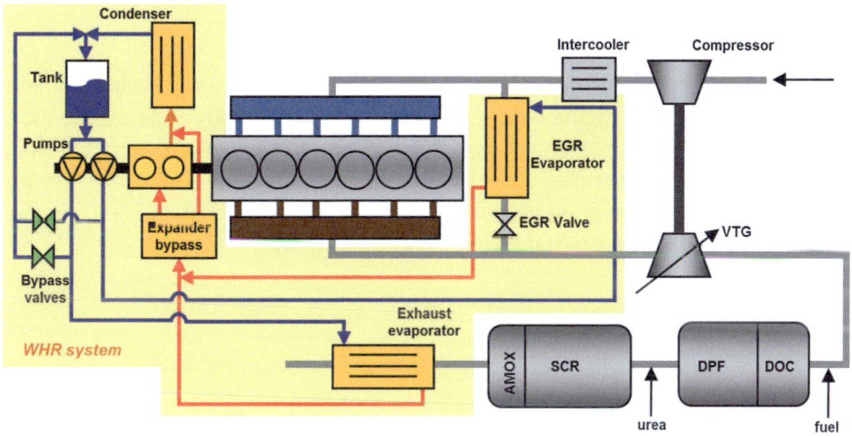

Fig. 14.1 Studied Euro-VI engine with WHR system

injected aqueous urea solution (tradename: AdBlue) in the hot exhaust gases. To avoid unacceptable NH_3 slip, an ammonia oxidation (AMOX) catalyst is installed.

The Euro-VI engine platform is extended with a Waste Heat Recovery (WHR) system. This system is based on a Rankine cycle and recovers thermal energy from both the EGR and exhaust gas flow using two evaporators. Note that the EGR evaporator replaces the original EGR cooler. The expander converts this recovered thermal energy into mechanical energy. It drives the two pumps and is directly connected to the crank shaft, such that the recovered energy can be transmitted to the engine. By using two bypass valves, the working fluid flow through the evaporators can be individually controlled, such that the working fluid is in vapor state downstream of the evaporators. Furthermore, an expander valve is present to bypass the expander in case no engine torque is requested (e.g., during braking or gear shifting) and to avoid damage of the expander (when the working fluid is in fluid or two phase state).

In what follows, a distinction is made between the simulation model, which aims at describing the main system characteristics, and the control model, which is a simplified version of the simulation model allowing for a real-time implementation of the control strategy. By using simplified control models, the proposed control strategy is assumed to be robust with respect to unmodeled dynamics.

14.2.1 Simulation Model

The simulation model, which is used in this chapter to model the powertrain, consists of models of an engine, a high-fidelity aftertreatment system and a waste heat recovery system. These models are described below in more detail.

14.2.1.1 Engine

To describe the behavior of the exhaust gas mass flow $\dot{m}_{exh}$, the exhaust gas temperature T_{exh}, and the engine out NO_x mass flow $\dot{m}_{NO_x}$, steady-state engine maps are used. These four-dimensional maps $f(N_e, \tau_e, u_{EGR}, u_{VTG})$ are constructed using a validated mean-value engine model. For varying combinations of EGR valve position u_{EGR} and VTG position u_{VTG}, the fuel mass $\dot{m}_f$ is varied such that the engine torque τ_e is realized (for constant engine speed N_e [rpm]).

14.2.1.2 Aftertreatment System

A high-fidelity aftertreatment model is implemented to simulate the DOC/DPF and SCR system. This modular model is built up using one-dimensional submodels of a pipe with urea decomposition, DOC, DPF, SCR, and AMOX catalyst. All catalyst models are based on first principle modeling and consist of mass and energy balances. By dividing the catalyst in various segments, these validated models describe the spatial distribution of pressure, temperature and chemical components. More details on the model approach and accuracy can be found in [12].

14.2.1.3 Waste Heat Recovery System

Assuming an ideal low-level WHR control system, the WHR system dynamics are described by a first-order model with constant overall efficiency η_{WHR}. This is inspired by the observed thermal dynamics in the studied engine. The recovered thermal energy from both the EGR and exhaust gas flow is given by:

$$\frac{d\dot{Q}_{WF}}{dt} = \frac{1}{\alpha_{WHR}} \left(\dot{Q}_{EGR,g} + \dot{Q}_{exh,g} - \dot{Q}_{WF}\right) \tag{14.1}$$

and the EGR and exhaust gas heat flows are defined by, respectively:

$$\dot{Q}_{EGR,g} = \dot{m}_{EGR} \cdot c_{p,EGR} \cdot (T_{EGR,in} - T_{EGR,out}) \tag{14.2}$$

$$\dot{Q}_{exh,g} = \dot{m}_{exh} \cdot c_{p,exh} \cdot (T_{SCR} - T_{tp}) \tag{14.3}$$

This thermal energy is finally converted into mechanical power at the expander shaft:

$$P_{WHR} = \tau_{WHR} \cdot \omega_{WHR} = \begin{cases} \eta_{WHR} \cdot \dot{Q}_{WF} & \tau_{d,req} > 0 \\ 0 & \tau_{d,req} \leq 0 \end{cases} \tag{14.4}$$

with total required torque $\tau_{d,req}$ and WHR output torque τ_{WHR} [see also Eq. (14.7)]. In case no power is requested from the WHR system, the expander bypass is activated, such that $P_{WHR} = 0$.

Table 14.1 Control model parameters

Constant	Unit	Definition	Value
c_1	kg^{-1}	$\frac{c_{p,exh}}{C_{DOC}}$	0.1163
c_2	kg^{-1}	$\frac{c_{p,exh}}{C_{SCR}}$	0.0512
c_3	$s \times kg^{-1}$	$\frac{h}{C_{SCR}}$	7.692×10^{-4}
c_4	$Jg^{-1}\ K^{-1}$	$\eta_{WHR} \cdot c_{p,EGR}$	100
c_5	$Jg^{-1}\ K^{-1}$	$\eta_{WHR} \cdot c_{p,exh}$	100

Based on engine dynamometer data, a constant overall efficiency $\eta_{WHR} = 0.10$, a time constant $\alpha_{WHR} = 60$ s, and a constant post-WHR exhaust gas temperature $T_{tp} = 110\,°C$ are chosen. Moreover, the values of $T_{EGR,in}$ and $T_{EGR,out}$ follow from four-dimensional engine maps.

14.2.2 Control Model

This section presents the control model that is used in the optimal control strategy of Sect. 14.3. For real-world implementation, this simplified model has to represent the main system characteristics and has to be evaluated in real-time. Compared to the simulation model, the main difference lies in the description of the aftertreatment system. More precisely, the control model uses a simplified aftertreatment model, which will be discussed below. Identical engine maps and a identical WHR model as in the simulation model are used.

The thermal behavior of the total DOC-DPF-SCR-AMOX system is described by two coupled differential equations, see Eq. (14.6). For the SCR conversion efficiency η_{SCR}, a combination of three stationary maps is used, which are determined for different pre-SCR concentration ratios $C_{NO_2}/C_{NO_x} = [0, 0.5, 1.0]$ and a specified ammonia slip level. The individual SCR efficiency maps depend on the average SCR catalyst temperature T_{SCR}°C and space velocity SV 1/h:

$$SV = 3600 \frac{\dot{m}_{exh}}{\rho_{exh} V_{cat}} \tag{14.5}$$

with normal condition exhaust gas density ρ_{exh} [g/m^3] and SCR catalyst volume V_{cat} [m^3]. Using the predicted C_{NO_2}/C_{NO_x} ratio from a stationary DOC efficiency map, the NO_x conversion efficiency η_{SCR} is computed by interpolation. Finally, the control model incorporates the WHR dynamics, which are given by Eqs. (14.1)–(14.4).

In summary, the control model is written in state space form $\dot{x} = f(x, u, t)$:

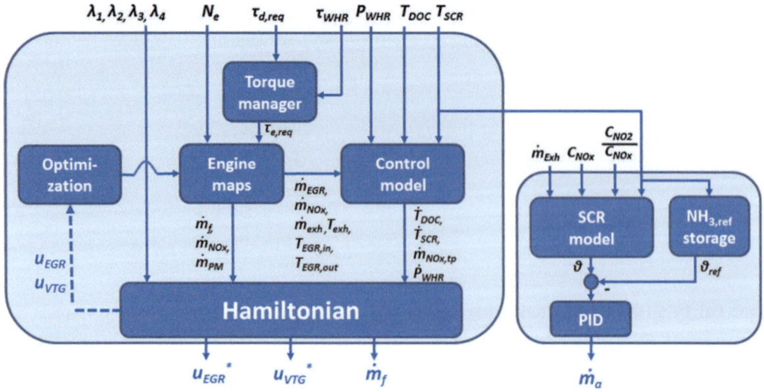

Fig. 14.2 Scheme of the engine control system

$$\dot{x} = \begin{bmatrix} c_1\,\dot{m}_{exh}\big(T_{exh} - T_{DOC}\big) \\ c_2\,\dot{m}_{exh}\big(T_{DOC} - T_{SCR}\big) - c_3\big(T_{SCR} - T_{amb}\big) \\ \dot{m}_{\mathrm{NO}_x}\big(1 - \eta_{SCR}(T_{SCR}, SV, C_{\mathrm{NO}_2}/C_{\mathrm{NO}_x})\big) \\ \frac{1}{\alpha_{WHR}}\big(c_4\,\dot{m}_{EGR}(T_{EGR,in} - T_{EGR,out}) + c_5\,\dot{m}_{exh}(T_{SCR} - T_{tp}) - P_{WHR}\big) \end{bmatrix} \tag{14.6}$$

with state variables $x = [T_{DOC}\ T_{SCR}\ m_{\mathrm{NO}_{x,tp}} P_{WHR}]^T$, in which T_{DOC} and T_{SCR} denote the DOC/DPF and SCR temperatures, respectively, $m_{\mathrm{NO}_x,tp}$ denotes the cumulative tailpipe $\mathrm{NO_x}$ mass, and P_{WHR} denotes the WHR power. The model parameter values that are used in this chapter are specified in Table 14.1.

14.3 Control Strategy

Figure 14.2 shows a scheme of the proposed engine control system. The main objective of this control system is to determine the settings for the inputs $\dot{m}_f$, $\dot{m}_a$, u_{EGR} and u_{VTG}, such that fuel consumption is minimized within the constraints set by emission legislation. By assuming ideal torque management, the requested engine torque is determined from:

$$\tau_{e,req} = \tau_{d,req} - \tau_{WHR}. \tag{14.7}$$

In this control system, two subsystems can be distinguished: a supervisory controller, which is described in this section, and a low-level SCR controller. Details about this model-based ammonia storage controller can be found in [11].

To select suitable values for the aforementioned inputs, the Integrated Powertrain Control (IPC) approach, which was first introduced in [13], is used. This model-based approach integrates energy and emission management by exploiting the synergy between the engine, WHR and aftertreatment system. The developed IPC strategy is compared with a baseline engine control strategy and with the optimal solution to the control problem, which serves as a benchmark. The examined control strategies are described below.

14.3.1 An Optimal Control Approach to IPC

Following the IPC approach, the studied control problem is formulated in the optimal control framework. We propose to minimize the total operational costs associated with fuel and AdBlue consumption and active DPF regeneration, which can be expressed as follows:

$$\min_{u \in \mathcal{U}} \int_0^{t_e} \pi_f \dot{m}_f + \pi_a \dot{m}_a + \pi_{PM} \dot{m}_{PM} \, \mathrm{d}t \tag{14.8}$$

subject to Eqs. (14.6) and (14.7). In Eq. (14.8), $u = [u_{EGR} \;\; u_{VTG}]^T$ is the vector with EGR and VTG valve positions, and $\mathscr{U}$ is the set of its allowable values. Moreover, the following end point constraint on tailpipe NO_x emission is imposed:

$$\int_0^{t_e} \dot{m}_{NO_{x,tp}} \, \mathrm{d}t \leq Z_{NO_x} \int_0^{t_e} \frac{1}{3.6 \times 10^6} P_d \, \mathrm{d}t \tag{14.9}$$

In these expressions, the diesel price $\pi_f = 1.34 \times 10^{-3}$ [Euro/g], AdBlue price $\pi_a = 0.50 \times 10^{-3}$ [Euro/g], and fuel costs associated with active DPF regeneration per gram of accumulated soot $\pi_{PM} = 7.10 \times 10^{-2}$ [Euro/g] are used. By explicitly taking into account the operational costs, it is easy to deal with variations in fuel and AdBlue prices. Moreover, the AdBlue mass flow $\dot{m}_a$ is determined by assuming that all injected urea decomposes in ammonia and is available for NO_x conversion. As such, the desired AdBlue dosage $\dot{m}_a$ g/s in Eq. (14.8) is given by:

$$\dot{m}_a = 2.0067 \cdot \eta_{SCR}(T_{SCR}, SV, C_{NO_2}/C_{NO_x}) \cdot \dot{m}_{NO_x} \tag{14.10}$$

Finally, the engine-out NO_x emission $\dot{m}_{NO_x}$, the fuel mass flow $\dot{m}_f$, and the (equivalent) cost of active DPF regeneration $\dot{m}_{PM}$ are determined by four-dimensional steady-state maps.

14.3.2 Optimal IPC Trategy

The optimal control problem presented in Sect. 14.3.1 can be solved by applying Pontryagin's Minimum Principle, see, e.g., [14]. This principle relies on a Hamiltonian, which entails the (integrand of the) objective function from Eq. (14.8), augmented with Lagrange multipliers λ and the state dynamics $f(x, u, t)$ from Eq. (14.6):

$$H(x, \lambda, u, t) = \pi_f \dot{m}_f(u, t) + \pi_a \dot{m}_a(x, u, t) + \pi_{PM} \dot{m}_{PM}(x, u, t) + \lambda^T f(x, u, t) \tag{14.11}$$

The Pontryagin's Minimum Principle gives two necessary (and, under some assumptions on the objective function, also sufficient) conditions for $u^\star \in \mathcal{U}$ to be the optimal control input along the optimal state trajectory $x^\star$. In particular, the optimal $u^\star \in \mathcal{U}$ satisfies:

$$H(x^\star, \lambda^\star, u^\star, t) \leq H(x^\star, \lambda^\star, u, t), \qquad \text{for all } u \in \mathcal{U} \tag{14.12}$$

where $\dot{\lambda} = -\frac{\partial H}{\partial x}$ subject to $\lambda^T(t_e)\delta x(t_e) = 0$. Note that Eq. (14.12) can be solved by minimising Eq. (14.11) over all $u \in \mathcal{U}$. The solutions to λ satisfy:

$$\dot{\lambda}_1 = c_1 \dot{m}_{exh} \lambda_1 - c_2 \dot{m}_{exh} \lambda_2 + \frac{d\eta_{SCR}}{dT_{DOC}} \dot{m}_{\mathrm{NO}_x} \left(\lambda_3 - c_6 \pi_a\right) \tag{14.13a}$$

$$\dot{\lambda}_2 = (c_2 \dot{m}_{exh} + c_3)\lambda_2 + \frac{d\eta_{SCR}}{dT_{SCR}} \dot{m}_{\mathrm{NO}_x} \left(\lambda_3 - c_6 \pi_a\right) - \frac{c_5 \dot{m}_{exh}}{\alpha_{WHR}} \lambda_4 \tag{14.13b}$$

$$\dot{\lambda}_3 = 0 \tag{14.13c}$$

$$\dot{\lambda}_4 = \frac{1}{\alpha_{WHR}} \lambda_4 - \frac{d\dot{m}_f}{dP_{WHR}} \pi_f - \frac{d\dot{m}_a}{dP_{WHR}} \pi_a - \frac{d\dot{m}_{PM}}{dP_{WHR}} \pi_{PM} \tag{14.13d}$$

subject to $\lambda_1(t_e) = \lambda_2(t_e) = \lambda_4(t_e) = 0$. From Eq. (14.13a)–(14.13d), it can be observed that the dynamics of λ_1, λ_2 and λ_4 are unstable. Moreover, they have end-point constraints. These two facts make the solution to this optimal control problem difficult to implement in practice, as it requires the entire drive cycle to be known *a priori*. Still, the optimal solution, given by Eq. (14.12) and (14.13a)–(14.13d), will be studied in this work to benchmark the approximate and real-time implementable solution that we will propose below. Note that λ_3 is an undetermined constant (as $\dot{\lambda}_3 = 0$, while no end-point constraint is given), which has to be tuned over a drive cycle to ensure that Eq. (14.9) is satisfied.

14.3.3 Real-Time IPC Strategy

As mentioned before, the optimal solution to the control problem is difficult to implement in real time (as the drive cycle has to be known *a priori*). In particular, the main difficulty lies in solving Eq. (14.13a)–(14.13d), not in solving Eq. (14.12). Namely, the latter is a minimisation problem that can be solved using standard nonlinear programming routines. Therefore, a real-time implementable IPC strategy is proposed that is based on approximating Eq. (14.13a)–(14.13d), such that it can be computed using real-time available information.

In the approximate solution that we propose in this chapter, the expression for Eq. (14.13c) is kept and its value is tuned over representative drive cycles (with cold and hot starts), and the solutions to Eqs. (14.13a), (14.13b), (14.13d) are approximated. In particular, the expression for Eq. (14.13d) subject to $\lambda_4(t_e) = 0$ is approximated by a suitably chosen (possibly nonzero) constant, which is also tuned

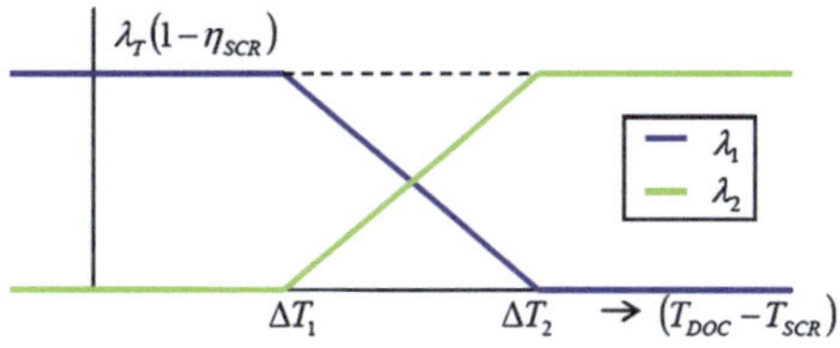

Fig. 14.3 Heuristic rule for λ_1 and λ_2 [15]

over representative drive cycles. The expressions for Eqs. (14.13a), (14.13b) are replaced by the heuristic, postulated rule that was proposed in [15]. This rule is illustrated in Fig. 14.3 and is parameterized by λ_T, ΔT_1 and ΔT_2. It will be shown in Sect. 14.5 that these approximations, which are essential for arriving at a real-time implementable IPC strategy, will cause only minor loss of optimality.

The rationale of the parametrization of Fig. 14.3 is that the effort taken to heat up the aftertreatment system should be proportional to the SCR conversion inefficiency. In case the DOC temperature is not significantly higher than the SCR temperature, it seems better to invest in raising the engine-out exhaust temperature rather than promoting heat convection from the DOC/DPF to the SCR system (which corresponds to a large λ_1). The converse holds when the DOC/DPF temperature is significantly higher than the SCR temperature (which corresponds to a large λ_2).

14.3.4 Baseline Strategy

In order to compare the results from the optimal control formulation, a baseline engine control strategy is proposed that mimics a state-of-the-art air management strategy for a standard Euro-VI engine configuration (without WHR system). As in [11, 15], this strategy is characterized by switching between two control modes:

1. **Thermal management mode (M1)** for rapid heat-up of the aftertreatment system ($T_{SCR} < 200$ °C);
2. **Low** NO_x **mode (M2)** for normal operation ($T_{SCR} \geq 250$ °C).

A fundamental difference with IPC is that the baseline strategy relies on fixed control settings u for each engine operating point (N_e, τ_e). For both modes, these settings are pre-determined in an off-line optimization procedure, which is often based on stationary test conditions.

As we want to use the same control structure for both strategies in simulations, two different sets of constant λ are used for the control modes (see Table 14.2). As engine calibration is mainly optimized using steady state-measurements, anticipated steady-state T_{DOC} and T_{SCR} values from the engine maps are used in the Hamiltonian to evaluate the SCR efficiency maps.

Table 14.2 Selected control parameters

Control strategy	Control parameters				
	$-\lambda_{1,M1}$	$-\lambda_{1,M2}$	λ_2	λ_3	λ_4
Baseline(-WHR)	1.40×10^{-3}	0	0	0.218	0
Recal-WHR	1.36×10^{-3}	0	0	0.200	0
	ΔT_1	ΔT_2	$-\lambda_T$	λ_3	$-\lambda_4$
Real-time	123	142	8.07×10^{-4}	1.13×10^{-2}	2.97×10^{-6}
	$\lambda_{3,cold}$	$\lambda_{3,hot}$			
Optimal	2.76×10^{-2}	8.53×10^{-3}			

14.4 Control Design

This section discusses the calibration procedure of the optimal and real-time IPC strategies. An overview of the selected control parameters is given in Table 14.2. For details on the calibration of the baseline strategies, the interested reader is referred to [11, 15]. Note that the baseline strategy will be applied to three cases:

- Conventional powertrain without WHR system (*Baseline*);
- Powertrain with WHR system using the same control strategy (*Baseline-WHR*);
- Powertrain with WHR system using a recalibrated control strategy (*Recal-WHR*).

As a result, two different sets of calibration parameters are used.

14.4.1 Optimal IPC Strategy

The optimal IPC strategy is completely determined by Eq. (14.12) and (14.13a)–(14.13d). Because of this, and because $\lambda_1(t_e) = \lambda_2(t_e) = \lambda_4(t_e) = 0$, only a value for λ_3 needs to be determined, such that Eq. (14.9) is satisfied for a given value for Z_{NO_x}. This can be done, e.g., by performing a line search resulting in a cost-NO_x tradeoff curve (see also left-hand graph of Fig. 14.5). Note that, contrary to the real-time IPC strategy, the optimal IPC strategy is not restricted to have the same λ_3 for both the cold and hot WHTC. The λ_3 values that are given in Table 14.2 are tuned, such that the Euro-VI target of $Z_{NO_x} = 0.41$ g/kWh is achieved for both the cold and hot WHTC.

Despite the fact that the optimal solution does not require much tuning, numerically obtaining a solution to the optimal control problem is more difficult. This is due to the fact that the dynamics for λ_1, λ_2 and λ_4 are unstable and have end-point constraints. Numerically finding solutions to unstable differential equations is a tedious task, as not just the solution to the differential equation grows, but also the approximation error of the numerical integration schemes. Therefore, the optimal solution is computed using an iterative procedure: the so-called forward–backward sweeping method, which is adopted from [16]. For a given λ_3 value, this iterative procedure consists of repeatedly executing the following two steps:

(1) For given $\lambda_1, \lambda_2, \lambda_4$, solve Eq. (14.6) with Eq. (14.12) from $t = 0$ to $t = t_e$
(2) For the x and u resulting from the previous step, solve Eq. (14.13a)–(14.13d) from $t = t_e$ to $t = 0$

The procedure has to be initialised with some $\lambda_1, \lambda_2, \lambda_4$. Here, they are chosen to be equal to zero. Under certain conditions, see [16], the algorithm converges, which means that the optimal solution has been found. Note that the numerical complexity of this iterative procedure is large, which is another reason why the optimal IPC strategy is difficult to implement in real-time.

14.4.2 Real-Time IPC Strategy

The real-time implementable IPC strategy is tuned using a numerical programming routine. In this chapter, we use the Nelder-Mead simplex algorithm. The objective is to find the control parameters $\Delta T_1, \Delta T_2, \lambda_T$ and λ_3, such that the operational costs are minimised over the hot WHTC. Simultaneously, the specified NO_x engineering target of 0.41 g/kWh has to be met for the weighted WHTC (where the cold and hot cycle are weighted by 16 % and 84 %, respectively).

14.5 Simulation Results

To evaluate the performance of the controllers, simulations are done for the World Harmonized Transient Cycle (WHTC). This test cycle specifies the requested engine speed N_e and torque $\tau_{d,req}$, see Fig. 14.4. Three parts can be distinguished: urban driving conditions (0–900 s), rural driving conditions (900–1380 s), and highway driving conditions. As we focus on Euro-VI legislation, results have to be generated for cold as well as hot cycle conditions. In case of a cold cycle, the initial SCR catalyst temperature is set to 20 °C, whereas 200 °C is used at the start of the hot cycle; engine and WHR system heat up are not modeled yet.

In this study, we focus on the results of the optimal and real-time IPC strategy. For reference, also the results for the baseline strategies are presented. All cases are compared with the standard Euro-VI engine without WHR system (*Baseline*). The main results are summarized in Table 14.3.

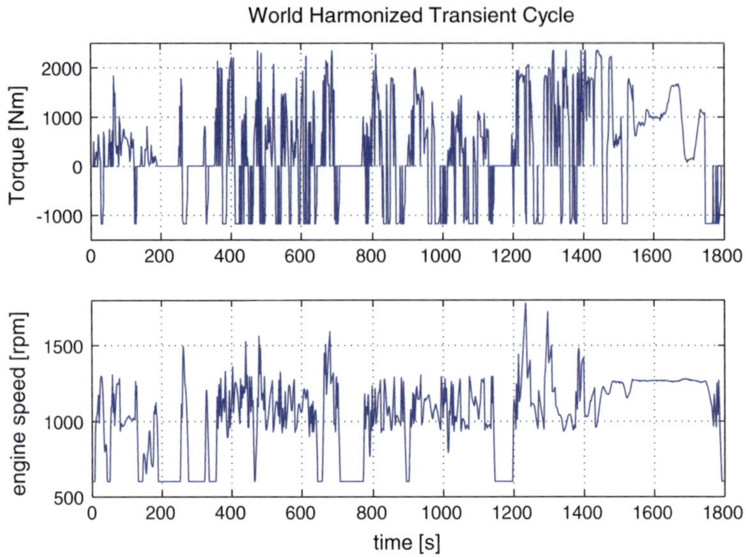

Fig. 14.4 World Harmonized Transient Cycle: specified torque $\tau_{d,req}$ and engine speed N_e

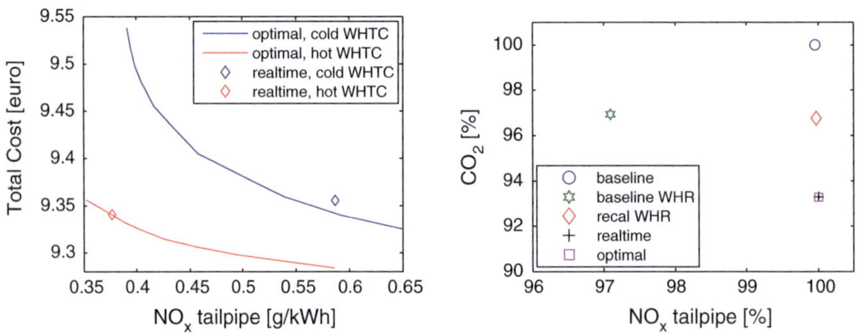

Fig. 14.5 Total operational cost-$NO_{x,tp}$ tradeoff for optimal IPC strategy (*left*) and CO_2–$NO_{x,tp}$ tradeoff for weighted WHTC (*right*)

14.5.1 Overall Powertrain Results

Following the procedure described in Sect. 14.4.1, the total operational cost-$NO_{x,tp}$ trade-off curves for the optimal IPC strategy are determined (see left-hand graph of Fig. 14.5). In the remainder of this section, simulation results are shown that correspond to a tailpipe NO_x emission of 0.41 g/kWh for both cold and hot WHTC. The diamonds in this figure indicate the results after tuning the real-time IPC strategy.

For the weighted WHTC results, the CO_2–$NO_{x,tp}$ tradeoff is shown in the right-hand graph of Fig. 14.5. All results are compared with the baseline (without

Table 14.3 Overview of WHTC results

Quantity	Control strategy				
	Baseline	Baseline WHR	Recal WHR	Real-time WHR	Optimal WHR
$NO_{x,eo}$ (g/kWh)					
Hot	3.67	3.55	3.63	5.42	5.62
Cold	3.80	3.67	3.74	5.42	4.61
Weighted	3.69	3.57	3.65	5.42	5.46
$NO_{x,tp}$ (g/kWh)					
Hot	0.353	0.342	0.354	0.376	0.411
Cold	0.708	0.691	0.702	0.586	0.410
Weighted	0.410	0.398	0.410	0.410	0.410
$NH_{3,max}$ (ppm)					
Hot	2	2	2	2	1
Cold	2	2	2	2	2
CO_2 %					
Hot	100	96.9	96.7	93.2	93.0
Cold	100	97.0	97.0	93.4	94.5
Weighted	100	96.9	96.8	93.3	93.3
Hot WHTC costs %					
Fuel	97.1	94.1	93.9	90.5	90.3
AdBlue	1.1	1.1	1.1	1.8	1.9
PM	1.8	1.8	1.7	0.8	0.8
Total	100.0	97.0	96.7	93.1	93.0

WHR). This figure shows that simply adding a WHR system to the existing engine (*Baseline-WHR*) results in 3.1 % CO_2 reduction. However, only a minor performance improvement is possible by exploiting the margin in tailpipe NO_x emission (*Recal-WHR*): additional 0.1 % CO_2 reduction, which is due to the reduced fuel consumption over the hot cycle (see Table 14.3). Both the real-time and optimal IPC strategy are able to reduce CO_2 emissions by an additional 3.5 %. From these results, it is concluded that the real-time strategy is a promising practical implementation of the optimal strategy. In total, the CO_2 saving potential of the WHR system sums up to 6.7 %.

14.5.2 Cold Cycle Results

To better understand the differences between the studied strategies, time traces of the results are shown in Fig. 14.6. We focus on the cold cycle, since this cycle is more challenging and the differences are more distinct here. For the baseline strategies, EGR valve and VTG settings are fixed for each operating point (N_e, τ_e) and each engine mode. As can be seen from the SCR temperature plot, the controller switches

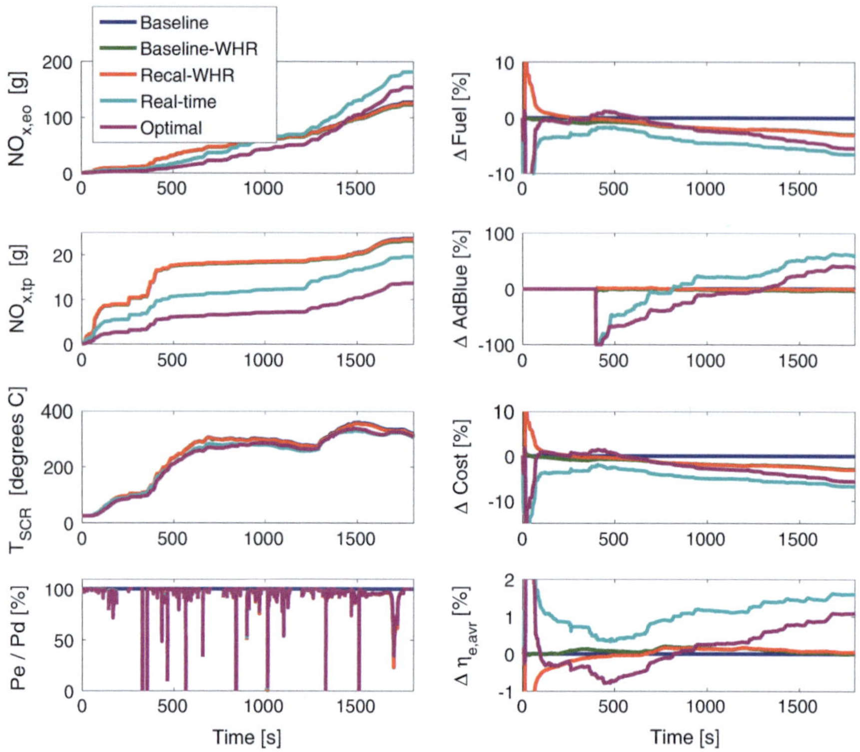

Fig. 14.6 Cold WHTC results. Cumulative emission results and SCR temperature (*left*). Operational costs, fuel and AdBlue consumption and engine efficiency relative to the baseline (*right*)

from thermal management to low NO_x mode around $t = 550$ s, when $T_{SCR} \geq 250$ °C (see also λ_1 in Fig. 14.7). This leads to similar cumulative engine-out and tailpipe NO_x emissions for the three baseline strategies.

Based on the actual state of the engine, aftertreatment and WHR system, the adaptive IPC strategy selects on-line the EGR valve and VTG settings that gives the lowest operational cost. In the first part of the cold WHTC, this results in low engine out NO_x emissions due to relatively high EGR rates. After $t = 600$ s and especially during the highway part, high SCR conversion efficiencies can be realized, so engine-out $NO_{x,eo}$ emissions can be relaxed by reducing EGR flow: so-called *EGR-SCR balancing* (see upper left-hand graphs in Fig. 14.6). This leads to a considerable increase in engine out $NO_{x,eo}$ emissions, but the tailpipe $NO_{x,tp}$ emission target is still met. This effect is even more pronounced for the optimal strategy.

In the bottom left-hand graph of Fig. 14.6, the engine power-to-total power ratio P_e/P_d is plotted. Although EGR and exhaust mass flows and temperatures differ for the studied control strategies, the observed differences are small. As expected, the WHR contribution is largest in the high way part: average power output of $P_{WHR} =$

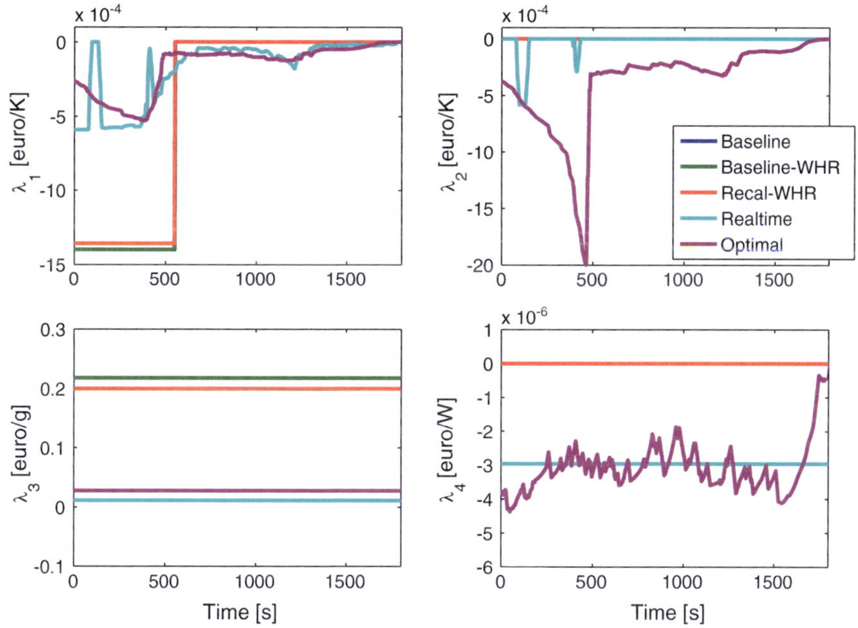

Fig. 14.7 Lagrange multipliers (cold WHTC)

4.9 kW for the IPC strategy. During the urban and rural part, the WHR system is seen to be switched off frequently, according to Eq. (14.4).

The right-hand graphs of Fig. 14.6 show the corresponding fuel and AdBlue consumption as well as the resulting operational costs over time. These results are given relative to the baseline. At every time instant t_k, the relative fuel consumption is determined by:

$$\Delta Fuel(t_k) = 100 \times \frac{\int_0^{t_k} \dot{m}_f - \dot{m}_{f,Baseline}\ \mathrm{d}t}{\int_0^{t_k} \dot{m}_{f,Baseline}\ \mathrm{d}t} \tag{14.14}$$

In a similar way, the relative AdBlue and total operational costs are computed. As seen from this figure, AdBlue injection starts around $t = 400$ s, when $T_{SCR} > 180$ °C. In the bottom right-hand graph, the moving average of the difference between engine efficiency and baseline engine efficiency is shown:

$$\Delta\eta_{e,avg}(t_k) = \eta_{e,avg}(t_k) - \eta_{e,avg,Baseline}(t_k) \tag{14.15}$$

where:

$$\eta_{e,avg}(t_k) = \frac{\int_0^{t_k} P_e\, dt}{\int_0^{t_k} P_{fuel}\, dt}$$

with $P_{fuel} = \dot{m}_f \times Q_{LHV}$ and the lower heating value Q_{LHV} of diesel.

As shown in the upper right-hand graphs of Fig. 14.6, implementation of the WHR system reduces fuel consumption in all studied cases. For the baseline strategies, this is achieved with only small changes in AdBlue consumption. The IPC strategies can further reduce fuel consumption and total operational costs. By on-line adaptation of the EGR valve and VTG settings, a cost optimal balance is found; although AdBlue consumption is significantly increased, it only has a small (absolute) contribution to the total operational cost due to its relatively low mass flow and cost. The corresponding high SCR efficiencies η_{SCR} allow for high engine-out NO_x emissions, which are associated with reduced fuel consumption. Consequently, the reduced fuel (and PM) costs compensate for the increased AdBlue costs. Results for the engine efficiency illustrate that the adapted EGR-VTG settings in the IPC strategy lead to 1.0–1.5 % increase in engine efficiency compared to the baseline. This effect is most dominant during the highway part. For the baseline strategies with WHR system, efficiency improvements are negligible over the cold WHTC.

Figure 14.7 shows the resulting Lagrange multipliers for the examined strategies. For the three baseline strategies, these values are chosen to be (switching) constants. Note that $\lambda_4 = 0$ for the baseline strategies, as the WHR dynamics are neglected in these cases. Comparison of the IPC strategies learns that the chosen λ_4 value in the real-time strategy is a good approximation of the co-state dynamics associated with the optimal case. This also holds for the λ_1 dynamics. However, the implemented heuristics for λ_2 are not capable to approximate the corresponding optimal co-state dynamics. In the optimal strategy, $\lambda_2 > \lambda_1$, which means that heat convection is promoted to heat up the aftertreatment system. This only holds for short periods in the real-time strategy; focus is on raising the exhaust gas temperature for most of the time. Based on these results, this Lagrange multiplier is believed to have only a minor effect on overall powertrain performance.

14.6 Conclusions and Future Work

Optimal control of a Euro-VI heavy-duty diesel engine with Waste Heat Recovery (WHR) system is challenging due to the high number of sub systems, interactions and requirements set by emission legislation. To optimize overall performance, an integrated approach is required which combines energy and emission management. Based on the Integrated Powertrain Control (IPC) approach, an optimal control strategy is presented, which minimizes total operational costs and explicitly deals with the tailpipe NO_x emission constraint. Following Pontryagin's Minimal Principle, the optimal solution for a cold and hot WHTC is numerically determined using a forward–backward sweeping method. Alternatively, a real-time implementable strategy is proposed. From simulation results over the WHTC, it is concluded that the optimal IPC strategy can be replaced by the proposed real-time strategy with negligible loss of optimality. This IPC strategy outperforms the current baseline engine control strategy: an additional 3.5 % CO_2 reduction within the tailpipe NO_x limit.

Current research concentrates on the robustness of the IPC strategy performance for model uncertainties and varying duty cycles. Tests will be performed on an engine dynamometer to demonstrate the potential of the proposed control strategy.

References

1. Park T, Teng H, Hunter GL, van der Velde B, Klaver J (2011) A Rankine cycle system for recovering waste heat from HD diesel engines-experimental results. In SAE World Congres, 2011
2. Bredel E, Nickl J, Bartosch S (2011)Waste heat recovery in drive systems of today and tomorrow, MTZ Worldwide Edition
3. Horst T, Rottengruber H, Seifert M, Ringler J (2013) Dynamic heat exchanger model for performance prediction and control system design of automotive waste heat recovery systems. Appl Energy 105:293–303
4. Hou G, Sun R Hu G, Zhang J (2011) Supervisory predictive control of evaporator in Organic Rankine Cycle (ORC) system for waste heat recovery. In International Conference on Advanced Mechatronic Systems, pp 306–311
5. Quoilin S, Aumann R, Grill A, Schuster A, Lemort V, Spliethoff H (2011) Dynamic modeling and optimal control strategy of waste heat recovery organic Rankine cycles. J Appl Energy 88(6):2183–2190
6. Howell T, Gibble J, Tun C (2011) Development of an ORC system to improve HD truck fuel efficiency. In: Directions in energy-efficiency and emission research DEER Conference
7. Tona P, Peralez J, Sciarretta A (2012) Supervision and control prototyping for an engine exhaust gas heat recovery system based on a steam Rankine cycle. In: IEEE/ASME international conference on advanced intelligent mechatronicsMechatronics, pp 695–701
8. Hounsham S, Stobart R, Cooke A, Childs P (2008) Energy recovery systems for engines. In: SAE World Congress
9. Merz F, Sciarretta A, Dabadie J-C (2012) On the optimal thermal management of hybrid-electric vehicles with heat recovery systems. Oil Gas Sci Technol Rev IFP Energies Nouvelles 67(4):601–612
10. Willems F, Kupper F, Cloudt R (2012) Integrated powertrain control for optimal CO_2–NOx trade-off in an Euro-VI diesel engine with waste heat recovery system. In: Proceeding of American control conference, Montreal, pp 1296–1301
11. Willems F, Kupper F, Cloudt R (2012) Integrated energy and emission management for heavy-duty diesel engines with waste heat recovery system. In: Proceedings of the IFAC workshop on engine and powertrain control, simulation and modeling (ECOSM'12) Rueil-Malmaison, pp 203–210
12. Cloudt R, Saenen J, van den Eijnden E, Rojer C (2010) Virtual exhaust line for model-based diesel aftertreatment development. SAE paper 2010-01-0888
13. Willems F, Foster D (2009) Integrated powertrain control to meet future CO_2 and Euro-6 emission targets for a diesel hybrid with SCR-deNOx system. In: Proceeding of American Control Conference, pp 3944–3949
14. Bertsekas D (2007) Dynamic programming and optimal control. Athena Scientific, Belmont
15. Cloudt R, Willems F (2011) Integrated emission management strategy for cost-optimal engine-aftertreatment operation. SAE Int J Engines 4(1):1784–1797
16. McAsey M, Mou L, Han W (2012) Convergence of the forward-backward sweep method in optimal control. Computational Optimization and Applications 53(1):207–226

Part IV
Optimization of the Engine Operation

Chapter 15
Learning Based Approaches to Engine Mapping and Calibration Optimization

Dimitar Filev, Yan Wang and Ilya Kolmanovsky

Abstract In this chapter we consider a class of optimization problems arising in the process of automotive engine mapping and calibration. Fast optimization algorithms applicable to high fidelity simulation models or experimental engines can reduce the time, effort and costs required for calibration. Our approach to these problems is based on iterations between local model identification and calibration parameter (set-points and actuator settings) improvements based on the learned surrogate model. Several approaches to algorithm implementation are considered. In the first approach, the surrogate model is defined in a linear incremental form and its identification reduces to Jacobian Learning. Then quadratic programming is applied to adjust the calibration parameters. In the second approach, we consider a predictor-corrector algorithm that estimates the change in the minimizer based on changing operating conditions before correcting it. Case studies are described that illustrate the applications of algorithms.

15.1 Introduction

The chapter is concerned with automating the process of internal combustion engine mapping and calibration using real-time optimization techniques. As the engine complexity is growing to meet more stringent fuel economy and emissions regulations and increasing customer demands for improved drivability, the interest in techniques

---

D. Filev (✉) · Y. Wang
Research and Advanced Engineering, Ford Motor Company, Dearborn, MI, USA
e-mail: dfilev@ford.com

Y. Wang
e-mail: ywang21@ford.com

I. Kolmanovsky
Department of Aerospace Engineering, The University of Michigan, Ann Arbor, MI, USA
e-mail: ilya@umich.edu

H. Waschl et al. (eds.), *Optimization and Optimal Control in Automotive Systems*, Lecture Notes in Control and Information Sciences 455,
DOI: 10.1007/978-3-319-05371-4_15,

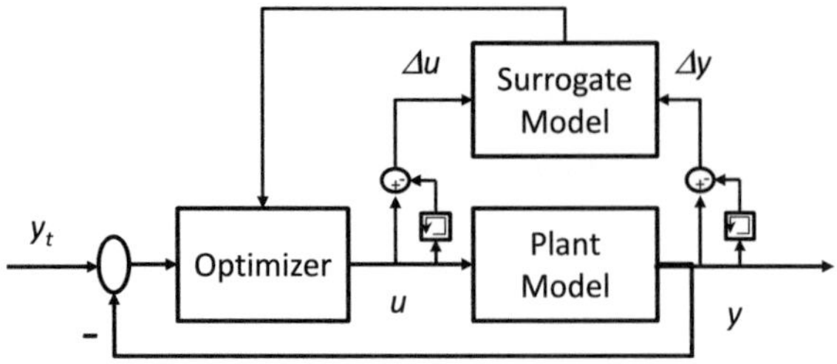

Fig. 15.1 The surrogate model based learning and optimization algorithm

for reducing engine calibration time and effort has been rapidly increasing [4, 11–13, 18–21, 25].

The engine mapping process involves engine characterization to identify component static response and calibration optimization which refers to determining optimal set-points and actuator settings given current engine operating conditions. The set-points are stored in look-up tables in the strategy or incorporated into the regression models.

Several approaches have been proposed to address engine calibration optimization. In more traditional approaches, the data are first collected from the engine, then regressed and the optimization is performed on the resulting regression models to find the optimal set-points and actuator settings. Extremum seeking, based on various types of optimization algorithms running in real-time on the engine, has been exploited in [11, 13, 20, 21, 25]. Finally, techniques where transient data are rapidly collected to identify a dynamic model and this dynamic model is used for static calibration optimization have been recently developed [3]. Besides automotive engines, calibration optimization of aircraft gas turbine engines has been also considered [17].

In this chapter we discuss another approach to engine calibration which is based on iteratively combining the local identification and local optimization steps. This approach follows [7] (see also [8, 23]). Specifically, we employ real-time learning of a surrogate model to locally characterize engine response around the current operating point and then we perform an optimization update to improve the parameters and actuator settings with respect to the identified surrogate model, see Fig. 15.1.

If the surrogate model is defined as a linear incremental form, the surrogate model identification reduces to estimating the Jacobian and we employ the Jacobian Learning (JL) technique proposed in [7]. See also [8] for a related approach.

The approach of [7] has not been previously applied to engine calibration and optimization problems. In this chapter, we present its extension to the case when there are constraints, and we demonstrate its capability to rapidly find minimizers based on a high fidelity GT-Power engine model.

A further extension of the algorithm is presented to achieve high accuracy in filling the calibration tables when operating conditions are (slowly) varying. This extension is based on treating the calibration optimization as a parameter-dependent optimization problem [2, 6, 9, 19, 22], and applying a predictor-corrector approach. The predictor part of the algorithm feed-forward compensates for changes in the minimizer with changing engine operating conditions while the corrector part improves on the predicted minimizer. As an illustration, the application of the predictor-corrector algorithm to a series HEV model is considered.

The chapter is organized as follows. In Sect. 15.2, we highlight the mathematical problem formulation. In Sect. 15.3, the JL-based optimization algorithm of [7] is described. The results of applying this approach to a high fidelity engine model in GT-Power using the co-simulation approach are presented in Sect. 15.4. Another case study is considered in Sect. 15.5, where engine fuel consumption minimization is performed on-board of a simulated series hybrid vehicle when the generator power is maintained at a constant value during the learning and optimization phase. In Sect. 15.6, we present the predictor-corrector algorithm and in Sect. 15.7 we illustrate its application to engine fuel consumption minimization on-board of a simulated series hybrid vehicle when the generator power is slowly varying. Concluding remarks are presented in Sect. 15.8.

15.2 Mathematical Problem Formulation

From a calibration perspective, a typical engine can be considered as a MIMO nonlinear system with adjustable actuator settings and/or modes of operation as inputs, and performance characteristics as outputs. The steady-state input-output relationship at different steady state operating conditions can be described as

$$y = F(u, p), \tag{15.1}$$

where the input vector u includes adjustable actuator settings—spark retard, intake valve opening/closing (IVO/IVC) and exhaust valve opening/closing (EVO/EVC) settings of the VCT actuator, throttle and wastegate settings, exhaust gas recirculation valve opening (EGR), air-to-fuel ratio, etc.; output vector y consists of engine performance characteristics—brake torque (equivalently, engine load), brake specific fuel consumption (BSFC), Crank Angle 50 (CA50), etc. and p is a vector of operating conditions, e.g. engine speed and target engine brake torque (equivalently, target engine load).

Many engine-related calibration problems can be reduced to minimizing a certain function with respect to a part of variables:

$$\text{Minimize } Q(u, p) \text{ with respect to } u, \tag{15.2}$$

and often the function $Q(u, p)$ can be specified as

$$Q(u, p) = ||F(u, p) - y_t||_{\Omega}^2 = (F(u, p) - y_t)^{\mathrm{T}}\Omega(F(u, p) - y_t), \tag{15.3}$$

where y_t is the vector of target values for the output vector, y, $\Omega = \Omega^{\mathrm{T}} \geq 0$ is weighting matrix, and for a vector z, $||z||_{\Omega}^2 = z^{\mathrm{T}}\Omega z$.

In filling the calibration tables, the minimizer in (15.2) is sought in the form dependent on p, i.e., in the form,

$$u = u^*(p). \tag{15.4}$$

For instance, Q can represent a weighted sum of BSFC squared (for which the corresponding BSFC target is zero) and squares of the deviation between some of the components of the output vector y and a corresponding set of target values of engine load, e.g. engine load and its target value, CA50 and its target, etc. We note that we do not assume that the targets are achievable.

There are several distinct characteristics of the optimization problem (15.2) for static engine calibration. Firstly, if the optimization is performed directly on an experimental engine, the measurements of the function Q are uncertain due to sensor measurement noise and due to deviations of engine operation from steady-state. The latter source of noise is exacerbated if it is desirable to move rapidly through the operating conditions, p, without waiting for engine to completely reach the steady-state. This latter source of noise is also present if the optimization is performed on an engine model, such as a high fidelity GT-Power engine model, where in addition the evaluation of the function Q can be expensive due to slow model simulations. Consequently, special care is required in the application of the optimization algorithms and techniques to engine-related optimization problems (15.2).

15.3 Jacobian Learning Based Optimization Algorithm

In this section we focus on the problem of engine mapping at steady state, for given operating conditions, i.e. assuming $p = const$ in (15.1). In this case, F is assumed to be nonlinear but smooth, the output vector, y, represents a set of performance variables, e.g., Brake Specific Fuel Consumption (BSFC), engine load, Crank Angle 50 (CA50), etc. The engine mapping problem is decomposed to a set of local mappings obtained at specific operating conditions, p,

$$y = \tilde{F}(u) = F(u, p), \tag{15.5}$$

where u is an r-dimensional vector, and y is a q-dimensional vector. We are interested in an approach for solving problem (15.2) that exhibits fast convergence so that it can be repeatedly applied at different p.

We consider a local surrogate model based on a linearized time-varying (Jacobian) approximation of the nonlinear input-output mapping $\tilde{F}$ in (15.5),

$$\Delta y(k) = J(k)\Delta u(k) \tag{15.6}$$

where

$$\Delta u(k) = u(k) - u(k-1), \quad \Delta y(k) = y(k) - y(k-1),$$

and $J(k)$ is the $q \times r$ Jacobian matrix defined by

$$J_{s,j}(k) = \frac{\partial y_s}{\partial u_j}(k), \quad 1 \le j \le r, 1 \le s \le q.$$

The surrogate model takes the form,

$$\hat{y}(k) = y(k-1) + \hat{J}(k-1)(u(k) - u(k-1)), \tag{15.7}$$

where $\hat{J}(k)$ is the estimate of the Jacobian of $\tilde{F}$ at the time instant $(k-1)$. The optimization algorithm solves the following Quadratic Programming (QP) problem,

$$u(k) = argmin(||y_t - \hat{y}(k)||^2_{\Omega} + ||u(k) - u(k-1)||^2_{\Gamma}), \tag{15.8}$$

subject to (15.7) and the constraints on the range of actuator settings,

$$u_{min} \le u(k) \le u_{max}. \tag{15.9}$$

Note that (15.8) includes a penalty on the increments of u which prevents the optimization algorithm from taking too large steps that render the Jacobian approximation $\hat{J}(k-1)$ inaccurate. In our numerical experiments, we typically used $\Omega = I$ and $\Gamma = \gamma I$, where γ is a tunable parameter in (15.8).

We note that the problem (15.8) with the simple box-like constraints (15.9) is a low dimensional quadratic programming problem that can be easily solved using any quadratic programming solver. In our numerical experiments, we use the functions *quadprog.m* and *lsqlin.m* of Matlab. We also note that (15.8) *without* the constraints on the ranges (15.9) is easily solvable and leads to an update of the form

$$\begin{aligned} u(k) &= u(k-1) + \left(\hat{J}^{\mathrm{T}}(k-1)\Omega\hat{J}(k-1) + \Gamma\right)^{-1} \hat{J}^{\mathrm{T}}(k-1)\Omega(y_t - y(k-1)) \\ &= u(k-1) + K(k-1)(y_t - y(k-1)). \end{aligned} \tag{15.10}$$

Since

$$\hat{J}^{\mathrm{T}}(\hat{J}\hat{J}^{\mathrm{T}} + \rho I_{q\times q})^{-1} = (\hat{J}^{\mathrm{T}}\hat{J} + \rho I_{r\times r})^{-1}\hat{J}^{\mathrm{T}},$$

the updates in (15.10) are similar to the one studied in [7] for the case $r \ge q$, which had the form,

$$\begin{aligned} u(k) &= u(k-1) + \hat{J}^{\mathrm{T}}(k-1)H\left(\hat{J}(k-1)\hat{J}^{\mathrm{T}}(k-1) + \rho I_{q\times q}\right)^{-1}(y_t - y(k-1)) \\ &= u(k-1) + \tilde{K}(k-1)(y_t - y(k-1)), \end{aligned} \tag{15.11}$$

where $\rho > 0$ and H is a diagonal $q \times q$ gain matrix. Note that both (15.10) and (15.11) are variants of Levenberg-Marquardt optimization algorithm. The diagonal matrix $\Gamma > 0$ in (15.10), typically chosen as $\Gamma = \rho I_{r \times r}$, where $I_{r \times r}$ is the $r \times r$ identity matrix, or the term $\rho I_{q \times q}$ in (15.11), where I is the $q \times q$ identity matrix, play the role of the Tikhonov regularization matrices [24]. Their use improves the numerical conditioning of the underlying inverse problem.

We employ the Kalman filtering for recursively learning $\hat{J}$. Specifically, let J_s denote the sth row of the Jacobian, J, $1 \leq s \leq q$. The following model is assumed as a basis for estimating $\hat{J}_s$,

$$\begin{aligned} J_s(k+1) &= J_s(k) + w_s(k), \\ \Delta y_s(k) &= J_s(k)\Delta u(k) + v_s(k), \end{aligned} \tag{15.12}$$

where $w_s(k)$ is the process noise representing the imprecision of the linearized model with zero mean and the covariance matrix, $Q_s \geq 0$, and $v_s(k)$ is the measurement noise with zero mean and covariance matrix, $R_s > 0$. The Kalman filter updates take the following form,

$$\begin{aligned} \hat{J}_s(k) &= \hat{J}_s(k-1) + L_s(k)\Big(\Delta y_s(k) - \hat{J}_s(k-1)\Delta u(k)\Big), \\ L_s(k) &= P_s(k-1)\Delta u(k)\Big(R_s + \Delta u^{\mathrm{T}}(k)P_s(k-1)\Delta u(k)\Big)^{-1}, \\ P_s(k) &= P_s(k-1) - L_s(k)\Delta u^{\mathrm{T}}(k)P_s(k-1) + Q_s. \end{aligned} \tag{15.13}$$

The matrix Q_s is a drift factor that is analogous to the forgetting factor in the commonly used form of the Recursive Least Squares (RLS) [1] and can be estimated from the expected changes in the Jacobian. The advantage of using the drift factor versus the exponentially forgetting factor is in the cases when the system is not excited [7, 16]. It forces the covariance matrix, P_s (which essentially controls the variable learning rate of the Kalman filter) to grow linearly rather than exponentially.

The advantage of (15.13) versus alternative techniques for estimating the Jacobian (e.g., based on the center differences) is that it can be used with general input excitation sequences; in many practical cases, the Jacobian can be estimated from the optimization algorithm iterates. The use of nonlinear surrogate model rather than (15.6) can be advantageous in specific problems; however, we found that (15.6) represents an effective choice, in general. Further comments on exploiting nonlinear surrogate models are made in Sect. 15.6.

15.4 Case Study 1: Application to Engine Mapping

In order to reduce the development time and costs, engine models are used more and more frequently to evaluate engine designs without building the actual hardware. A two step procedure is typically used to compare different engine designs: (1) optimal settings of all actuators at different engine conditions (speed/load) are determined that result in the best performance (e.g., minimum fuel consumption) at each condition; (2) the overall performance is compared as quantified by a weighted sum of performances at different engine conditions where the weighting factors may reflect the time spent at a particular operating condition over a drive cycle. For Step 1, Design of Experiments (DoE) based sweeping of actuator settings, followed by off-line data processing is employed to find the optimal actuator settings at each engine conditions. This approach is not very efficient, especially as new engine technologies are introducing significant numbers of new actuators. At the same time, more detailed engine models that more accurately predict the response of the new engines are becoming more complex and take increasingly longer time to simulate. Even though improvements in computing hardware and parallel computing are alleviating the problem to some extent, fast and smart searching/optimization methods applicable to simulation models are highly desired to speed up this process.

For our case study, the JL-based algorithm described in Sect. 15.3 has been applied to rapidly determine the actuator settings that are optimal for each steady-state operating condition based on an engine model implemented in GT-Power. The GT-Power is a well established engine CAE tool provided by Gamma Technologies and is a popular modeling software package used by engine designers and system developers. Our algorithm has been implemented in Simulink and co-simulated with the engine model running in GT-Power. The co-simulation environment has been setup so that the engine model runs continuously, and model steady-state output data are sent to the optimization algorithm, which then updates the Jacobian estimate (i.e., surrogate model), and, consequently, the actuator settings as discussed in Sect. 15.3.

In this case study, the nonlinear vector function, F in (15.1), represents Brake Specific Fuel Consumption (BSFC), Engine Load, and $CA50$ (defined as the crank angle after top dead center at which 50 % of fuel is burnt). The input vector u represents throttle position, spark timing, intake cam timing, and exhaust cam timing. The target vector y_t for y prescribes the target of 0 for BSFC, the target load the engine should be running at, and the target value for $CA50$ based on the standard correlation between $CA50$ location and best fuel efficiency. The parameter vector p comprises engine speed and target engine load.

Table 15.1 summarizes the results of an exemplar algorithm application. Our engine model is running at fixed speed of 3,000 rpm. The $CA50$ target at this engine speed is 5.54°. The target engine load values are 0.2, 0.5, and 0.8, which change every time algorithm convergence is achieved. The targets, found optimal actuator settings, achieved values of engine load, $CA50$ and BSFC, and the number of optimization iterations are given for three load targets in Table 15.1.

Table 15.1 Calibrated engine variables for 3,000 rpm and target engine loads of 0.2, 0.5 and 0.8

Rpm	Load des	CA50 des	Throttle	Spark Spark	Intake cam	Exhaust cam	Load	CA50	BSFC	Iterations
3,000	0.2	5.54	7.15	−75.00	27.00	34.88	0.20	5.66	1106.74	16
3,000	0.5	5.54	15.62	−39.91	9.00	39.00	0.50	5.63	285.62	4
3,000	0.8	5.54	90.00	−26.95	−32.30	39.00	0.80	5.51	253.77	10

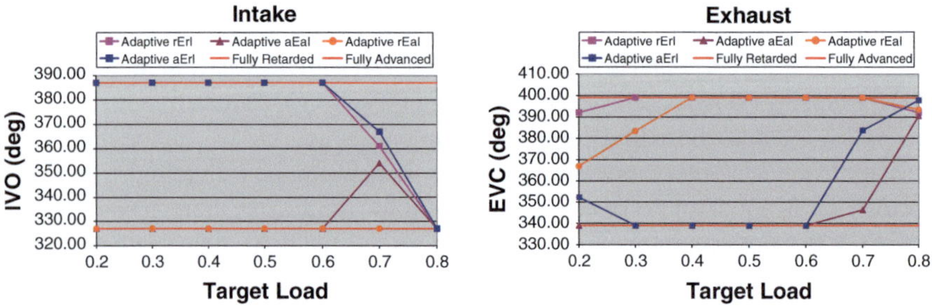

Fig. 15.2 Optimized intake and exhaust cam timings at different loads, when initialized at four different combinations (both retarded, both advanced, and one retarded/one advanced)

Table 15.2 The intake and exhaust cam timings that deliver the best BSFC at different loads

Load	0.2	0.3	0.4	0.5	0.6	0.7	0.8
Intake	387	387	327	327	327	366.8997	327
Exhaust	391.9614	399	399	399	399	383.7112	391.9786

We note that the cam timing optimization is challenging because the effects of cam timing on the BSFC are not monotonic, and dependent on the engine torque outputs, as shown in [14]. Therefore, applying the algorithm with different initial conditions of the cam timings will yield local minimizers (that may not necessarily be global minimizers). The conventional approach is to run a sweep of the intake/exhaust cam timings to find the best BSFC, which is very time consuming. In this work, we applied our JL-based algorithm but initialized the cam timings at several different values; we then selected the actuator settings that gave the best BSFC over these runs. Figure 15.2 shows the best IVO and EVC at different loads all at 3,000 rpm, when initialized at four different initial values.

Combining all the results, the best intake and exhaust cam timings at each loads are summarized in Table 15.2.

The results show that the optimal exhaust cam should be fully retarded or close to being fully retarded for all the cases. The optimal intake cam is retarded at lower loads, and advanced at higher loads. These results are consistent with [14].

The algorithm was applied next to the same engine model to complete a full map that finds the best throttle/spark/intake cam timing/exhaust cam timing settings. The

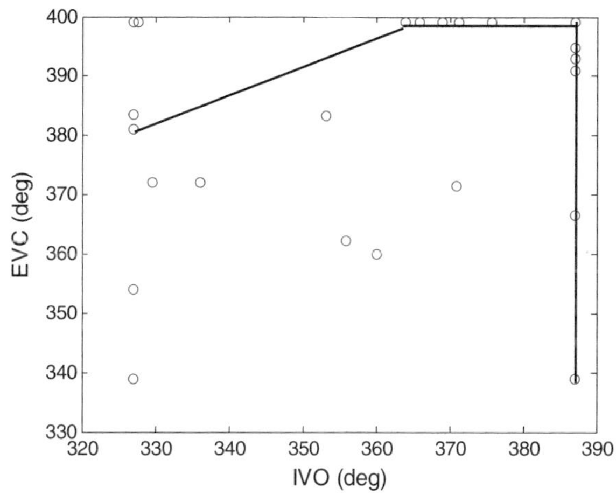

Fig. 15.3 Identifying a smooth CAM track through clustering the optimized CAM settings

algorithm runs fast and converges at all conditions, except for a high speed low load target that is not a feasible operating point for this engine. It takes on average 12 iterations at each point.

The results of optimal CAM settings across all the speed/load points can be summarized in Fig. 15.3. By clustering the CAM settings, we can obtain the CAM track that can be implemented in production strategy so the actual CAM positions are not commanded to change significantly during normal operations.

15.5 Case Study 2: On-board Fuel Consumption Optimization in Series HEV

As another case study, we consider finding online engine operating points that minimize fuel consumption for the specified constant engine/generator power request, P_{eng}^{req} in a series hybrid electric vehicle. We note that in hybrid electric vehicles, the battery power can complement engine power during such a learning phase to ensure the requested time-varying electric motor power at the wheels; thus learning of optimal operating points can be performed on-board without causing driver disturbance. As in [10], an estimate of the engine power can be used that is generated by an input observer based on the measurements of the engine speed and generator power, where the latter is estimated based on the measurements of the current in the generator. The battery power output is controlled as to ensure that the wheel power matches the driver demand by complimenting the estimated generator power. After learning is completed, the optimal operating points can be used in the power smoothing strategy proposed in [5, 15].

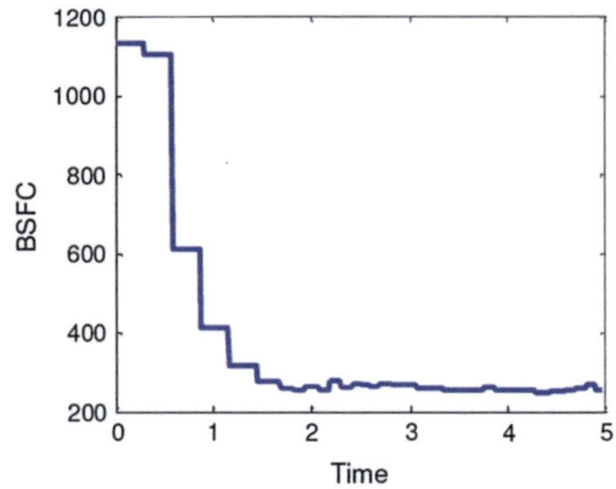

Fig. 15.4 The time history of BSFC improvement in a series HEV

We consider the case when the engine has to meet the target of fixed engine power, P_{eng}^{req} while minimizing the BSFC. The vector u consists of throttle angle, spark timing, engine speed, intake valve opening and exhaust valve closing timing. Note that the engine speed is varied by the optimization algorithm, and therefore, the actual torque target varies with it, as shown in Fig. 15.5. With the assumption of constant P_{eng}^{req}, the application of the Jacobian Learning based optimization algorithm of Sect. 15.3 yields the responses shown in Figs. 15.4, 15.5, 15.6, 15.7. The plots show the power target is tracked and BSFC is minimized, while all actuators converge to the optimal values.

In order to reduce battery State of Charge fluctuations during on-board learning and reduce learning time, it is desirable to track the fuel consumption minimizer in slow transients, i.e., when P_{eng}^{req} is slowly-varying. To realize such a functionality, predictor-corrector algorithms are considered next.

15.6 Predictor-Corrector Algorithm

The updates in Sect. 15.3 can be generalized to the case of the parameter-dependent optimization (15.2), (15.3) when parameters are slowly varying. With the surrogate model based on estimated Jacobians, we assume that

$$\Delta y = \hat{J}_u \Delta u + \hat{J}_p \Delta p. \tag{15.14}$$

This leads to the following update in the case of no control bounds,

$$\begin{aligned} u(k) = u(k-1) + \left(\hat{J}_u^{\mathrm{T}}(k-1)\Omega \hat{J}_u(k-1) + \Gamma \right)^{-1} \hat{J}_u^{\mathrm{T}} \Omega \\ \times \Big(- \hat{J}_p(k-1)(p(k) - p(k-1)) + y_t - y(k-1) \Big). \end{aligned} \tag{15.15}$$

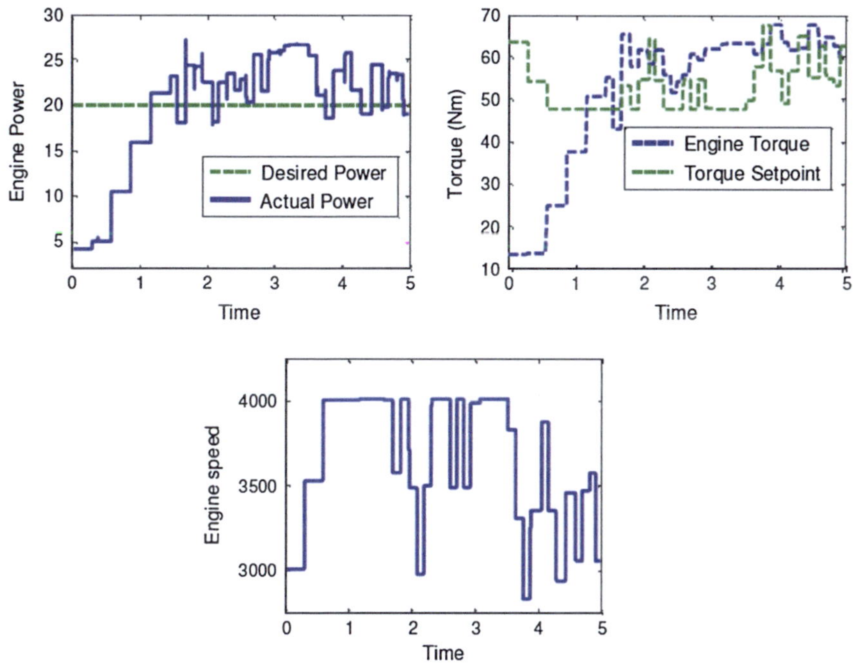

Fig. 15.5 The time history of engine power (*top*, *left*), engine torque and torque set-point (*top*, *right*), and engine speed (*bottom*). Engine torque set-point is generated to achieve desired power

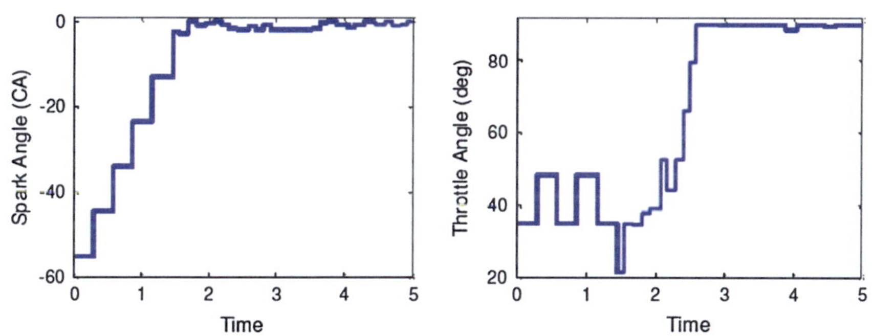

Fig. 15.6 The time history of spark timing and throttle position

Note that the update (15.15) incorporates prediction for the change in the solution due to the parameter change which is due to the term, $\hat{J}_p(k-1)(p(k)-p(k-1))$. The Jacobian $\hat{J}_p$ is estimated with the Kalman filter using the same approach as for $\hat{J}_u$ in Sect. 15.3.

In the case of constraints,

$$y_c = g(u, p) \leq 0, \tag{15.16}$$

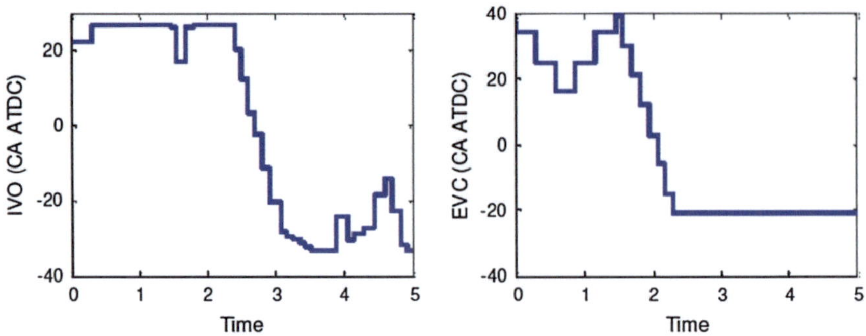

Fig. 15.7 The time history of intake valve opening and exhaust valve closing

Jacobian Learning can be applied also to constraint learning based on a surrogate constraint model,

$$\Delta y_c = \hat{g}_u \Delta u + \hat{g}_p \Delta p. \tag{15.17}$$

By denoting the vector of active constraints by superscript a, the update is based on solving a system of linear algebraic equations,

$$\begin{aligned}
&\hat{J}_u^{\mathrm{T}}(k-1)\Omega(y(k-1)-y_t) + \hat{J}_u^{\mathrm{T}}(k-1)\Omega \hat{J}_p(k-1)(p(k)-p(k-1))\\
&\quad + (J_u^{\mathrm{T}}(k-1)\Omega J_u(k-1) + \Gamma)(u(k)-u(k-1))\\
&\quad + (\hat{g}_u^a(k-1))^{\mathrm{T}}(\lambda^a(k)-\lambda^a(k-1)) = 0\\
&\hat{g}^a(u(k-1), p(k-1)) + \hat{g}_u(k-1)(u(k)-u(k-1))\\
&\quad + \hat{g}_p^a(k-1)(p(k)-p(k-1)) = 0,
\end{aligned} \tag{15.18}$$

with respect to $u(k)$ and the vector of Lagrange multipliers of the active constraints, $\lambda^a(k)$. Care needs to be taken to handle changes in the set of active constraints. The change in the set of active constraints can be detected if as a result of the update (15.18) some of the components of the vector $\lambda^a(k)$ become zero or negative or if some of the constrained outputs predicted according to (15.17) change sign. In such a case, (15.18) is applied with $\tilde{p}(k) = \alpha p(k) + (1-\alpha)p(k-1)$ replacing $p(k)$, where $0 < \alpha < 1$ corresponds to the first predicted change in the sign of any of Lagrange multipliers or constrained outputs (15.16); the corresponding $\tilde{\lambda}^a(k)$ and $\tilde{u}(k)$ are determined based on (15.18); (15.18) is re-configured for a different set of active constraints; and the process is repeated with $p(k-1)$ replaced by $\tilde{p}(k)$, $\lambda^a(k-1)$ replaced by $\tilde{\lambda}^a(k)$ and $u(k-1)$ replaced by $\tilde{u}(k)$.

As another extension, the surrogate model (15.14) can be replaced by a more general nonlinear model and the minimization problem (15.2) can involve an arbitrary, sufficiently smooth function $Q(u, p)$, i.e., not necessarily (15.3). A more general form of the predictor-corrector algorithm for minimizing the function $Q(u, p)$ with respect to u in the unconstrained case can be based on Newton's method,

$$u(k) = u(k-1) + \hat{Q}_{uu}^{-1}(k-1)(-\hat{Q}_u(k-1) - \hat{Q}_{up}(k-1)(p(k) - p(k-1))), \quad (15.19)$$

where the partial derivatives are evaluated at $u(k-1)$ and $p(k-1)$. The update (15.19) requires that the Hessians of Q be estimated on-line. Under appropriate assumptions and if gradients and Hessians of Q are accurately known, constants $\alpha > 0$ and $\beta > 0$ can be found such that the tracking error of the minimizer satisfies a relation of the form,

$$||u(k+1) - u^*(p(k+1))|| \leq \alpha ||u(k) - u^*(p(k))||^2 + \beta ||p(k+1) - p(k)||. \quad (15.20)$$

This relation suggests strong, quadratic convergence properties to the minimizer if $p(k)$ stays constant and strong tracking properties of the minimizer if the parameter $p(k)$ is slowly-varying. In the constrained case, the Newton's method can be applied to the root finding problem resulting from KKT conditions augmented with an extra equality constraint. We discuss a case study where this approach was followed in the next section and leave it to future work to explore its details and properties further.

15.7 Case Study 2 (Cont'd): On-board Fuel Consumption Optimization in Series HEV

The application of (15.19) to identification of Optimal Operating Points (OOP) line, i.e., engine speed and engine torque values that minimize BSFC for a given P_{gen}^{req}, is shown in Figs. 15.8, 15.9. Here u is the engine speed and $p = P_{gen}^{req}$ is slowly varying as shown in Fig. 15.8. The advantage of handling slowly varying P_{gen}^{req} during learning is that battery SoC fluctuations are reduced. The function Q is the BSFC and the surrogate model for it is assumed to be second order in the two variables (engine speed u and $p = P_{gen}^{req}$). Small amplitude random excitation has been added to $u(k)$ and $p(k)$ to facilitate the identification of surrogate model parameters. The algorithm is able to rapidly converge to OOL and follow it as the generator power is varied. See [10] for further details. We note that a version of HEV model in Simulink was used for experimentation with (15.19) differently from GT-Power model which we used for the results in Sect. 15.5. The OOL is required for implementation of the power smoothing strategy [5, 15], and as our results demonstrate, it can be learned on-board without creating driver disturbance.

15.8 Concluding Remarks

In this chapter we have considered an iterative optimization approach that combines the steps of real-time learning of a surrogate model and then performing optimization with respect to the identified model. In the case when the surrogate model is given in a

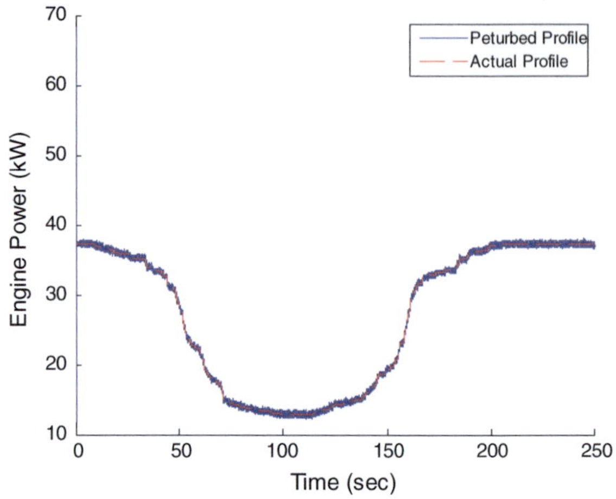

Fig. 15.8 The time history of engine/generator power

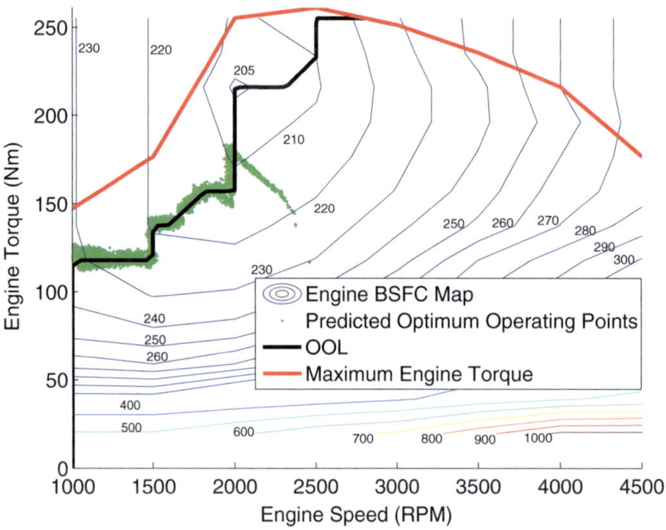

Fig. 15.9 The engine speed and torque superimposed on engine map

linear incremental form, we refer to such a method as Jacobian Learning (JL)-based. The algorithm derivation has been presented and we demonstrated its application to a case study of mapping an engine represented by a detailed GT-Power simulation model. Results have shown that the proposed algorithm can be used as an efficient, robust, and generic tool for fast engine calibration optimization in the case of fixed operating conditions providing the optimal actuator settings in less than 20 iterations for the engine model considered. Since the surrogate model is learned from data and no assumption on the type of engine model is made, the methodology can be applied to virtually any type of engine, including diesel engines and aircraft gas turbine engines.

We have also presented a case study showing that such an approach can be applied on-board in an HEV, and considered its various extensions, including a potentially faster predictor-corrector form of the algorithm, and the treatment of more general types of surrogate models. These will be further considered in the future work.

Acknowledgments The authors wish to thank Rohit Gupta and Heyongjun Park of the University of Michigan and John Michelini of Ford Research and Advanced Engineering for their contributions to research reflected in this chapter.

References

1. Astrom K, Wittenmark B (1989) Adaptive control. Addison-Wesley, Reading
2. Buskens C, Maurer H (2001) Sensitivity analysis and real-time optimization of parametric nonlinear programming problems. In: Grotschel M, Krumke M, Rambau J (eds) Online optimization of large scale systems. Springer, Berlin
3. Cameo Powertrain Calibration Tool (2013). https://www.avl.com/calibration
4. Chen W-K (1993) Linear networks and systems. Wadsworth, Belmont, pp 123–135
5. Di Cairano S, Liang W, Kolmanovsky IV, Kuang ML, Phillips AM (2013) Power smoothing energy management with applications to a series hybrid powertrain. IEEE Trans Control Syst Technol 21(6):2091–2103
6. Fiacco AV (1983) Introduction to sensitivity and stability analysis in nonlinear programming. Mathematics in science and engineering, vol 165. Academic Press, New York
7. Filev D, Larsson T, Ma L (2000) Intelligent control for automotive manufacturing—Rule based guided adaptation. In: Proceedings of IEEE conference on IECON'00, Nagoya, Japan, pp 283–288
8. Gorinevsky D (1997) An approach to parametric nonlinear least square optimization and application to task-level control. IEEE Trans Autom Control 42(7):912–927
9. Guddat J, Vasquez FG, Jongen HT (eds) (1990) Parametric optimization: singularities, path following and jumps. Wiley, New York
10. Gupta R, Kolmanovsky IV, Wang Y, Filev DP (2012) On-board learning-based fuel consumption optimization in series hybrid electric vehicles. In: Proceedings of 2012 American control conference, Montreal, Canada, pp 1308–1313
11. Haskara I, Ozguner U, Winkelman J (2000) Extremum control for optimal operating point determination and set-point optimization via sliding modes. ASME J Dyn Syst, Meas Control 122:719–724
12. Hrovat D, Jankovic M, Kolmanovsky I, Magner S, Yanakiev D (2010) Powertain control. In: Levine W (ed.) IEEE Controls Handbook, 2nd edn. CRC Press, Boca Raton
13. Jankovic M, Magner S (2004) Optimization and scheduling for automotive powertrains. In: Proceedings of the 2004 American control conference, pp 4054–4059
14. Jankovic M, Magner S (2006) Fuel economy optimization in automotive engines. In: Proceedings of the 2006 American control conference, Minneapolis, MN, pp 1191–1196
15. Konev A, Lezhnev L, Kolmanovsky IV (2006) Control strategy optimization for a series hybrid vehicle. SAE, World Congress, Detroit, MI
16. Liung L (1987) System identification—Theory for user. Prentice Hall, Englewood Cliffs
17. Lyantsev OD, Breikin TV, Kulikov GG, Arkov VY (1993) On-line optimization of aero engine control system. Automatica 39:2115–2121

18. Malikopoulos A, Papalambros PY, Assanis DN (2007) A learning algorithm for optimal internal combustion engine calibration in real-time. In: Proceedings of the ASME 2007 international design engineering technical conference and computers and information in engineering conference, IDETC/CIE, Las Vegas, Nevada
19. Malikopoulos A, Papalambros PY, Assanis DN (2007) Real-time, self-learning optimization of diesel engine calibration. In: Proceedings of 2007 fall technical conference of the ASME internal combustion engine division, ICEF2007-1603, Charleston, South Carolina, USA, 14–17 Oct 2007
20. Pan Y, Ozguner U (2004) Extremum seeking control of a variable valve timing engine. In: IFAC symposium on advances in automotive control, Salerno, Italy, pp 173–178
21. Popovic D, Jankovic M, Magner S, Teel A (2006) Extremum seeking methods for optimization of variable cam timing engine operation. IEEE Trans Control Syst Technol 14(3):398–407
22. Quoc TD, Savorgnan C, Diehl M (2012) Adjoint-based predictor-corrector sequential convex programming for parametric nonlinear optimization. SIAM J Optim 22(4):1258–1284
23. Ryan J, Speyer J (2010) Peak-seeking control using gradient and hessian estimates. In: Proceedings of American control conference, pp 611–616
24. Tychonoff AN, Arsenin VY (1977) Solution of Ill-posed problems. Winston and Sons, Washington
25. Zhang C, Ordonez R (2006) Non-gradient extremum seeking control of feedback linearizable systems with application to ABS design. In: Proceedings of the 45th IEEE conference on decision and control, San Diego, CA, pp 6666–6671

Chapter 16
Online Design of Experiments in the Relevant Output Range

Nico Didcock, Andreas Rainer and Stefan Jakubek

Abstract Nonlinear system identification requires informative data obtained from experiments in order to parameterise a model of the underlying process. As an example for the automotive industry, good models for NO_x and smoke emissions are required to effectively calibrate modern combustion engines. With continuously increasing complexity in terms of the number of variation channels available in the engines the experimental effort provides a growing challenge for efficient calibration. Design of Experiments (DoE) refers to optimal excitation of the system in order to maximise the knowledge gained for a process under investigation from a limited amount of measurements. We introduce a methodology that omits measurements unrelevant for calibration via pre-specification of restrictions on the inputs *as well as the outputs*. Output restriction to a certain *target region* is obtained via a supervising online model that is trained during the workflow. The distribution of input samples obtained via this method is non-uniform over the pre-image of the target region. The effectiveness of this concept is demonstrated for the modeling of NO_x and smoke emissions of a diesel engine.

N. Didcock · S. Jakubek
Christian Doppler Laboratory for Model Based Calibration Methodologies, Institute of Mechanics and Mechatronics Division of Control and Process Automation, Vienna University of Technology, Wiedner Hauptstrasse 8-10/325/A5, 1040 Vienna, Austria
e-mail: nico.didcock@tuwien.ac.at

S. Jakubek
e-mail: stefan.jakubek@tuwien.ac.at

A. Rainer (✉)
AVL List GmbH, Hans-List-Platz 1, 8020 Graz, Austria
e-mail: andreas.rainer@avl.com

H. Waschl et al. (eds.), *Optimization and Optimal Control in Automotive Systems*, Lecture Notes in Control and Information Sciences 455,
DOI: 10.1007/978-3-319-05371-4_16,

16.1 Introduction

Stationary optimisation of combustion engines is an important task especially at the beginning of a calibration project. For heavy-duty engines it at the same time provides the basis for stationary legislative testruns (e.g. ESC, the European Stationary Cycle). For transient emission cycles representative stationary operating points form the base for dynamic modeling procedures.

The increasing number of control parameters brings along new challenges for data collecting procedures. Test-plans for the system exciting parameters are required to minimise the amount of measurements while maximising the informative value of the experiments. The goal of these experiments is usually to gain knowledge on emissions such as NO_x and smoke, or other variables, e.g. fuel consumption. Legislative constraints however usually restrict emissions below some bounds, making large amounts of combinations of control variables redundant for the experiment. In this context online DoE is considered to be a very effective method in order to accelerate the workflow of engine calibration. The term online refers to an adaptive process where the experiment is planned during the workflow according to the current knowledge obtained by the measurements. More specifically, an online evolving model is used as it has been proposed in [1]. Here, the workflow is accompanied with the online training of efficient methods to estimate the input-output relation, see Fig. 16.1 for an illustration. This enables us to restrict a number of outputs to predefined, *custom output ranges* (COR) [2]. In each iteration step we invert the currently trained model to find the pre-image of the COR regions. The design for future experiments is then chosen to some distribution over this pre-image. Maximin-Euclidian-distance designs would converge to a uniform distribution. Designs according to the input–output distance lead to a distribution that is less dense for flat output behavior. The model is re-trained after each measurement, thereby improving the model quality in areas that are most relevant for calibration, see [3].

The procedure is demonstrated on verification results for NO_x and smoke emissions as well as fuel consumption. In Sect. 16.2 we introduce the problem we faced using the state of the art DoE approaches. We give a precise description of our methods in Sects. 16.3 and 16.4. Sections 16.5 and 16.6 demonstrate model improvements after applying the so called COR DoE procedure.

16.2 State of the Art Development Approach

To assess the benefits of the new method, which will be described in the following sections, consider the following example of the a test plan evaluation of a modern passenger car diesel engine in a single operating point. One standard optimisation task on such common-rail engines is to optimise the fuel consumption while keeping the legislative limits for NO_x and soot emissions within the prescribed driving cycle.

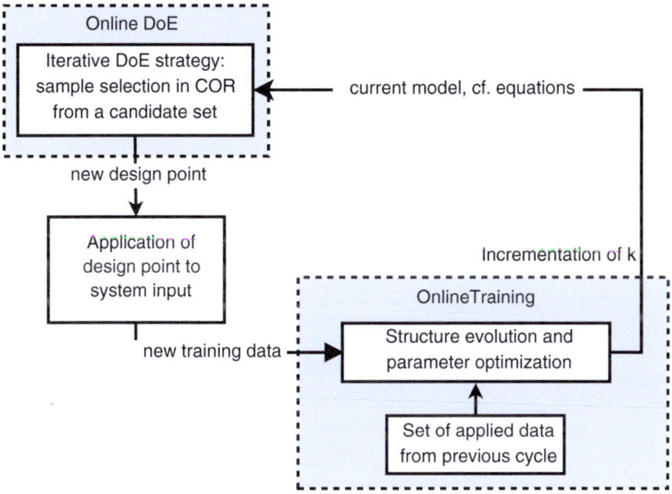

Fig. 16.1 Interaction between online DoE and online training for COR

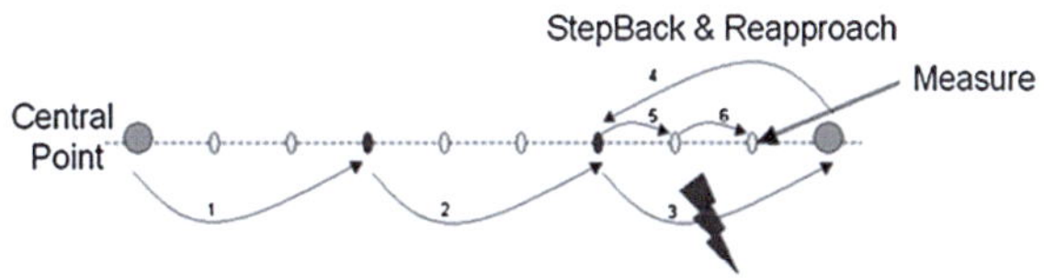

Fig. 16.2 AVL CAMEO™ DoE screening procedure

The optimisation was performed w.r.t. to five varying control parameters (variation parameters henceforth) within the ECU (engine control unit):

- Rail Pressure
- Main Injection Timing
- Boost Pressure
- EGR (Exhaust Gas Recirculation)
- Swirl Position.

To take into account constraints on the drivability, such as component temperatures, cylinder peak pressure, combustion stability etc., we use the DoE-Screening technique [4]. Here, the engine starts from a stable center point and changes the ECU-settings step by step in starlike directions until system limits are reached or a limit violation occurs. In case that a limit occurs, the strategy offers different reactions and measures the variation point as near as possible to this limit, see Fig. 16.2 for an illustration. The ecu variations within the convex hull of the drivable points and possibly within some external input restrictions then constitute the candidate input set. The task for the DoE strategies is now to find the ecu variations within this set that have maximum informative value for our purposes. Our measurements are

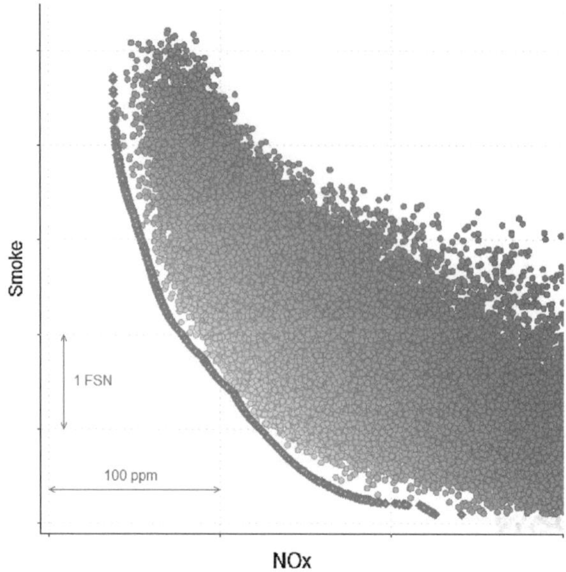

Fig. 16.3 NO_x and soot output estimation for random and pareto optimal inputs

double-checked via repetition measurements. This guarantees that we can assure the quality of our measurement devises and a stable workflow of the engine.

As state of the art DoE method a D-optimal design (see e.g. [5]) was calculated for the five variation parameters and measured using AVL CAMEO™, see [6]. Using these 59 variation points (plus 8 repetition points) output models were built and the output estimates for NO_x and soot were calculated, see Fig. 16.3. The figure shows how the output behaves for random sampling of input variations. Obviously, we are most interested in the pareto optimal values in the bottom left area. A large amount of input variations is therefore negleglible for calibration purpose and should be avoided during the test run.

We performed additional measurements to verify the model quality for the optimised input variations. As can be seen in Figs. 16.5 and 16.6, the fuel consumption and NO_x emissions for the verification variations (solid, gray points) can be predicted with the same quality the training data (circles). The achieved residuals are at the same time comparable with the variance of the measured repetition points (solid, black points) which indicates the model error is mainly measurement error. However, evaluating the smoke emissions, it can be seen in figure Fig. 16.6 that for some optimisation points the model underestimates the smoke values significantly. Since these regions are highly relevant for calibration the model quality needs to be particularly improved here. This motivated the development of a strategic DoE that aims to improve model quality for relevant output areas. We performed additional measurements to verify the model quality for the optimised input variations. As can be seen in Figs. 16.5 and 16.4, the fuel consumption (BSFC) and NO_x emissions for

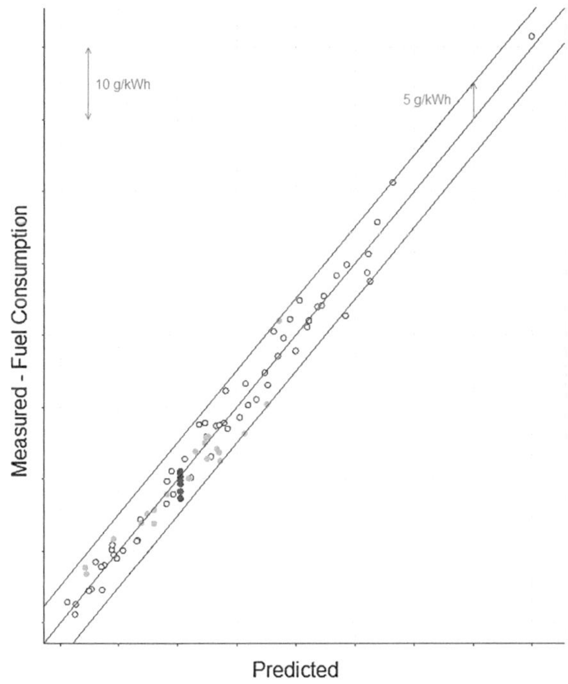

Fig. 16.4 Measured versus predicted plot for break specific fuel consumption (*BSFC*) after a D-optimal design

the verification variations (solid gray points) can be predicted with the same quality the training data (circles). The achieved residuals are at the same time comparable with the variance of the measured repetition points (full, black points) which indicates the model error is mainly measurement error. However, evaluating the smoke emissions, it can be seen in figure Fig. 16.6 that for some optimisation points the model underestimates the smoke values significantly. Since these regions are highly relevant for calibration the model quality needs to be particularly improved here. This motivated the development of a strategic DoE that aims to improve model quality for relevant output areas.

16.3 Mathematical Background of the COR Design

We now present a general framework for model based DoE where the estimated model is to be kept as general as possible. We commence with the task to draw a finite sample from an input space $\mathscr{U}$. We are interested in the estimation of the map $f : u \mapsto y$ for some one dimensional output such as NO_x , soot or BSFC. Let u_k

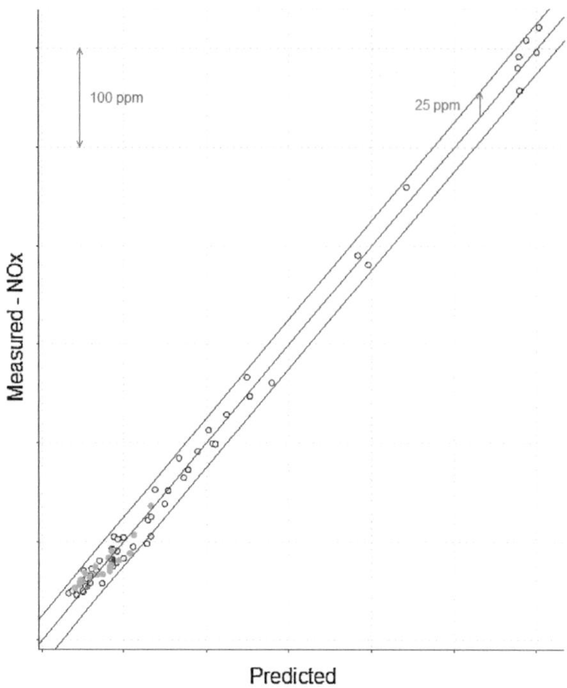

Fig. 16.5 Measured versus predicted plot for NO_x after a D-optimal design

and y_k denote the measurements of inputs and outputs, respectively. We assume that the output has the form

$$y_k = y(u_k) = f(u_k, \theta) + \varepsilon(u_k) = f_k + \varepsilon_k$$

Let f_k denote the *deterministic component*, depending on the parameters θ, and ε_k the non-systematic measurement error. The parametric form of the model component has to be chosen appropriately for the data. Examples are the linear-in-parameters regression model $f_k = \varphi(u_k)^T\theta = \varphi_k^T\theta$ for some regression vector φ and parameters θ, neural networks or local linear models, see e.g. [7]. Noise is assumed to be independent and identically distributed with zero mean and equal variance σ^2. The model component is estimated from a sample $S = [s_1 \dots s_n]$ where n denotes the number of samples and s_k denotes the index for the kth sample. The error covariance depends heavily on the input design $U_S = [u_{s_1} \dots u_{s_n}]$ where we use the equivalent notations y_S and ε_S. The output estimator $\hat{y}_k = f(u_k, \hat{\theta})$ depends on the parameter estimator $\hat{\theta}$. The parameter variance and therefore the model variance depend heavily on the inputs u_{s_k}. In the OLS/GLS model the parameter covariance depends exclusively on the inputs.

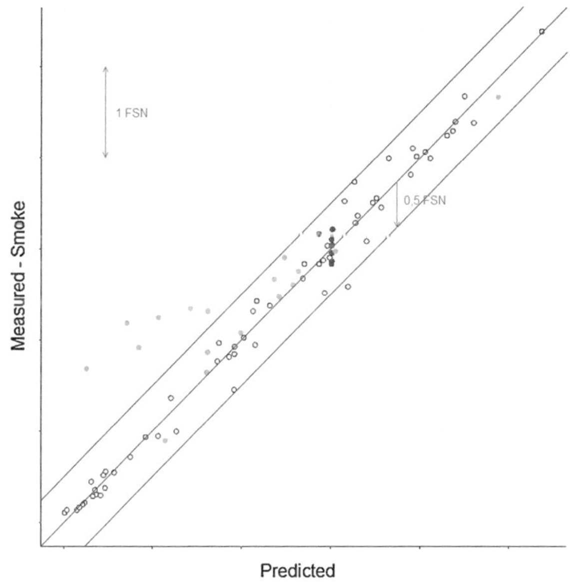

Fig. 16.6 Measured versus predicted plot for smoke emissions after a D-optimal design

16.3.1 A Local Model Architecture

We briefly present a local model architecture that we have used for the estimation of the mappings in the application example presented. We use a Takagi-Sugeno fuzzy model architecture as in [8]. Locally, in a region indexed by j, the output for u_k is modeled as a full quadratic function of the input, the local estimate shall be denoted $\hat{y}_{j,k} = \hat{y}_j(u_k)$. The validity function $\eta_{j,k} = \eta_j(u_k)$ denotes the proportion of the local model estimate relative to the overall estimate which is then composed as the weighted sum of I local outputs,

$$\hat{y}_k = \sum_{j=1}^{I} \eta_{j,k} \hat{y}_{j,k}$$

The coefficients θ_j according to the regressor of the jth local model can be estimated as the WLS estimator weights η_j. The weights correspond to a partitioning of the input space in regions that are dominated by one local model each. With a growing number of observations the parameters of the model are updated and, if necessary, the number of local models is increased incrementally, see [8] for details. Note, however, that our proposed concepts are not restricted to this specific type of model.

16.3.2 State of the Art Designs

Ideal designs of input excitation signals maximise the information content of our experiments. However it is not obvious what should be regarded as the information content. Model based DoE maximises the information gain w.r.t. the precision of the estimated parameters of some model structure, see [9] and [10] for details. Model based DoE is basically related to Fisher's concept of information. The Fisher Information matrix in its general form is defined as

$$I_F(\theta, S) = \mathcal{E}\left\{\left(\frac{\partial \log p(Y_S|U_S,\theta)}{\partial\theta}\right)^T \left(\frac{\partial \log p(Y_S|U_S,\theta)}{\partial\theta}\right)\right\}$$

Where $Y_S = [y_{s_1}...y_{s_n}]$ denotes the sample output. Its inverse, the Cramer-Rao-Bound, is an upper bound for the covariance of any unbiased parameter estimator. Optimal designs are based on optimising some functional on the Fisher matrix $S_F = \arg\max_S J\,(I\,(\theta, S))$ such as the determinant (D-optimality), the trace (A-optimality) or the maximum value of the diagonal entry (G-optimality). In [11] it is pointed out that D-optimal designs tend to lie on boundaries which we find mischievous since the boundaries of our inputs tend to become undrivable. In comparison, so called maximin designs use a distance measure to the design space as the informative value or the experiment, e.g. the Euclidian distance,

$$d_2(u_k, u_j) = \sqrt{(u_k - u_j)^T (u_k - u_j)}$$

Maximin designs then fulfill

$$S_{\max\min} = \arg\max_S \min_{u_k \neq u_l \in U_S} d_2(u_k, u_l)$$

See [11] for a general introduction to maximin designs. In particular it should be noted that Euclidian maximin distance designs converge to a uniform distribution over the input space. In [12] the equivalence between D-optimality and maximin designs is shown when the distance measure is chosen appropriately.

16.3.3 Online Procedures

Our method differs substantially from these state-of-the-art DoE procedures. We are not primarily interested in the training a model or a uniform coverage of the input space. Our aim is to collect testbed data that is relevant for application and we therefore need a procedure that incorporates knowledge on outputs such as NO_x emissions or exhaust temperature that are, more likely than inputs, constrained to lie in some region of interest. In [13–16], DoE methods w.r.t. relevant output levels are presented. Here, one is interested in a certain (single) level set of the output, e.g. where some system failure occurs. Model accuracies are improved for this critical value of interest. We do not restrict our target outputs to one specific value but to an applicable set of values. Moreover, the distribution of the design points in the input space is not asymptotically uniform, but depends on the output as well. With our method we obtain measurements that are more dense where the output is less flat.

We train our model with measurements from a standard initial design such as a central composite designs (CCF) and a *DoE screening* procedure. These methods are useful for determining the input space as the drivable area of the engine. After the initial design we run an online design procedure where the measurements influence the position of the following design points. At each time step we augment the sample S_t by a number of observations such that $S_t \subset S_{t+1}$ and $U_{S_t} \subset U_{S_{t+1}}$. The proposed designs in Sect. 16.3.2 do not take into account the measurements Y_S since the distance criteria in general depend on the inputs only. We find it important for our design to leave out regions that are of no particular interest. Take for instance the modeling of smoke particles then we are less interested in regions where the emission rate is low but model accuracy in high emission areas becomes crucial.

16.4 Design Strategies

Although maximin designs in the input space are easy to calculate they suffer from the drawback of entirely ignoring the structure of the model. Our work involves the estimation of various output components of combustion engines where we found this strategy inefficient due to the existence of trivial output areas that need not be explored exhaustively. In order to reduce the number of observations to a minimum we want to reject experiments that yield little information. We focus on two extensions to state of the art methods. First, we want to omit design points in areas where the output becomes trivial to avoid unnecessary costs. When the output is comparably flat we need fewer observations than when the mapping is more complex. Second, bearing legislative constraints in mind we constrain the experiment to yield measurements in a relevant region. In the following we discuss the two extensions in detail.

16.4.1 A Distance Criterion in the Product Space

We modify the distance criterion according to [17]. We use a distance design with respect to the inputs as well as the output. The method proved to measure the output dynamics more efficiently than other concepts. Let X denote the product space of inputs and estimated outputs $U \times \hat{y}$ with elements $x_k = (u_k, \hat{y}_k) \in X$ and X_S the set which first components are the sample inputs. We define the *input–output distance* as

$$d_{IO}(x_k, x_j) = \sqrt{(x_k - x_j)^T (x_k - x_j)}$$

and the input output maximin design as

$$S_{IO-\max\min} = \arg\max_{S} \min_{x_k \neq x_l \in X_S} d_{IO}(x_k, x_l)$$

This simple extension helps to concentrate the design in areas where the output is less flat. Take for example an input region where the output grows fast with respect to the input, then the distances will become comparably large. Therefore maximin distances will concentrate on regions with higher output escalation, that is, where the output either increases or decreases relatively fast. By reducing the number of observations made in trivial regions we raise the relevant model performance relative to the number of measurements.

16.4.2 The Custom Output Region (COR)

Second, we try to force the output measurement to lie in a given region of interest. The calibration of combustion engines is subject to fulfill legislative constraints and we therefore find high estimation precision required in the drivable areas with relevant emission rates. Again, the design procedure depends on the estimated output since we need to have an idea where the relevant emissions are located. Running an online strategy we use the latest trained model to identify the inverse image of the output region of interest. We force the design to be placed in this region in the next design step. We commence from the candidate in put set $u_k \in U_{\text{cand}}$ e.g. from the previously described screening procedure, a model estimate $\hat{y}_k$ as well as the Custom Output Region, for example an interval COR $= [y_{\min}, y_{\max}]$. Next, we compute the candidate set that is expected to lie in the COR and result in the final candidate set $U_{\text{cand,COR}} = \{u_k | u_k \in U_{\text{cand}}, \hat{y}_k \in \text{COR}\}$. Finally, we calculate an input-output-maximin design strategy over this set.

16.4.3 The iDoE Strategy

The methodology is illustrated on a simple academic example in Fig. 16.7. Large parts of the input space map to a constant output that—by assumption—is not of interest. In contrast, the relatively small areas where there is more fluctuation in the output are considered highly relevant. A standard Maximin DoE Procedure is compared to the IO-Maximin DoE procedure after the online application of ten design points. The standard space filling approach fails to measure in the areas where the output shows its characteristic behavior. Contrarily, if we apply a COR DoE and define the COR region between 0.2 and 1 the points are distributed more in the non-trivial areas, providing the model is good enough to estimate a precise pre-image of the COR.

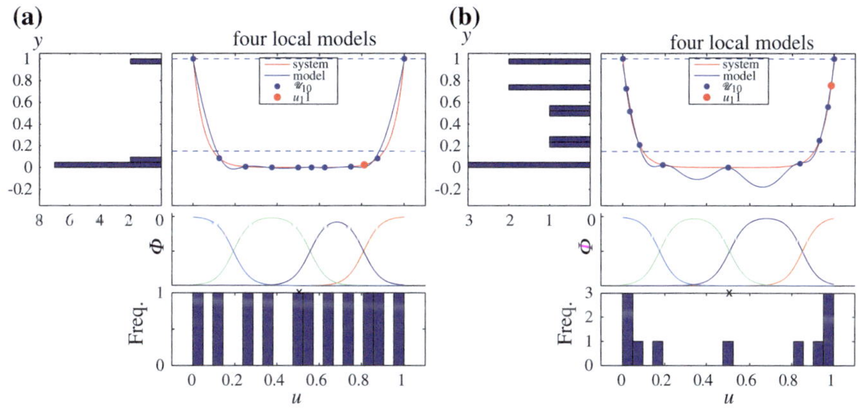

Fig. 16.7 Comparison of model and process after the online application design points with different DoE

What can also be observed is that the distribution of the applied points is not uniform in the input space. It is due to the modified distance criterion that design points are applied evenly on the *output graph*.

We call the combination of these two extensions the iterative DoE (iDoE) strategy, see [18]. The term iterative stresses the fact that the design sequence is updated after every measurement through the estimation update of the output $\hat{y}$. Although both extensions - the augmented distance criterion as well as the COR procedure—work independently, we combine the tools to generate a powerful strategy that

- avoids unnecessary repetition points,
- favors areas with less flat output behavior
- generates outputs in a desired output range.

16.5 Improved Development Approach using the COR Design

We now demonstrate the usefulness of our methodology using the same diesel engine as in Sect. 16.2. Since model building and therefore our online procedure requires data, the method needs an initial design. A CCF (Central Composite Faced) design which consisted of 42 variation points (plus 7 repetition points) was chosen. This initial design was measured using the screening strategy, giving one measurement at the variation point or near the drivability limit plus a second measurement on the way back to the center point. Based on these 91 measurements a model was calculated and online re-iterated during the iDoE test. The necessary measurement time for the automated test was approximately 10 % higher compared to the state of the art procedure. see Table 16.1. But after the application of only ten iterative design points an acceptable model quality can be achieved for all relevant output regions.

Table 16.1 Measurement time comparison for D-optimal design and COR DoE: additional time spent on the online procedure can be seen as negligible

	# Variation points	Time/measurement (min)	Total measurement time (min)
D-optimal	59	4	236
Repetition points	8	1	8
			244
Startdesign	42	4	168
2nd measurement	42	1	42
Repetition points	7	1	7
iDoE	10	4	40
2nd measurement	10	1	10
			267

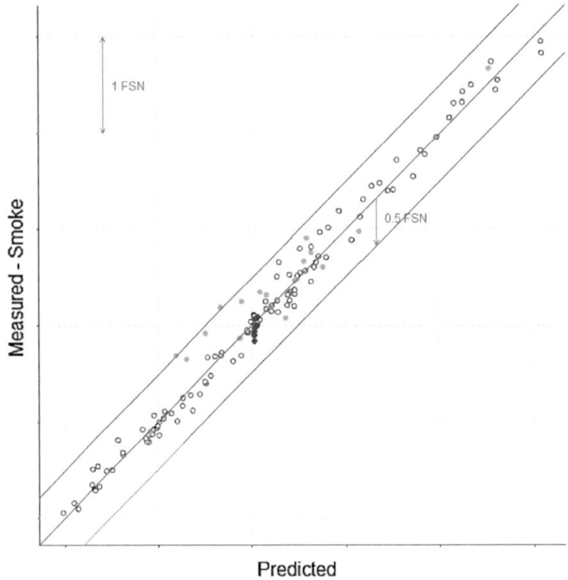

Fig. 16.8 Measured versus predicted plot for smoke emissions after a COR design

Using the state of the art D-optimal design the verification measurements of the optimised points resulted in residuals of up to 1.5 FSN (see Fig. 16.6). Using iDoE and COR design, the highest residuals of the same verification measurements could be reduced by 66 % to a maximum of 0.5 FSN (see Fig. 16.8) whereas the repetition measurement of the (more stable) center point showed a measurement deviation of ±0.2 FSN.

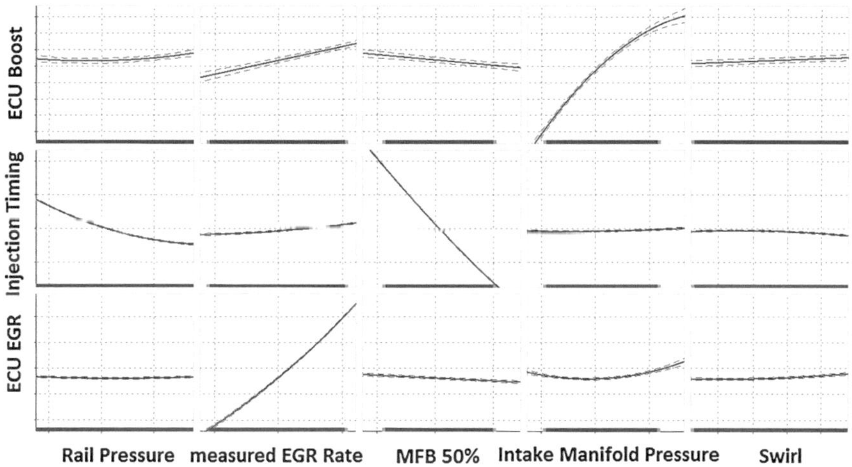

Fig. 16.9 Modeled ECU variation parameters

16.6 Further Improvement

Further improvement of the model accuracy for all models can be achieved when a D-optimal initial design is chosen. For this example 47 measurements were performed which is the suggested amount of points for 5 variation parameters expecting a 3rd oder polynomial behavior.

Model accuracy can be improved if the ECU variation parameters are not directly used as input channels for model estimation. Instead, we use channels that are, from a physical understanding of the combustion process, more likely to explain the response variables. E.g. the estimated ECU channels for boost-pressure, injection timing and EGR were exchanged by the measured intake-manifold pressure, the MFB 50 % (mass fraction burned 50 %) which can be measured with an indicating device (AVL IndiCom™) and the EGR rate, measured with an emission bench. The necessary values for the according ECU channels are still available after the optimisation in case models are calculated for them. This means that these original ECU variation parameters are not seen as variation channels any more but as response channels during the optimisation. Usually already simple 2nd order polynomial model approaches can fit their behavior well, this is illustrated in Fig. 16.9.

Finally, the model quality can be increased continuously when adding additional iterative design points as described in Sect. 16.5. As a measure of model fit we calculate a *leave one out* estimator, denoted R^2 pred (predicted). During the testrun, the R^2 pred does not change significantly which shows that the model structure is flexible enough for fitting the measurement data, see Fig. 16.10. A second measure, the root mean squared error on verification points, denoted RMSE*, decreases the more iterative points are measured. We see that the model quality in the relevant area

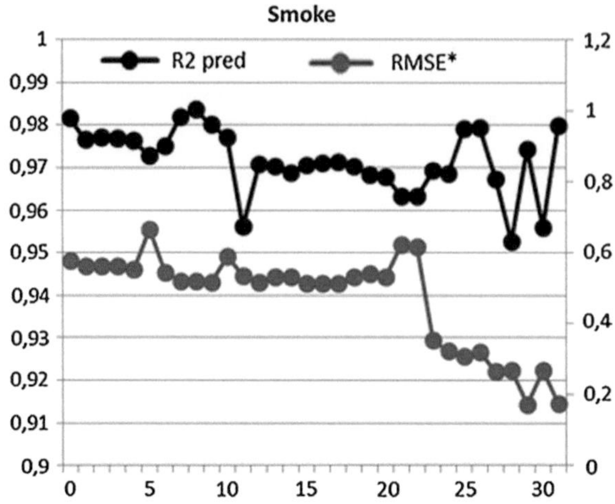

Fig. 16.10 R^2 and RMSE* for modeled smoke calculated for different numbers of iDoE points

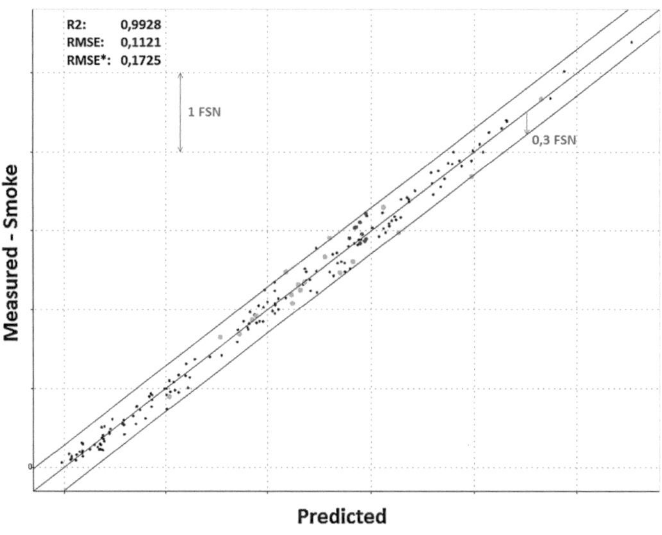

Fig. 16.11 Model quality for smoke using physical variation parameters and 30 iterative design points

becomes more accurate during the test. In this example, 30 iterative design points would be sufficient to decrease the RMSE* to less than 0.2 FSN.

The measured versus predicted plot is shown in Fig. 16.11. The residual for the worst verification point in this case is 0.3 FSN which means an improvement by 80 % compared to the state of the art procedure. The necessary measurement time increases by $\approx$ 60 % (see Table 16.2) which is still reasonable and so 9 stationary

Table 16.2 Measurement time comparison for D-optimal design and COR DoE with 30 iDoE points

	# Variation points	Time/measurement (min)	Total measurement time (min)
D-optimal	59	4	236
Repetition points	8	1	8
			244
Startdesign	47	4	188
2nd measurement	47	1	47
Repetition points	7	1	7
iDoE	30	4	120
2nd measurement	30	1	30
			392

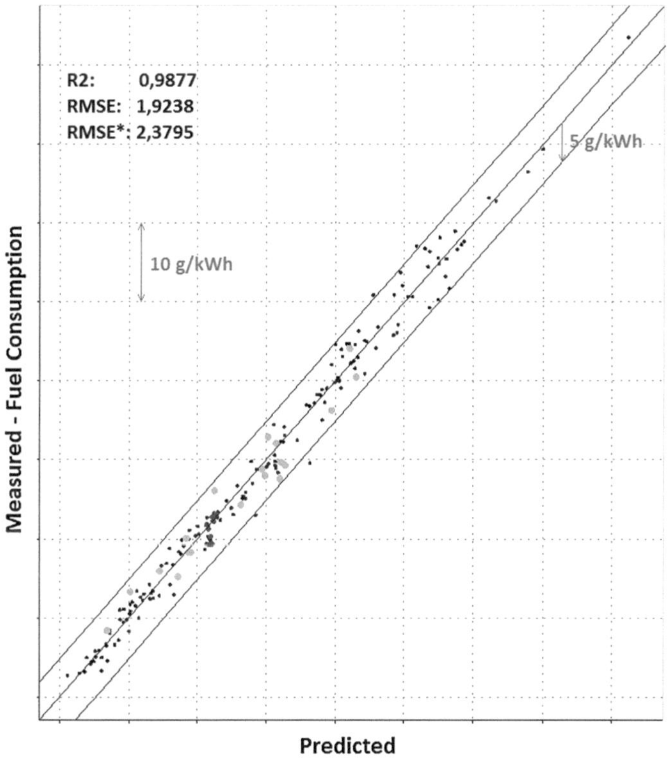

Fig. 16.12 Model quality for BSFC using physical variation parameters and 30 iterative design points

operating points (with 5 variation parameters) can be measured fully automatically during one weekend.

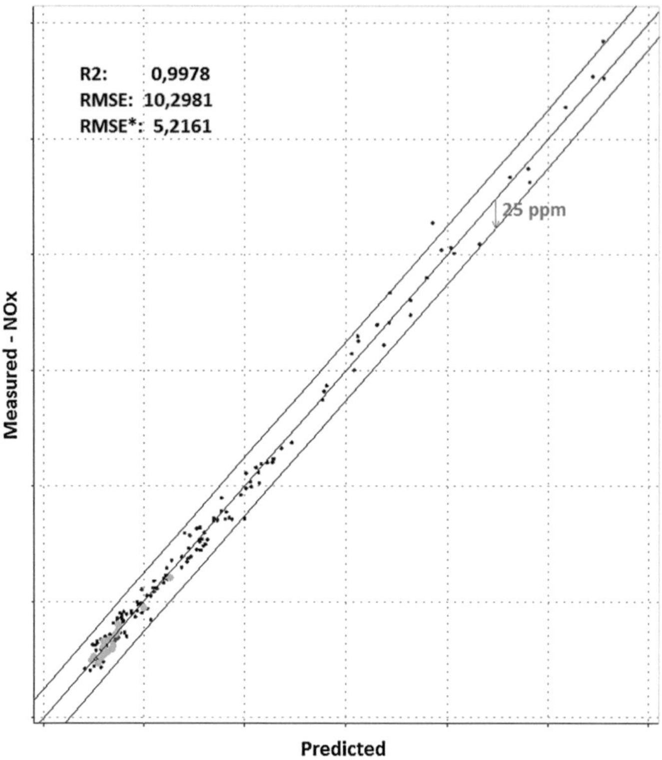

Fig. 16.13 Model quality for NO_x using physical variation parameters and 30 iterative design points

At the same time also all other models get improved in the relevant area by adding iterative variation points. The result for the fuel consumption and NO_x emission is shown in Figs. 16.12 and 16.13.

16.7 Conclusion

We presented a procedure that calculates the optimal design of an experiment online during the workflow. It proved especially useful for the calibration of combustion engines where legislative restrictions require certain output values to be bounded. The training of online estimators of these outputs allows us to improve model quality in areas that are relevant for further steps. Pre-knowledge of the calibration engineer can be entered in the early stage of the test design for achieving an adequate measurement point distribution in critical areas which is a prerequisite for gaining realistic models reflecting the real engine behavior.

Another technique for improving the model quality is the usage of physical variation parameters which can be modeled with much higher accuracy. At the same time all models like fuel consumption, emissions, pressures, temperatures, etc. are improved and therefore very accurate estimations can be expected as result of the optimisation. The original ECU variation parameters are treated as response channels so that they are also outputs of the optimisation and can directly be used e.g. for the calibration of ECU maps.

References

1. Hametner C, Stadlbauer M, Deregnaucourt M, Jakubek S (2013) Incremental optimal process excitation for online system identification based on evolving local model networks. Mathematical and computer modelling of dynamical systems. Taylor & Francis, UK
2. Hametner C, Stadlbauer M, Deregnaucourt M, Jakubek S (2013) Methode zur ermittlung eines modells einer ausgangsgröße eines technischen systems NULL
3. Hametner C, Jakubek S (2013) Local model network identification for online engine modelling. Info Sci. 220:210–225. http://dx.doi.org/10.1016/j.ins.2011.12.034. Accessed from 20 January, 2013
4. Pfluegl H, Riel A, Gschweitl K (2006), Model-based on-line optimization, European Patent EP1150186, 30 Aug 2006, year filed 2003
5. Atkinson A, Donev AN, Tobias RD (2007) Optimum experimental designs, with SAS. Oxford University Press, Oxford
6. AVL CAMEO^{TM} (2013). http://www.avl.com/cameo
7. Nelles O (2002) Nonlinear system identification. Springer, Heidelberg
8. Hametner C, Jakubek S (2011) Combustion engine modeling using an evolving local model network. In: Proceedings of the 2011 international conference on fuzzy systems
9. Pronzato L (2008) Optimal experimantal design and some related control problems. Automatica 44:303–325
10. Franceschini G, Macchietto S (2008) Model-based design of experiments for parameter precision: state of the art. Chem Eng Sci 63:4846–4872
11. Johnson ME, Moore LM, Ylvisaker D (1990) Minimax and maximin distance designs. J Stat Planning Infer 26:131–148
12. Kiefer J, Wolfowitz J (1960) The equivalence of two extremum problems. Can J Math 12:363–366
13. Ranjan P, Bingham D, Michailidis G (2008) Sequential experiment design for contour estimation from complex computer codes. Technometrics 50:527–541
14. Vazquez E, Bect J (2009) A sequential bayesian algorithm to estimate a probability of failure. In: Proceedings of the 15th IFAC symposium on system identification
15. Arenbeck H, Missoum S, Basudhar A, Nikravesh P (2010) Reliability-based optimal design and tolerancing for multibody systems using explicit design space decomposition. J Mech Des 132(2):021010
16. Picheny V, Ginsbourger G, Roustant O, Haftka RT, Kim NH (2010) Adaptive designs of experiments for accurate approximation of a target region. J Mech Des 132(7):071008
17. Stadlbauer M, Jakubek S (2013) Publication forthcoming
18. Hametner C, Stadlbauer M, Deregnaucourt M, Jakubek S, Winsel T (2013) Optimal experiment design based on local model networks and multilayer perceptron networks. Eng Appl Artif Intel 26:251–261

Chapter 17
Optimal Control of HCCI

Per Tunestål

Abstract HCCI (Homogeneous Charge Compression Ignition) is a very control-intensive combustion concept which has been studied for over a decade because of its favorable combination of high efficiency and low emissions. Various optimal control methods have been applied to HCCI and this chapter gives an overview of them. Optimal control of HCCI can be divided into model based and non-model based where MPC is an example of model based and extremum seeking control is an example of non-model based control. The model-based methods can be divided based on whether they use physics based or black box models. Finally a division can be made based on whether the control aims for optimal set-point tracking of e.g. combustion timing or whether it attempts to optimize an overall design criterion such as fuel consumption. This chapter presents and characterizes a number of published methods for optimal HCCI control and characterizes them according to the above criteria.

17.1 Introduction

HCCI (Homogeneous Charge Compression Ignition) combustion has been studied intensely for more than a decade because of its ability to combine high efficiency with low emissions of particularly nitric oxides (NOx) and soot. One great difficulty with HCCI is however, that it lacks direct control of ignition. Unlike spark ignition combustion which is ignited by a spark and diesel combustion where ignition is triggered by fuel injection, HCCI combustion has spontaneous ignition of a homogeneous charge which means that the charge conditions have to be very accurately controlled in order to assure ignition at the right time.

P. Tunestål
Lund University, P.O. Box 118, 221 00 Lund, Sweden
e-mail: per.tunestal@energy.lth.se

H. Waschl et al. (eds.), *Optimization and Optimal Control in Automotive Systems*,
Lecture Notes in Control and Information Sciences 455,
DOI: 10.1007/978-3-319-05371-4_17,

The problem with controlling HCCI combustion timing has been recognized by many researchers [1, 2] and many different control methods have been devised. A majority of the published solutions have applied linear control e.g. PID [3] or linear state feedback [4] with simple, low-level, control objectives such as tracking a desired combustion timing trajectory. The main merits with these linear control methods are low complexity, robustness and tunability.

Perfect tracking of combustion timing is however, not interesting in itself but rather a tool to achieve other goals. Such goals could be low fuel consumption and/or low emissions. This is recognized by some of the optimal control methods where the optimality criterion is specified directly in terms of e.g. fuel consumption. Another shortcoming of the linear control methods is constraint handling. Constraints are nonlinear artifacts and as such can not be handled by truly linear controllers. For this reason special constraint handling is added e.g. integrator windup protection, often with less than satisfactory results.

Constraint handling is thus another reason to apply optimal control methods such as MPC (model predictive control) which is essentially online constrained optimization. With MPC the constraints can be explicitly taken into account in the optimization and thus there is no need to add separate constraint handling. This chapter presents examples of extremum seeking control as well as MPC control to illuminate the issues mentioned above.

17.2 Optimal Control of HCCI

17.2.1 Multi-output MPC of HCCI

In [4], Bengtsson et al. show the first example of MPC (Model Predictive Control) applied to control of HCCI combustion. The modeling approach is system identification of cylinder individual MIMO models using the subspace identification method. Excitation was provided by individually designed PRBS (Pseudo-Random Binary Sequence) signals on each input.

The test engine had a dual-fuel port injection system capable of injecting individual quantities of ethanol and n-heptane to each cylinder. It also had a cylinder-individual VVT (variable valve timing) system capable of changing the (IVC) intake valve closing angle from cycle to cycle which was not used in this control implementation.

The input/output selections for the HCCI cylinder models are illustrated in Fig. 17.1. The inputs selected for the cylinder models were fuel mass per cycle (W_f), fraction of ethanol (R_f), inlet temperature (T_i) and engine speed (n). The outputs were combustion timing (α_{50}), load ($IMEP_n$) and maximum pressure derivative ($dp/d\theta$). $dp/d\theta$ represents the combustion noise and a reasonable limit for the heavy duty test engine in the study is 15 bar/CAD (crank angle degree).

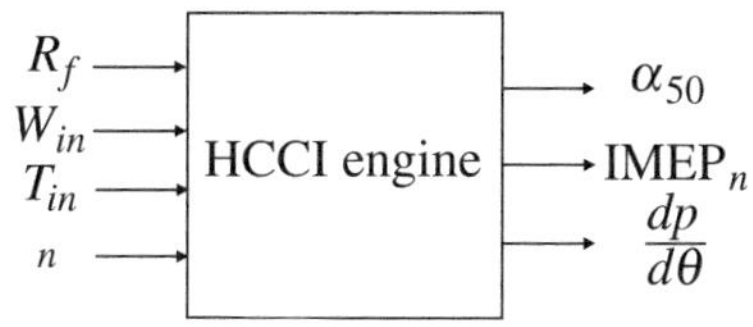

Fig. 17.1 The multi-input/multi-output HCCI engine cylinder model used for multi-output MPC control design [4]

The MPC design took a fairly simplistic approach where α_{50} was kept as close to TDC (top dead center) as possible in order to minimize HC (hydrocarbon) and CO (carbon monoxide) emissions. The tracking error of $IMEP_n$ was also included in the cost function. Hard constraints were applied on inputs and soft constraints on outputs, most importantly $dp/d\theta$ which was given a soft constraint of 15 bar/CAD.

Figure 17.2 shows an experiment where multiple stepwise load changes are applied. It can be seen that α_{50} is delayed when necessary to satisfy the constraint. When it is impossible to delay α_{50} further the $IMEP_n$ tracking is sacrificed by reducing the fuel mass.

17.2.1.1 Discussion

The strength of this approach is that it can minimize a cost function subject to multiple constraints which can be both simple input saturation constraints and output constraints. In this example the cost function was very simple. A more complicated cost function would have made the optimization problem more complex and the computation time would have increased substantially.

17.2.2 Physics-Based MPC of HCCI Combustion Timing

In [6] Widd et al. takes a physics based approach to MPC control of HCCI combustion. A central part is a sub-model describing the heat transfer between cylinder gas, cylinder walls and engine coolant. The continuous heat transfer is modeled as taking place at three specific time instances in each cycle: after intake/mixing, after combustion and after the exhaust stroke. The heat transfer model is illustrated in Fig. 17.3. The individual durations of the heat transfer events were tuning parameters.

Ignition was modeled using a simplified Arrhenius rate threshold model were the temperature was approximated by the TDC temperature. Compression and expansion were modeled as isentropic processes and $IMEP_n$ could be derived from cycle temperatures using ideal cycle analysis.

The inputs to the model were the inlet valve closing angle (θ_{IVC}) and T_i and the outputs were $IMEP_n$ and combustion timing (θ_{50}). The resulting model is of second order and a linearization was used for the MPC design. The control objective was θ_{50} tracking but a small weight was introduced on $\theta_{IVC}^r - \theta_{IVC}$, where θ_{IVC}^r is a reference

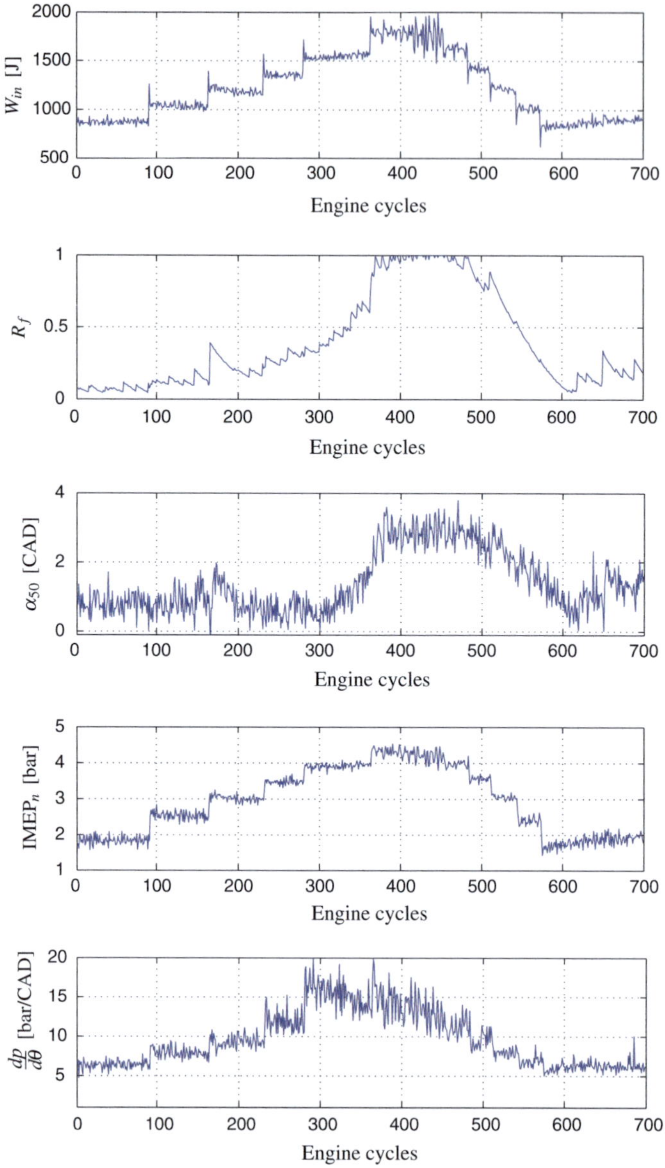

Fig. 17.2 Multiple load step changes illustrating the characteristics of the multi-output MPC controller [5]

crank angle in the middle of the controllable range of the inlet valve closing angle, in order to achieve a midranging [7] effect since T_i and θ_{IVC} are to some extent redundant. Midranging is a heuristic control design method that can be used when two control inputs affect the same output. If one of the control inputs has a high

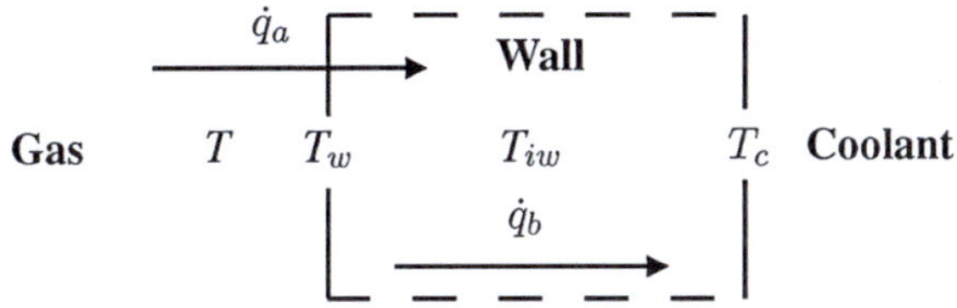

Fig. 17.3 Illustration of the heat transfer model used for physics based MPC control of HCCI combustion [6]

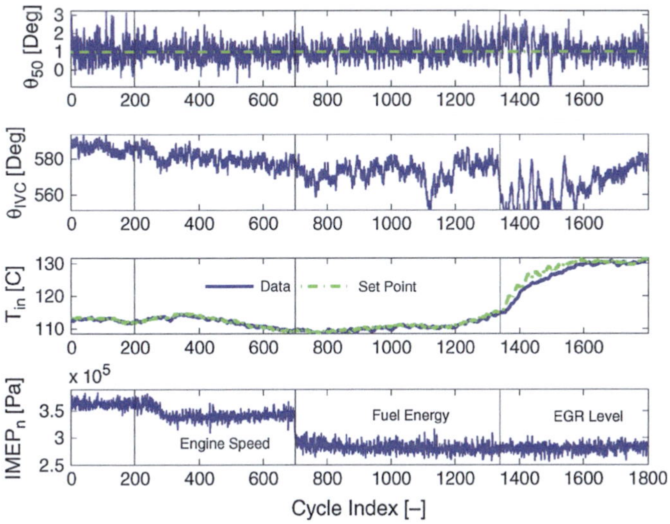

Fig. 17.4 Illustration of the disturbance rejecting characteristics of the physics-based MPC controller with respect to disturbances in engine speed, fuel enginery and EGR level [6]

bandwidth and the other one has a wide range, the slow control input can be used to push the fast one towards the middle of its range and thus make sure that high bandwidth control is always possible.

Due to the low model order short prediction and control horizons could be used which kept the computational load at a reasonable level. Figure 17.4 shows a disturbance rejection experiment with the physics based MPC controller which is able to reject disturbances in engine speed, fuel mass and EGR level.

17.2.2.1 Discussion

The physics-based approach to MPC is attractive since it provides modularity and a component-based structure. E.g. if material of the cylinder liner is changed in the presented example, only the heat-transfer part of the entire model is affected and everything else stays the same. For an identified black-box model, the entire identification would have to be repeated with the new hardware.

In this example the inlet valve closing timing and the inlet temperature are somewhat redundant in controlling the combustion phasing and then a midranging functionality can be obtained by adding a weak penalty to deviations in the intake valve closing timing from a reference value in the middle of its range, thus assuring maneuverability at all times.

17.2.3 Hybrid MPC of Exhaust Recompression HCCI

In [8] Widd et al. take a similar physics based modeling approach is in [6] but without the heat transfer model. The reason for omitting the heat transfer is that the engine used in this case operates with exhaust recompression with a considerable amount of burned gas retained from one cycle to the next. The heat transfer then has a minor influence on the charge temperature and instead focus is on the effect of NVO (negative valve overlap). NVO is the crank angle interval when both exhaust valves and inlet valves are closed around gas exchange TDC. By varying the NVO, the amount of retained burned gas and thus the charge temperature can be controlled. As in [6], the model is of second order. In [8] it is noted that the combustion timing behavior is quite different for early and late combustion timings respectively (see Fig. 17.5) with more cycle-cycle variation and less damping in the case of late combustion. For this reason different linearizations are used for early, mid and late combustion timings respectively in order to improve the control performance.

Tracking control of θ_{50} is implemented both using switching LQ design and using hybrid MPC and a comparison for a large setpoint change is shown in Fig. 17.6. It can be seen that the hybrid MPC controller handles the setpoint change significantly better and the reason is believed to be the fact that the hybrid MPC controller can anticipate the system behavior by using the correct linearization when jumping between early, mid and late combustion timing. The LQ controller can however, only use one linearization at a time based on the present combustion timing.

17.2.3.1 Discussion

The hybrid MPC is suitable for cases when the operating range can be partitioned into a small number of regions with similar system behavior within each region. The MPC can then perform nearly optimally throughout the operating range and even during transitions between regions. It can still be used for systems where the necessary number of regions is larger but the memory requirement as well as the identification effort will scale with the number of regions.

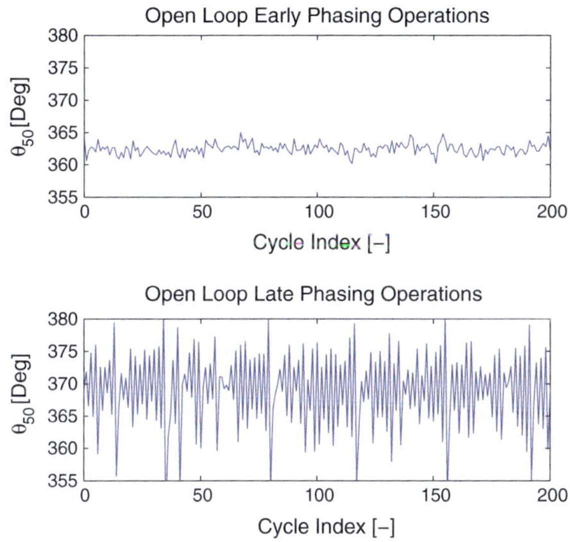

Fig. 17.5 Combustion timing behavior at early and late combustion timing respectively [8]

17.2.4 Optimizing Gains and Fuel Consumption of HCCI Using Extremum Seeking

In [9] a completely non-model based approach is taken where extremum seeking control is used for both tuning of controller gains for combustion timing control and subsequently for fuel consumption minimization by optimizing the combustion timing. The extremum seeking control is defined in Fig. 17.7 and minimizes the cost function $J(\theta)$ with respect to the parameter θ.

Extremum seeking calibration of the control parameters is achieved by defining the cost function as the tracking error and performing repeated positive and negative step changes of the combustion timing (CA50) setpoint. Figure 17.8 illustrates how PI parameters and a feed forward gain are optimized in 1600 s using this approach.

Using the calibrated CA50 controller extremum seeking control of CA50 was subsequently applied in order to minimize fuel consumption. Figure 17.9 shows how the fuel-optimal CA50 is found in approximately 2,000 s.

17.2.4.1 Discussion

Extremum seeking is attractive since it does not require a system model. It can also handle any type of cost function without local optima. The drawbacks with extremum seeking is that it usually requires artificial excitation and the excitation normally has to be of significantly lower frequency than the bandwidth of the system. Each

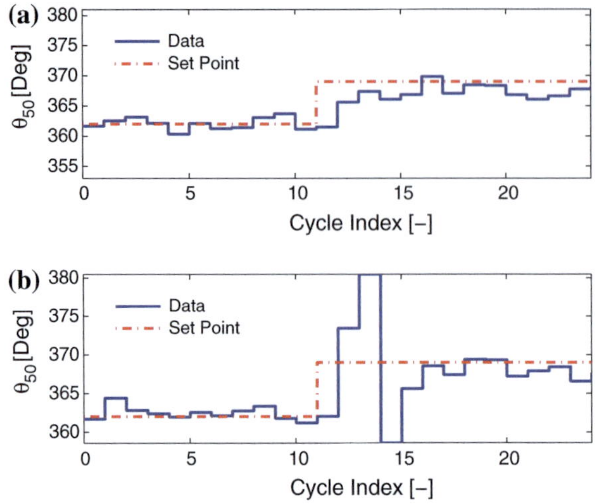

Fig. 17.6 Comparison of large setpoint changes for MPC and LQ controllers [6]

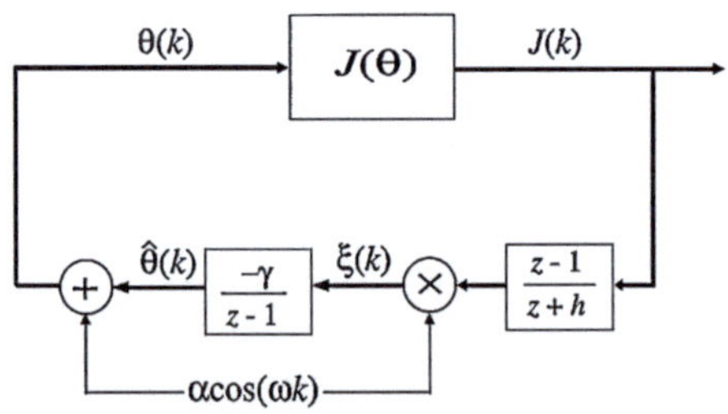

Fig. 17.7 Discrete extremum seeking control with sinusoidal excitation and optimization of the cost function $J(\theta)$ [9]

additional parameter to be optimized requires its own excitation frequency which means slower convergence.

17.3 Conclusions

Four different optimal HCCI control methods have been presented of which three are based on MPC. MPC is valuable for HCCI control mainly because of its ability to explicitly handle constraints. MPC can be applied both to black-box models based on system identification and to linearized physics-based models. When using piece-wise linear models MPC can anticipate model switching which can greatly improve the dynamic behavior for e.g. large setpoint changes. Second order models and relatively short prediction and control horizons have been sufficient for the presented cases and thus the resulting MPC designs have reasonable computational demands. Extremum

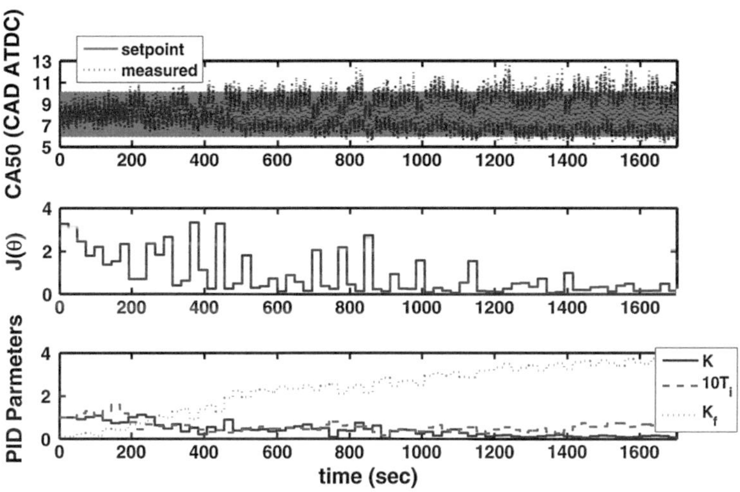

Fig. 17.8 Extremum seeking calibration of PI and feedforward gains [9]

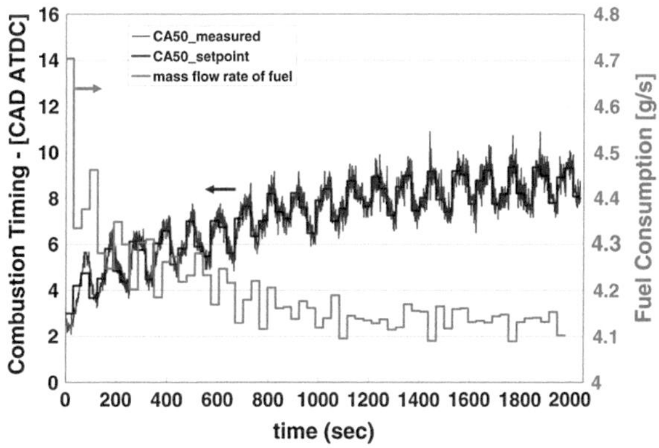

Fig. 17.9 Fuel consumption minimization using extremum seeking control of combustion timing [7]

seeking control provides a completely non-model based alternative. The advantage compared to MPC is that there is no need to derive and calibrate models but extremum seeking is essentially to be considered as a steady-state calibration method since the closed-loop bandwidth is a few orders of magnitude lower than for the presented MPC methods.

References

1. Hong Jo S, Shoda K, Do Jo P, Kato S (1979) Active thermo-atmosphere combustion (ATAC)—a new combustion process for internal combustion engines. SAE technical paper 790501, Society of Automotive Engineers
2. Najt P, Foster DE (1983) Compression-ignited homogeneous charge combustion. SAE technical paper 830264, Society of Automotive Engineers
3. Olsson J-O, Tunestål P, Johansson B (2001) Closed-loop control of an HCCI engine. SAE technical paper 2001–01-1031, Society of Automotive Engineers
4. Bengtsson J, Strandh P, Johansson R, Tunestål P, Johansson B (2006) Model predictive control of homogeneous charge compression ignition (HCCI) engine dynamics. In: Proceedings of the 2006 IEEE international conference on control applications, Munich
5. Pfeiffer R, Haraldsson G, Olsson J-O, Tunestål P, Johansson R, Johansson B (2004) System identification and LQG control of variable compression HCCI engine dynamics. In: Proceedings of the 2004 IEEE conference on control applications, Taipei, Taiwan
6. Widd A, Ekholm K, Tunestål P, Johansson R (2012) Physics-based model predictive control of HCCI combustion phasing using fast thermal management and VVA. IEEE Trans Control Syst Technol 20(3):688–699
7. Karlsson M, Ekholm K, Strandh P, Johansson R, Tunestål P, Johansson B (2007) Closed-loop control of combustion phasing in an HCCI engine using VVA and variable EGR. In: Proceedings of the 5th IFAC symposium on advances in automotive control, Monterey
8. Widd A, Liao H-H, Gerdes JC, Tunestål P, Johansson R (2014) Hybrid model predictive control of exhaust recompression HCCI. Asian J Control 16(2):1–12
9. Killingsworth NJ, Aceves SM, Flowers DL, Francisco E-L, Krstic M (2009) HCCI engine combustion-timing control: optimizing gains and fuel consumption via extremum seeking. IEEE Trans Control Syst Technol 17(6):1350–1361

Chapter 18
Optimal Lifting and Path Profiles for a Wheel Loader Considering Engine and Turbo Limitations

Vaheed Nezhadali and Lars Eriksson

Abstract Time and fuel optimal control of an articulated wheel loader is studied during the lift and transport sections of the short loading cycle. A wheel loader model is developed including engine (with turbo dynamics), torque converter, transmission and vehicle kinematics, lifting hydraulics and articulated steering. The modeling is performed with the aim to use the models for formulating and solving optimal control problems. The considered problem is the lift and transport section of the wheel loader that operates in the short loading cycle, with several different load receiver positions, while the considered criteria are minimum time and minimum fuel. The problem is separated into four phases to avoid solving a mixed integer problem imposed by the gearshifting discontinuities. Furthermore, two different load lifting patterns are studied one with the lifting free and one with the lifting performed only in the last 30 % of the transport. The results show that the optimal paths to the load receiver are identical for both minimum time and minimum fuel cycles and do not change when the loading lifting pattern is altered. A power break-down during the wheel loader operation is presented for the selected cycles of normal and delayed lifting where it is shown that the cycle time remains almost unchanged when lifting is delayed while the fuel consumption slightly decreases in minimum time transients.

18.1 Introduction

Wheel loaders (WL) consume significant amounts of fuel and usually operate in loading cycles where the same task is executed repeatedly. In such operations, optimizing the maneuvers as well as reducing the fuel consumption in every single

V. Nezhadali (✉) · L. Eriksson
Linköping University, SE-581 83 Linköping, Sweden
e-mail: Vaheed.nezhadali@liu.se

L. Eriksson
e-mail: larer@isy.liu.se

H. Waschl et al. (eds.), *Optimization and Optimal Control in Automotive Systems*, Lecture Notes in Control and Information Sciences 455,
DOI: 10.1007/978-3-319-05371-4_18,

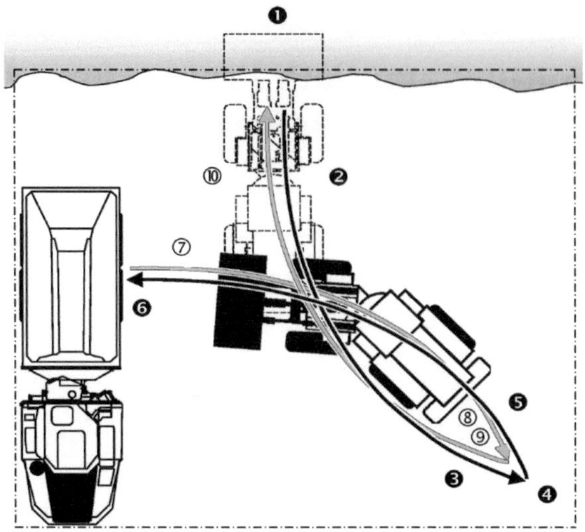

Fig. 18.1 Numbered sequence of actions in a short loading cycle between the loading point and the load receiver, *point 4* is called the reversing point, picture from Ref. [5]

cycle can significantly reduce the total production cost. This is therefore an interesting application where optimal control can guide engineers in the development of an efficient vehicle and users in how to utilize it efficiently.

A WL is a complex system, consisting of several subsystems with a power producing engine and several power consumers that compete about the power. Consumers are for example lifting system, powertrain with torque converter and gearbox for propulsion, as well as steering system which interacts with the powertrain in performing an efficient transport mission. Optimal control of such a system provides valuable knowledge about the optimum usage of the system dynamics. For example minimum fuel (Min M_f) and minimum time (Min T) operation solutions provide valuable information, and enables manufacturers to point out bottle necks as well as potential for improvements in the system design. Furthermore control algorithms and strategies of autonomous WL control systems can also be developed and improved by the optimal control results, see e.g. [7, 8, 11].

In this chapter, a WL is modeled as a nonlinear dynamic system with 9 states and 4 control inputs. The insight of modeling is to develop a model suitable for calculating the optimal controls of the WL in the short loading cycle. The short loading cycle, depicted in Fig. 18.1 is a frequent application of WLs where the optimal control analysis of the vehicle operation gives much insight into reducing fuel costs and cycle operation times. Different studies have been carried out to calculate the optimal WL working path profile (trajectory) in a loading cycle [1, 6, 18, 19, 21]. Another path is followed in [13] and [14] where optimal control is used for a fixed length short loading cycle where the major dynamics (engine, lifting, powertrain, and longitudinal motion) of the WL operation in the short loading cycle is studied. The results here extend the results in [13, 14] to also cover and solve the optimal WL path from loading point to load receiver. The emphasis in the modeling section is to model the components that have the largest power consumption during the WL operation, while still using models that are compact enough for utilization in optimal control problem formulation.

18.1.1 Outline

Section 18.2 describes the model of the wheel loader. The model is used, in Sect. 18.3, to formulate two optimal control problems Min M_f and Min T. These are formulated so that they can be solved to obtain the trajectory and transients of the WL system. The properties and requirements of the loading cycle and WL components are represented as boundary conditions and path constraints in the problem. The change of gear ratio during the WL operation introduces a discontinuity into the optimal control problem which is remedied by dividing the cycle into multiple phases with constant gear ratio.

In Sect. 18.4 the solutions from the multi-phase optimal control problem, for several different load receiver positions are presented. Optimal trajectories, with control inputs and states are analyzed for the loading cycles. The power consumption by different components during the WL operation is calculated and the power distribution between various components is analyzed. The effect of changing the load lifting strategy on the fuel consumption, cycle time and power distribution is also studied. Finally the conclusions are given in Sect. 18.5. The symbols, parameters and constants used in the following are summarized in the Tables 18.1, 18.2.

18.2 System Model

The WL model consists of three main sub systems namely powertrain, steering and lifting, see Fig. 18.2. Vertical bucket acceleration U_{ab} is selected as the control input to the lifting system, and correlates to the hydraulic forces in the lifting cylinders which are adjusted by the hydraulic valves. In the submodels for fuel injection, braking and steering the control inputs are selected as fuel mass injected per combustion cycle U_{mf}, braking torque U_b, and the derivative of steering angle U_{str}.

The state variables are selected to be engine speed ω_{ice}, intake manifold pressure P_{im}, bucket height H_{buc}, bucket lifting velocity V_{buc}, steering angle δ, vehicle speed V, heading angle θ and positions in X and Y directions determined by the differential equations in (18.1)–(18.5).

Table 18.1 List of symbols

Parameter	Description	Unit
ω_{ice}	Engine speed	rps
T_{ice}	Engine torque	Nm
U_{mf}	Injected fuel per combustion cycle	kg/cycle
U_{ab}	Vertical lifting acceleration	m/s^2
U_{str}	Derivative of steering angle	rad/s
U_b	Torque from service brakes	Nm
H_{buc}	Height of the bucket	m
V_{buc}	Vertical speed of the bucket	m/s
θ	Wheel loader heading angle	rad
δ	Steering angle	rad
V	Wheel loader speed	m/s
X	Wheel loader position in X	m
Y	Wheel loader position in Y	m
P_{im}	Intake manifold pressure	Pa
P_{str}	Power consumed for steering	W
P_{trac}	Power consumed for traction	W
P_{lift}	Power consumed for lifting	W
P_m	Intake manifold pressure model	Pa
F_{trac}	Traction force	N
F_{roll}	Rolling resistance force	N
$\dot{m}_a$	Air mass flow	kg/s
$\dot{m}_f$	Fuel mass flow	kg/s
T_{ig}	Combustion generated torque	Nm
T_{fric}	Engine friction torque	Nm
λ	Air to fuel ratio indicator	-
ϕ_λ	Smoke limit	-
ϕ	Speed ratio in torque converter	-
ω_{gb}	Gearbox rotational speed	rps
T_{pump}	Torque on pump side of torque converter	Nm
$T_{turbine}$	Torque on turbine side of torque converter	Nm
T_{gb}	Torque at gearbox input	Nm
T_w	Torque at wheels	Nm
R	Turning radius	m
σ	path curvature	1/m
F_{load}	Required force for load lifting	N
L_{cyl}	Length of lift cylinder	m
A_{cyl}	Cross section area of lift cylinder	m^2
P_{cyl}	Pressure in lift cylinders	Pa
$P_{cyl,max}$	Maximum pressure in lift cylinders	Pa
Q_{pump}	Hydraulic pump flow	m^3/s
$v_{cyl,max}$	Maximum lift cylinder speed	m^2
$V_{lift,max}$	Maximum lifting speed	m^2

(continued)

Table 18.1 (continued)

Parameter	Description	Unit
F_w	Boom weight	N
F_p	Applied force on lift pistons	N
α	Angular acceleration of boom	rad/s^2
M_o	Sum of torques applied on boom	Nm
M_f	Total mass of consumed fuel	kg
T	Cycle duration	s
t_i	Phase duration ($i \in \{1, 2, 3\}$)	s
$\dot{s}_i$	State derivative	-
X_{end}	Load receiver position in X	m
Y_{end}	Load receiver position in Y	m
$P_{loss,GB}$	Power losses in gearbox	W
$P_{loss,TC}$	Power losses in torque converter	W
$P_{eng,acc}$	Power required for engine acceleration	W

Table 18.2 Parameters and constants used in the model and optimal control problem formulation

Parameter	Description	Value	Unit
M_{buc}	Mass of bucket	10,000	kg
η_{lift}	Efficiency of lift pump	0.9	–
$\eta_{volumetric}$	Volumetric efficiency of the lift pump	0.98	–
$\eta_{cycl,l}$	Mechanical efficiency of the lift cylinder	0.95	–
η_{gb}	Efficiency of gearbox	0.9	–
c_r	Rolling resistance	0.03	–
λ_{min}	Minimum air to fuel ratio	1.2	–
$(A/F)_s$	Stoichiometric air to fuel ratio	14.57	–
c_p	Steering power parameter	3e4	–
$c_{p,1}$	Intake manifold pressure parameter	−0.328	–
$c_{p,2}$	Intake manifold pressure parameter	−121.519	–
$c_{p,3}$	Intake manifold pressure parameter	0.057	
$c_{p,4}$	Intake manifold pressure parameter	97,179.699	–
$c_{t,1}$	Time constant parameter	38.5857	–
$c_{t,2}$	Time constant parameter	−0.6869	–
R_a	Gas constant, air	287	J/kg K
T_{amb}	Ambient temperature	300	K
P_{amb}	Ambient pressure	101.57	kPa
$\eta_{vol,\,eng}$	Volumetric efficiency of diesel engine	0.9	–
M_{veh}	Mass of wheel loader	32,000	kg
J_w	Wheel inertia	100	kg m^2
r_w	Wheel radius	0.7	m
M_{wheels}	Mass of wheels	$4 * J_w/(r_w^2)$	kg

(continued)

Table 18.2 (continued)

Parameter	Description	Value	Unit
I_{ice}	Engine inertia	3	kg m^2
η_{gb}	Gearbox efficiency	0.9	–
n_{cyl}	Number of cylinders	6	–
V_d	Engine displacement volume	13e-3	m^3
γ	Gearbox gear ratio	[−60, 0, 60]	–
cfr_1	Engine friction coefficient	0.7196	–
cfr_2	Engine friction coefficient	−0.1414	–
cfr_3	Engine friction coefficient	0.3590	–
q_{lhv}	Heating value, diesel	42.9 e6	J/kg
n_r	Engine rev. per power cycle	2	–
$\eta_{ig,ch}$	Combustion chamber efficiency	0.6877	–
r_c	diesel engine compression ratio	17.3	–
γ_{cyl}	Specific heat capacity ratio of cyl. gas	1.35004	–
D_{pump}	Hydraulic pump displacement	220/1,900	L/round
r	boom length	2.9	m
r_1	boom length	1.7	m
x_c	dimension	0.19	m
y_c	dimension	−0.3	m
G	dimension	2.3	m
r_{pist}	lift piston radius	0.19/2	m
r_{rod}	lift rod radius	0.09/2	m
$P_{cyl,max}$	max lift cylinder pressure	34	MPa
$U_{str,min}$	lower limit on U_{str}	−1	rad/s
$U_{str,max}$	higher limit on U_{str}	1	rad/s
$U_{b,max}$	higher limit on U_b	2e5	Nm
$U_{mf,max}$	higher limit on U_{mf}	265	–
$U_{ab,min}$	lower limit on U_{ab}	−5	m/s^2
$U_{ab,max}$	higher limit on U_{ab}	5	m/s^2
$\omega_{ice,min}$	lower limit on ω_{ice}	57	rps
$\omega_{ice,max}$	higher limit on ω_{ice}	230.38	rps

$$\frac{d\omega_{ice}}{dt} = \frac{1}{J_{ice}}\Big(T_{ice}(U_{mf}, \omega_{ice}) - \frac{P_{lift}(U_{ab}, V_{buc}) + P_{str}(U_{str}) + P_{trans}(\omega_{ice}, V)}{\omega_{ice}}\Big) \tag{18.1}$$

$$\frac{dP_{im}}{dt} = \frac{1}{\tau_m(\omega_{ice})}(P_m(\omega_{ice}, T_{ice}) - P_{im}) \tag{18.2}$$

$$\frac{dH_{buc}}{dt} = V_{buc}, \quad \frac{dV_{buc}}{dt} = U_{ab} \tag{18.3}$$

$$\frac{dV}{dt} = \frac{sign(V)\Big(F_{trac}(U_b, \omega_{ice}) - F_{roll}\Big)}{M_{tot}} \tag{18.4}$$

$$\frac{d\delta}{dt} = U_{str}, \quad \frac{d\theta}{dt} = \frac{V}{R(\delta)}, \quad \frac{dX}{dt} = V\,\cos(\theta), \quad \frac{dY}{dt} = V\,\sin(\theta) \tag{18.5}$$

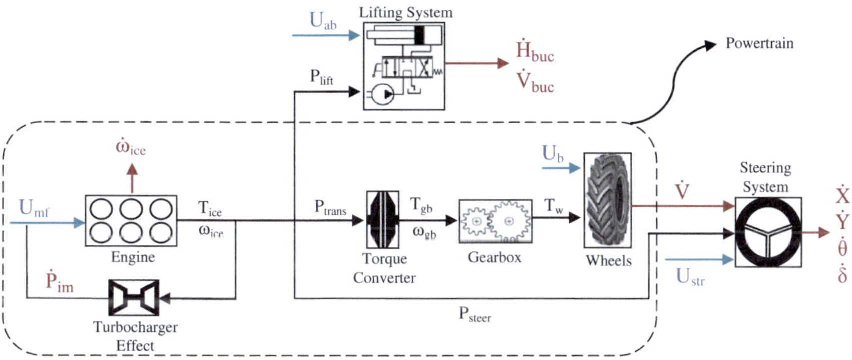

Fig. 18.2 Building blocks of the WL system model and the interdependence between components. States in *red*, control inputs in *blue*

In the following sections, the component models of the WL system are presented. Using the $sign$ function in (18.4), same differential equation can be used during the reversing ($sign(V) = -1$) and forwarding ($sign(V) = +1$) phases of the cycle.

18.2.1 Powertrain and Longitudinal Dynamics

The powertrain delivers the torque for vehicle traction to the wheels of the WL. The powertrain model consists of sub models for diesel engine, torque converter (TC), gearbox and the wheel that connects the powertrain to the longitudinal dynamics, see Fig. 18.2. The control inputs to the powertrain model are fuel injection per combustion cycle U_{mf} and braking torque U_b, the states of the components are engine speed ω_{ice}, intake manifold pressure P_{im} and vehicle speed V.

18.2.1.1 Diesel Engine

The engine model is a simplified version of the model in [22] which represents a 12 Liter 6 cylinder turbocharged diesel engine. The power required for lifting, steering and traction in the WL model is generated by the engine. The engine dynamics are represented by ω_{ice} and P_{im}, while the control input to the engine model being U_{mf}. The mass flows into the engine model stated in [kg/s] are air $\dot{m}_a$ and injected fuel $\dot{m}_f$

$$\dot{m}_a = \frac{\eta_{vol,eng} V_d \omega_{ice} P_{im}}{4\pi R_a T_{amb}}, \quad \dot{m}_f = \frac{10^{-6}}{4\pi} U_{mf} \omega_{ice} n_{cyl} \tag{18.6}$$

Engine gross indicated torque T_{ig} is calculated based on the fuel mass per combustion cycle and engine friction torque is modeled as a polynomial of engine speed.

$$T_{ig}(U_{mf}) = \frac{\eta_{ig}\, q_{hv}\, n_{cyl}\, U_{mf}\, 10^{-6}}{4\,\pi} \tag{18.7}$$

$$T_{fric}(\omega_{ice}) = \frac{V_d\, 10^5}{4\pi}\left(c_{fr1}\,\omega_{ice}{}^2 + c_{fr2}\,\omega_{ice} + c_{fr3}\right) \tag{18.8}$$

$$T_{ice}(U_{mf}, \omega_{ice}) = T_{ig}(U_{mf}) - T_{fric}(\omega_{ice}) \tag{18.9}$$

The increase in the intake manifold pressure depends on the mass flow over the compressor which is strongly dependent on the turbocharger speed. When the driver hits the accelerator pedal in a turbocharged diesel engine, the engine torque increases transiently due to slower speed dynamics of the turbocharger [3]. The time that it takes for the turbocharger speed build up is called turbo lag and it is accounted for by modeling the intake manifold pressure as a function of engine speed and torque including a variable time constant τ_m which depends on the engine speed.

$$P_m(\omega_{ice}, T_{ice}) = c_{p,1}\,\omega_{ice}^2 + c_{p,2}\,T_{ice} + c_{p,4}\,(T_{ice}\,\omega_{ice})^2 + c_{p,3} \tag{18.10}$$

$$\tau_m(\omega_{ice}) = c_{t,1}\,\omega_{ice}^{c_{t,2}} \tag{18.11}$$

where the tuning parameters are $c_{p,1,2,3,4}$, $c_{t,1}$ and $c_{t,2}$. The dynamics of the intake manifold pressure is then described by the following differential equation

$$\frac{dP_{im}}{dt} = \frac{1}{\tau_m(\omega_{ice})}(P_m(\omega_{ice}, T_{ice}) - P_{im}) \tag{18.12}$$

The relative air to fuel ratio λ is defined by

$$\lambda = \frac{(\dot{m}_a/\dot{m}_f)}{(A/F)_s} \tag{18.13}$$

In order to avoid division by zero in (18.13) when fuel is cut off ($U_{mf} = 0$), ϕ_λ is defined as

$$\phi_\lambda = \dot{m}_a - \dot{m}_f\,(A/F)_s\,\lambda_{min} \tag{18.14}$$

where the λ_{min} is set according to the smoke limits during the engine operation.

18.2.1.2 Torque Converter, Gearbox and Longitudinal Dynamics

TCs are used instead of using mechanical clutches to improve the drivability of vehicles at the cost of adding an additional efficiency into the powertrain. A TC transfers the engine torque to the wheels by means of a hydrodynamic coupling between pump and turbine sides where the stator blades multiply the magnitude of the transferred torque. When the TC transients are to be studied, differential equations are used to describe the TC dynamics [10], however, for powertrain control study only efficiency of the TCs is included in the models using efficiency look-up tables

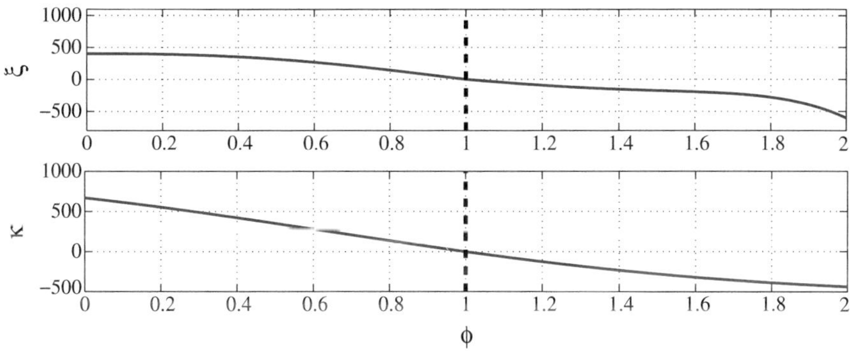

Fig. 18.3 TC characteristic curves ξ and κ define the operation mode depending on the speed ratio ϕ

which are simpler for controller design, [9] and [23]. TC characteristics depend on the speed ratio ϕ over the component, and here, functions are fitted to the experimental data to model the TC. The transferred torque from the TC to the gearbox $T_{turbine}$ is calculated as follows

$$\phi = \frac{\omega_{gb}}{\omega_{ice}} \quad \omega_{gb} = \frac{V\,\gamma}{r_w} \tag{18.15}$$

$$T_{pump} = \xi(\phi)\,(\frac{\omega_{ice}}{1{,}000})^2 \quad T_{turbine} = \kappa(\phi)\,(\frac{\omega_{ice}}{1{,}000})^2 \tag{18.16}$$

where $\xi(\phi)$ and $\kappa(\phi)$ are the TC characteristics depicted in Fig. 18.3. The TC is modeled to operate in two different modes depending on the ϕ value. Mode I during traction where the engine torque is transfered to the wheels for vehicle acceleration ($0 \le \phi \le 1$), and mode II where the turbine side overruns the pump side ($\phi > 1$). A rapid drop in the engine speed while the vehicle is moving ($\omega_{gb} \neq 0$) results in $\phi > 1$ meaning that the kinetic energy is transferred from the gearbox side to the engine side of the TC. A constant efficiency gearbox transfers the TC output torque to the wheels.

$$T_{gb} = T_{turbine} \quad T_w = T_{gb}\,\eta_{gb}\,\gamma \tag{18.17}$$

where γ is the gearbox gear ratio. Vehicle longitudinal acceleration is calculated by neglecting the aerodynamic resistive forces, due to low vehicle velocities, and considering only the rolling resistance forces while including the wheel inertia as an equivalent mass in the total mass of the vehicle.

$$F_{roll} = sign(V)\, c_r\, (M_{veh} + M_{buc})\, g, \quad F_{trac} = \frac{T_w - sign(V)\, T_b}{r_w}, \quad T_b = U_b \tag{18.18}$$

$$M_{tot} = M_{veh} + M_{buc} + \frac{4\, J_w}{r_w^2}, \quad \frac{dV}{dt} = \frac{sign(V)\,(F_{trac} - F_{roll})}{M_{tot}} \tag{18.19}$$

The $sign$ operator defines the direction of WL travel in the reversing and forwarding sections of the short loading cycle. The power required at the input to the TC, P_{trans}, in order to generate the tractive force at wheels is calculated according to

$$P_{trans} = T_{pump}\, \omega_{ice} \tag{18.20}$$

18.2.2 Steering and Ground Position

Articulated steering is used in WLs where the two main bodies of the vehicle are connected by a revolute joint such that the front and rear axles are equidistant to the articulation point. This type of steering is favorable in WLs as the front and rear wheels move over the same trajectory easing the movement of vehicle on muddy surfaces [18]. Figure 18.4 shows the geometry of the WL while turning. From the system and control points of view, several studies have been performed on path planning and trajectory optimization, e.g. [2, 12, 17], while many are specifically dedicated to WLs path planning and optimization [1, 18, 19, 21]. Here, in the steering system, WL position and heading angle are modeled using simple vehicle kinematics. (X, Y) and heading angle θ during load carrying from the loading point to the load receiver are determined based on the vehicle speed V and steering angle δ while the derivative of steering angle U_{str} is the control input to the model. The steering dynamics are then determined by

$$\frac{d\theta}{dt} = \frac{V}{R}, \quad \frac{dX}{dt} = V\, \cos\theta, \quad \frac{dY}{dt} = V\, \sin\theta \tag{18.21}$$

where R is the turning radius and (X, Y, θ) are calculated at point C, see Fig. 18.4, which remains equidistant from the front and rear axles during the loading cycle. For control purposes, the steering angle must remain continuous during the operation which is ensured by selecting the steering angular velocity U_{str} as the control input to the steering system and bounding it within limits.

$$\frac{d\delta}{dt} = U_{str}, \quad U_{str,min} < U_{str} < U_{str,max} \tag{18.22}$$

Vehicle turning performance is specified as a minimum turning radius, R_{min}. To ensure that the model fulfills this, the following connection between R and δ is used

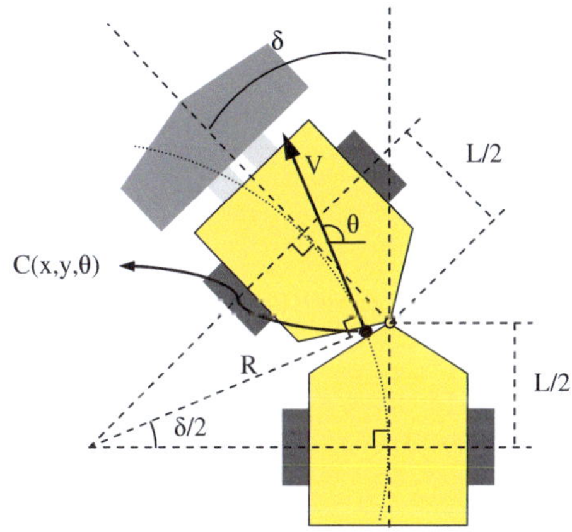

Fig. 18.4 The position of WL during the loading cycle is determined for point C

$$R = \frac{L}{2\ \tan(\frac{\delta}{2})}, \quad \sigma = \frac{1}{R} \tag{18.23}$$

where σ is the path curvature.

The hydraulic power required for steering P_{str} is modeled as a quadratic function of steering angular velocity meaning that there is no power demand for steering when steering angle is unchanged.

$$P_{str} = c_p U_{str}^2 \tag{18.24}$$

where c_p is the tuning parameter.

18.2.3 Lifting System

The lifting speed V_{buc} and bucket position H_{buc}, both in vertical direction, are determined in the lifting system model. The control input to the model is the vertical acceleration of the bucket U_{ab} and the required power for lifting P_{lift} is calculated as follows

$$F_{load} = M_{load}\,(g + U_{ab}), \quad P_{lift} = \frac{F_{load}\,V_{buc}}{\eta_{lift}} \tag{18.25}$$

where the load mass M_{load} is assumed to remain constant.

Lifting speed V_{buc} is dependent upon engine speed, as the hydraulic pumps in the lifting system rotate at the same speed as the engine until the flow in the pumps becomes saturated at 1,500 rpm. Also, the maximum lifting acceleration is limited by the maximum allowed pressure in the hydraulic system. In order to derive the constraints on the lifting speed and acceleration, the boom geometry is analyzed. While the boom is lifted, the displacement along the lift cylinders ΔL_{cyl} is multiplied

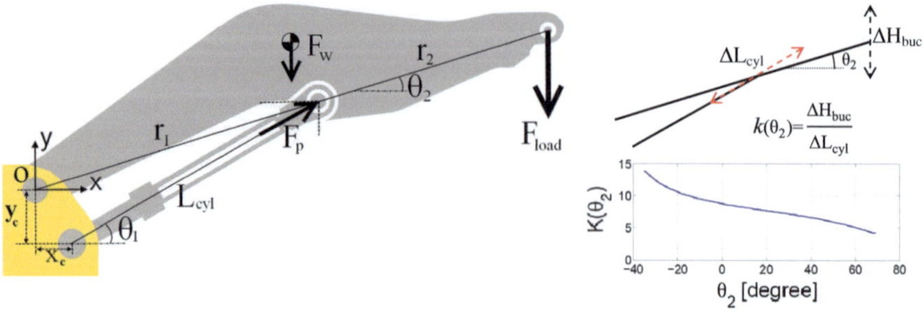

Fig. 18.5 The boom geometry and acting forces during lifting (*left*), and the multiplication factor between the lift cylinder and boom end displacement (*right*)

with a factor k resulting in vertical displacement $\Delta_{H_{buc}}$ at the end of the boom, see Fig. 18.5-right. k is calculated as a function of the boom angle θ_2 as follows

$$\theta_2 = \sin^{-1}(\frac{H_{buc} - G}{r}), \quad \theta_1 = \tan^{-1}(\frac{r_1 \cos(\theta_2) - x_c}{r_1 \sin(\theta_2) - y_c}) \tag{18.26}$$

$$L_{cyl} = \sqrt{\left(r_1 \cos(\theta_2) - x_c\right)^2 + \left(r_1 \sin(\theta_2) - y_c\right)^2} \tag{18.27}$$

$$k(\theta_2) = \frac{\Delta(r \ \sin(\theta_2))/\Delta\theta_2}{\Delta L_{cyl}/\Delta\theta_2}, \quad r = r_1 + r_2 \tag{18.28}$$

where G is the distance between the boom and body joint, point O, from the ground level and $r_1, r_2, \theta_{1,2}, x_c$ and y_c are illustrated in Fig. 18.5-left. Lifting is carried out by means of lift cylinders where two identical hydraulic pumps deliver the fluid to the system. The maximum displacement speed of the lift cylinders $v_{cyl,max}$ is determined by the maximum amount of fluid Q_{pump} pumped into them which is calculated as

$$A_{cyl} = \pi\,(r^2_{piston} - r^2_{rod}), \quad Q_{pump} = \min(\omega_{ice}, 157) \times D_{pump,max}\,\eta_{volumetric} \tag{18.29}$$

$$v_{cyl,max}(\omega_{ice}) = \frac{Q_{pump}\,\eta_{cyl,l}}{A_{cyl}} \tag{18.30}$$

using the k factor calculated in (18.28) the maximum possible bucket lifting speed at the end of the boom is determined as a function of the engine speed.

$$V_{lift,max}(\omega_{ice}) = k(\theta_2)\,v_{cyl,max}(\omega_{ice}) \tag{18.31}$$

Forces acting on the boom, during lifting are depicted in Fig. 18.5. In order to calculate the magnitude of the force exerted on the lifting cylinders, the torque equilibrium equation is solved around joint O.

$$F_w = M_{boom}\,(g + U_{ab}) \tag{18.32}$$

$$\sum M_o = I_{boom}\,\alpha \quad \Rightarrow \quad F_p = \frac{\frac{I_{boom}\,U_{ab}}{r} + F_{load}\,r\cos(\theta_2) + F_w\frac{r}{2}\cos(\theta_2)}{r_1\sin(\theta_1 - \theta_2)} \tag{18.33}$$

accordingly, the exerted pressure in the lifting system at any boom position and bucket acceleration is calculated as

$$P_{cyl}(U_{ab}, H_{buc}) = \frac{F_p(U_{ab}, H_{buc})}{A_{piston}} \tag{18.34}$$

It is noted that the bucket height H_{buc} is selected as a system state since it is required in (18.26), to determine the geometry of the boom θ_2, which is needed when calculating the maximum lifting speed and lift cylinder pressure.

18.3 Optimal Control Problem Formulation

The WL model developed in the previous section is utilized to formulate an optimal control problem which is solved in order to obtain the Min T and Min M_f transients of the WL system. In this section, first the loading cycle requirements are described in terms of boundary conditions of the optimization problem, then the path constraints are defined and finally the formulated optimal control problems is presented.

Gear shifts during the loading cycle introduce a discontinuous variable, gear ratio, into the problem. To avoid discontinuities in the problem and thus avoiding the need to solve a mixed integer problem, the loading cycle is divided into four separate phases where the gear ratio of the gearbox remains constant during each phase. The first phase starts when the WL leaves the loading point and reverse gear is selected to accelerate towards the reversing point, then in the second phase, the gearbox goes into neutral and service brakes are used to stop the vehicle. This process is repeated in the third and fourth phases with the difference that the WL moves in forward direction. The short loading cycle requirements are described in terms of boundary conditions in Table 18.3 where $\dot{s}_i$ are the system states determined by the differential equations (18.1)–(18.5) and $\dot{s} = 0$ at the end of the cycle ensures that the vehicle reaches the load receiver at stationary condition.

The final position of the WL (X_{end}, Y_{end}) depends on the configuration of the construction site. It is assumed that the WL starts reversing from the origin at the loading point and the load receiver is located down to the left of the origin. The working range of WL is constrained using the coordinates of the loading point and load receiver as follows

$$X \geq X_{end}, \quad Y \leq 0, \quad 0 \leq \theta \leq 2\pi \tag{18.35}$$

Table 18.3 The load carrying section of the short loading cycle is defined in terms of boundary conditions in the optimal control problem

Time	0	Phase 1 reversing $\gamma=-60$ $U_b=0$		Phase 2 reversing $\gamma=0$ $U_b\neq 0$		Phase 3 forwarding $\gamma=60$ $U_b=0$		Phase 4 forwarding $\gamma=0$ $U_b\neq 0$
		t_1^-	t_1^+	t_2^-	t_2^+	t_3^-	t_3^+	T
ω_{ice}	1,500 [rpm]	–	–	–	–	–	–	–
P_{im}	1.1 P_{amb}	–	–	–	–	–	–	–
H_{buc}	0.7 [m]	–	–	–	–	–	–	5 [m]
V_{buc}	0	–	–	–	–	–	–	0
V	0	–	–	0	0	–	–	0
U_{str}	0	–	–	–	–	–	–	0
δ	0	–	–	–	–	–	–	0
θ	90°	–	–	–	–	–	–	–
X	0	–	–	–	–	–	–	X_{end}
Y	0	–	–	–	–	–	–	Y_{end}
$\dot{s}_i$	–	–	–	–	–	–	–	0

The free variables t_1, t_2 and t_3 are the gear shifting times which will be optimized

In order to avoid exerting uneven forces on the boom structure, the WL must be perpendicular to the load pile (at $t=0$, $\theta=90°$) and load receiver at the beginning and end of the cycle, [18]. At the beginning of the first phase, it is assumed that the bucket is lifted as high as the wheel radius and since the engine has already been producing power for bucket filling, the initial engine speed and intake manifold pressure are set higher than the idling engine speed and ambient pressure respectively. When the bucket is loaded and raised, the vehicle should not brake harshly in order to avoid structural damages and this is ensured by adding the constraint on the vehicle speed derivative. The smoke limiter constraint on ϕ_λ and other component limitations according to the properties stated in [16] are defined as the following path constraints

$$\begin{aligned}
0 &\le \phi_\lambda & V_{buc} &\le V_{lift,max}(\omega_{ice})\\
R_{min} &\le R(\delta) & T_{ice} &\le T_{ice,max}(\omega_{ice})\\
\omega_{ice,min} &\le \omega_{ice} \le \omega_{ice,max} & |V| &\le V_{max}\\
P_{cyl}(U_{ab}, H_{buc}) &\le P_{cyl,max} & -0.18\,g &\le \frac{dV}{dt}\\
\delta_{min} &\le \delta \le \delta_{max} & P_{amb} &\le P_{im}\\
U_{ab,min} &\le U_{ab} \le U_{ab,max} & 0 &\le U_{mf} \le U_{mf,max}\\
U_{str,min} &\le U_{str} \le U_{str,max} & 0 &\le U_b \le U_{b,max}
\end{aligned} \tag{18.36}$$

In order to ensure the continuity of the states between the successive phases, phase connectivity constraints are applied to the problem. The bucket acceleration U_{ab} and steering angular velocity U_{str} are kinematic properties which cannot be discontinuous for the sake of mechanical stability, therefore, the connectivity constraints are applied on these control inputs as well.

$$U_{ab}, U_{str} \text{ and } s_i \text{ at start of phase } j+1 = U_{ab}, U_{str} \text{ and } s_i \text{ at end of phase } j \quad (18.37)$$
$$i \in \{1,2,3,4,5,6,7,8,9\},\ j \in \{1,2,3\}$$

The fuel consumption during the WL operation is calculated as

$$M_f = \int_0^{t_1} \dot{m}_f\,dt + \int_{t_1}^{t_2} \dot{m}_f\,dt + \int_{t_2}^{t_3} \dot{m}_f\,dt + \int_{t_3}^{T} \dot{m}_f\,dt \quad (18.38)$$

where $\dot{m}_f$ is obtained by (18.6). The optimal control problems formulated to calculate the Min T and Min M_f system transients and the phase shifting times t_1, t_2, t_3 are

$$\min_{u(\cdot)} M_f \quad \text{or} \quad \min_{u(\cdot)} T$$
$$\text{s.t: } \dot{s} = f(s,u) \text{ and constraints in (18.36), (18.37) and Table 18.1} \quad (18.39)$$

Due to the complexity of the problem and number of states and control inputs, solving the OCPs with methods such as Dynamic Programming or Pontryagin's Maximum Principle would require very large computational effort. Therefore, an optimal control solver PROPT [20] is used to solve the problem in (18.39). The solver employs pseudospectral collocation method [4] to solve the OCP where states, controls and cost function are described in terms of high order polynomials satisfying the constraints.

18.4 Results

The multi-phase optimal control problem in (18.39) subjected to the path constraints and boundary conditions in (18.35)–(18.37) and Table 18.3 is solved for thirty different load receiver positions (X_{end}, Y_{end}) which are evenly spread over a $20 \times 20\,\text{m}^2$ working area, and the Min M_f and Min T transients and optimal trajectory of the WL are calculated. The effects of imposing an alternative lifting pattern on the transients is analyzed and power distribution between the different components during the operation is presented.

18.4.1 Optimal WL Trajectory from Loading Point to the Load Receiver

The optimal WL trajectory from the loading point to the load receiver is found to be nearly identical at different points of the working area for Min M_f and Min T transients. The results show that the minimum traveling distance between the loading

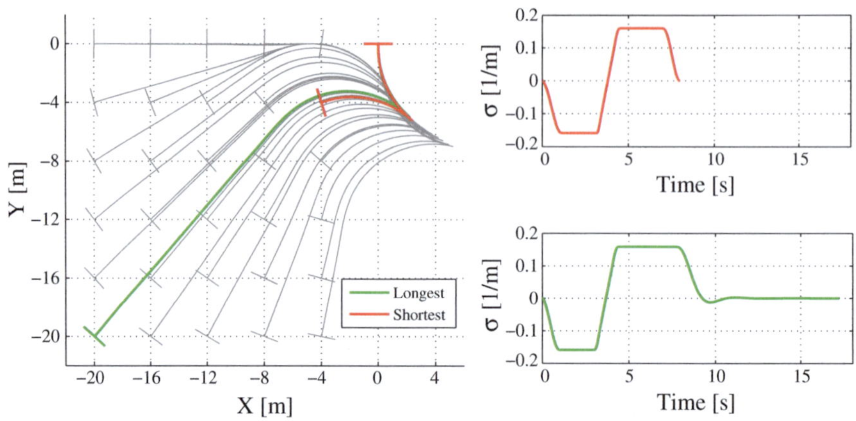

Fig. 18.6 The optimal WL trajectory in the short loading cycle from the loading point to the load receiver for Min M_f and Min T transients are identical and depend only on the load receiver position (*left*). The trajectory curvature remains continuous during the operation which is a necessity for trajectory controller design (*right*)

point and the load receiver is selected in both Min M_f and Min T solutions, although the Min M_f transients are almost twice as long as the Min T transients. Figure 18.6-left, shows the optimal trajectories of the WL, and final heading angles corresponding to the optimal orientation of the load receiver, at various load receiver positions. The shortest and the longest trajectories are highlighted and the system transients will be presented for them in the sequel. In the same figure, to the right, the trajectory curvature (σ) is presented for the highlighted cycles where it is seen that it remains continuous during the cycle. This is a necessity for trajectory controller design, [18, 19, 21], and is the result of using the constrained derivative of the steering angle U_{str} as the control input to the steering system. The curvature is zero at the beginning where the WL leaves the loading point and also at the end of the cycle when the vehicle reaches the load receiver.

18.4.2 Min M_f and Min T System Transients

Figures 18.7 and 18.8 show the Min T and Min M_f transients and the engine operating points for the short and long cycles, highlighted in Fig. 18.6. According to [15], a typical lift and transport operation consists of around 5 s reversing and 5 s forwarding (for the same range of distance as in the shorter cycle here), while the Min T transients are calculated to be only 8 s. The duration of Min M_f transients for the same cycle are nearly two times longer with 36 % lower fuel consumption. In the Min T case, at the beginning of the first phase when the WL starts to move, the increase in engine output torque is limited by the smoke limiter constraint which is caused by the turbocharger speed build up effect with insufficient P_{im} resulting in

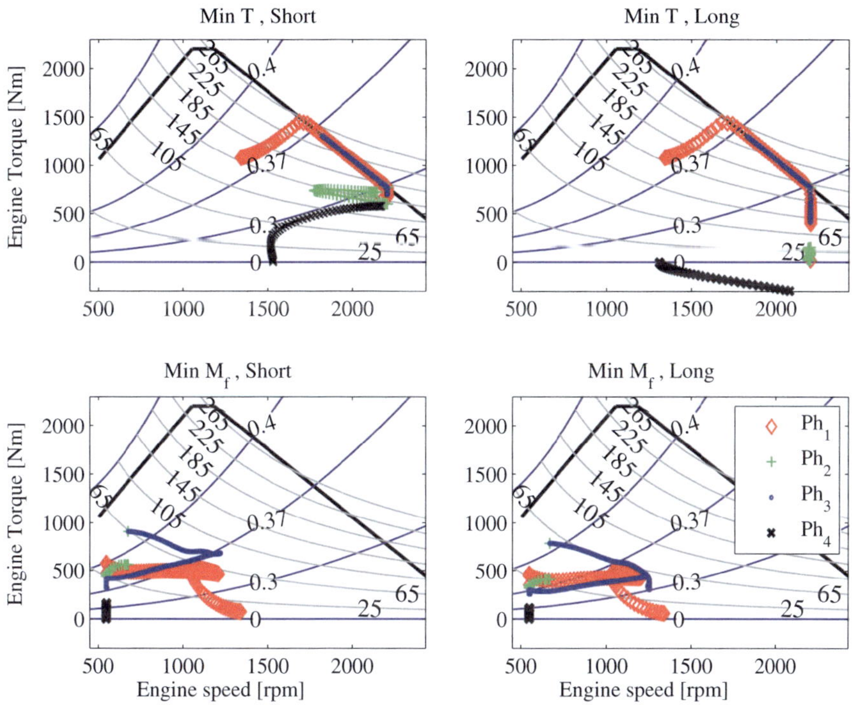

Fig. 18.7 Engine operating points in time and fuel optimal transients with respect to the engine torque limits for the short and long cycles (constant efficiency curves in *blue* and constant power [kW] curves in *gray*)

low $\dot{m}_a$. The bucket lifting starts a bit later in order to leave all of the engine power for faster vehicle acceleration and steering at the beginning. The steering angle rapidly decreases to the lowest allowed value by the minimum curvature radius constraint and remains unchanged until moments before reversing point where it goes towards zero.

In the second phase, the engine is decoupled from the wheels when the gear is shifted into neutral. All the engine power is used for lifting while higher engine speed at the end of this phase is desirable, since the kinetic energy of the engine is going to be used for faster vehicle acceleration at the beginning of the next phase via a rapid drop in the engine speed.

In the Min T transients of the short cycle, half of the final bucket height is reached during reversing and the rest in the forwarding section. In case of the long cycle, the length of the reversing and forwarding sections of the cycle are unequal. In the reversing section, faster dynamics of the steering angle occur in order to position the vehicle quickly in the trajectory where less steering power would be required during forwarding. Most of the engine power is allocated to vehicle acceleration and steering without major load lifting during reversing. In the section with forward movement, except short intervals at the beginning where vehicle is

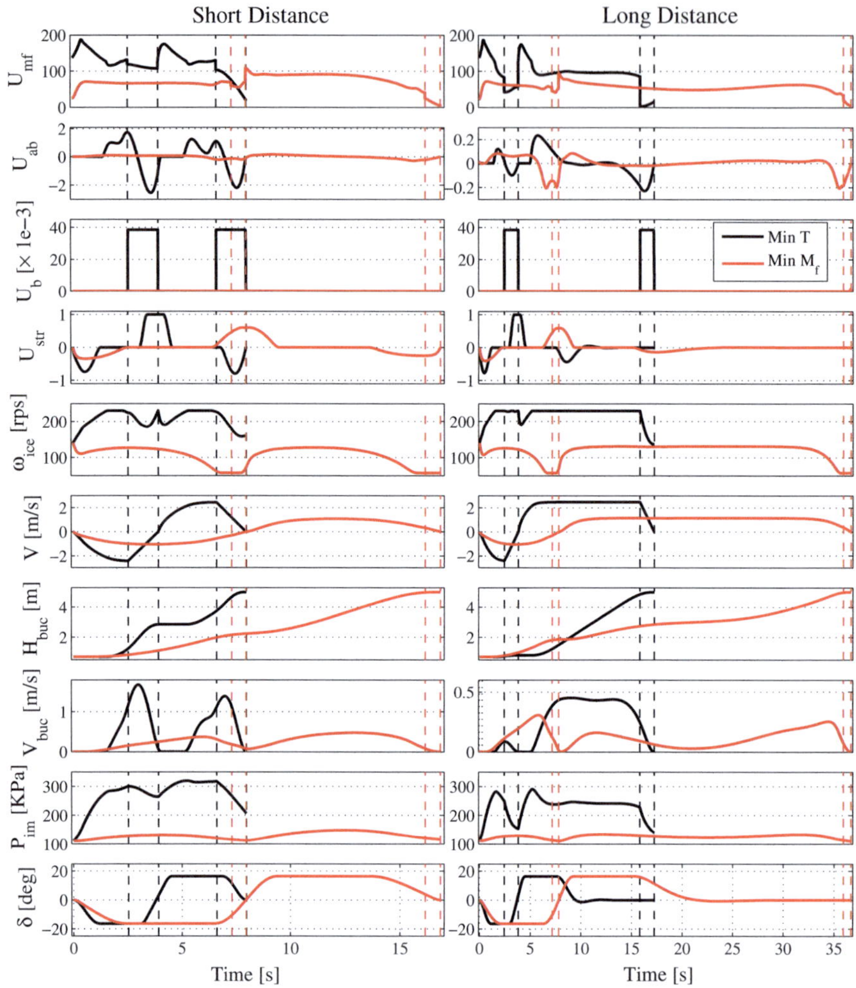

Fig. 18.8 The Min T and Min M_f transients of the short (*left*) and long (*right*) cycle. *Vertical lines* are the phase boundaries

accelerated, the WL travels mostly on a straight line (no steering power required) with a constant speed (small traction power required) where most of the engine power remains untouched and available for fast load lifting.

In case of the short cycle and for both Min M_f and Min T transients, the third phase has similar dynamics as the first one in the sense that first the vehicle is accelerated to high speeds and then the lifting starts.

The Min M_f transients are similar for both short and long cycles in the sense that high engine speeds are avoided and less rapid changes occur in U_{mf}. But still U_{str}

has fast dynamics in order to ensure that the shortest trajectory to the load receiver is traveled.

18.4.3 Delayed Lifting

Considering the uneven surfaces where WLs operate on, the risk of losing vehicle stability increases when the loaded bucket is raised during the period that the vehicle is moving. The results in Fig. 18.8 show that the load lifting begins early in the reversing phase and the bucket is raised from its initial position during a long period of the cycle. In this section, the problem formulated in (18.39) is solved with an additional constraint limiting the bucket lifting to only the last 30 % of the cycle duration ensuring that the bucket remains mostly on lower height levels. To study the effects of the lifting pattern on the system transients and power distribution among different components, the Min T and Min M_f transients and power break down between various components are analyzed.

18.4.3.1 Effects of Delayed Lifting on Min T and Min M_f Transients

The optimal trajectory from the loading point to the load receiver in different positions remains the same as that of the normal lifting pattern and looks as depicted in Fig. 18.6. The Min T and Min M_f transients of the WL for the short and long cycles are illustrated in Fig. 18.9 where lifting is delayed. The reversing section remains similar to the previous case with the difference that after accelerating to high speeds, no power is required for lifting and less engine torque is generated late in the reversing section. The major difference happens, as expected, late in the forwarding section where the vehicle is controlled to reach high speed before start of lifting in order to leave most of the engine power for lifting in the rest of the cycle duration. The fuel injection level increases to produce larger torque as the lifting starts and the intake manifold pressure rises to deliver more air to the combustion chamber so that more fuel can be injected without reaching the smoke limit.

18.4.4 Power Break Down

The power generated by the diesel is consumed for vehicle traction P_{trac}, load lifting P_{lift} and vehicle steering P_{str}. TC losses $P_{loss,TC}$ also constitute a major portion of power consumption and are calculated separately while the power loss due to the efficiency of the lifting system is included in P_{lift}. The diesel engine power is required for traction when the traction force F_{trac} is positive.

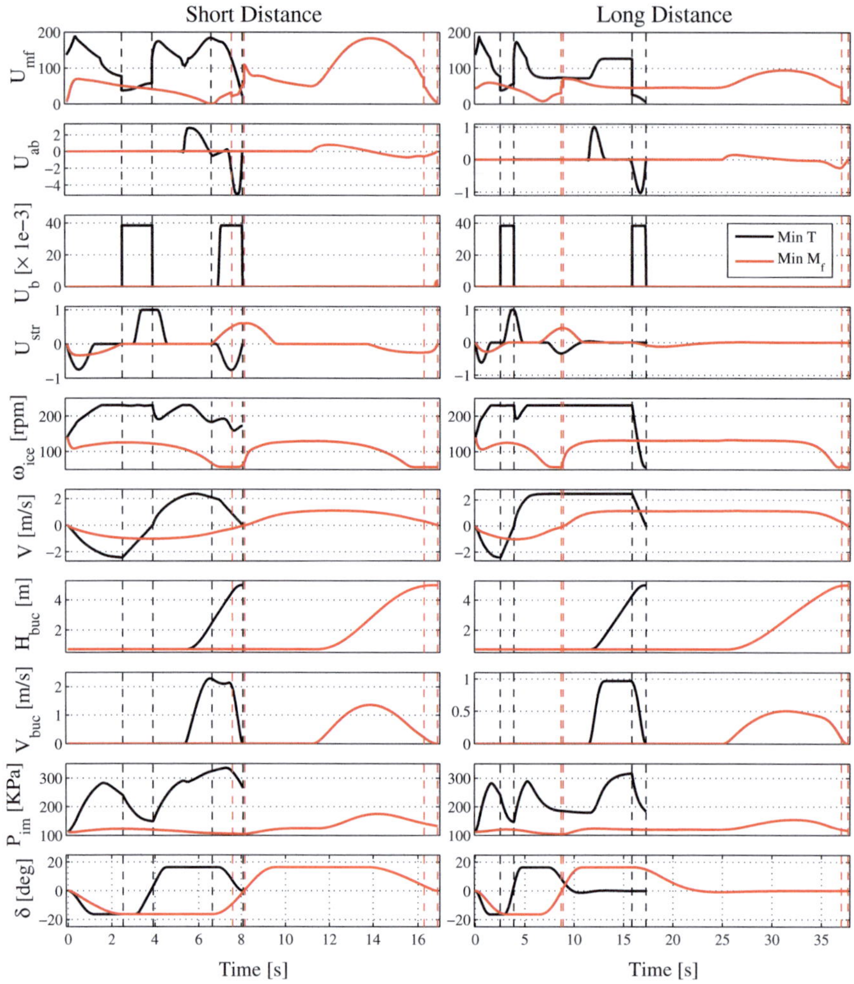

Fig. 18.9 The Min T and Min M_f transients of the short (*left*) and long (*right*) cycle when the lifting is delayed to the last 30 % of the cycle duration. *Vertical lines* are the phase boundaries

$$F_{trac} = M_{tot}\frac{dV}{dt} + F_{roll}, \quad P_{trac} = F_{trac}\,V \text{ when } F_{trac} > 0 \tag{18.40}$$

The power loss in the gearbox is calculated as follows

$$P_{loss,GB} = \frac{1-\eta_{gb}}{\eta_{gb}}\,P_{trac} \tag{18.41}$$

The loss of power in the TC is calculated as the difference between the input and output powers of the TC as

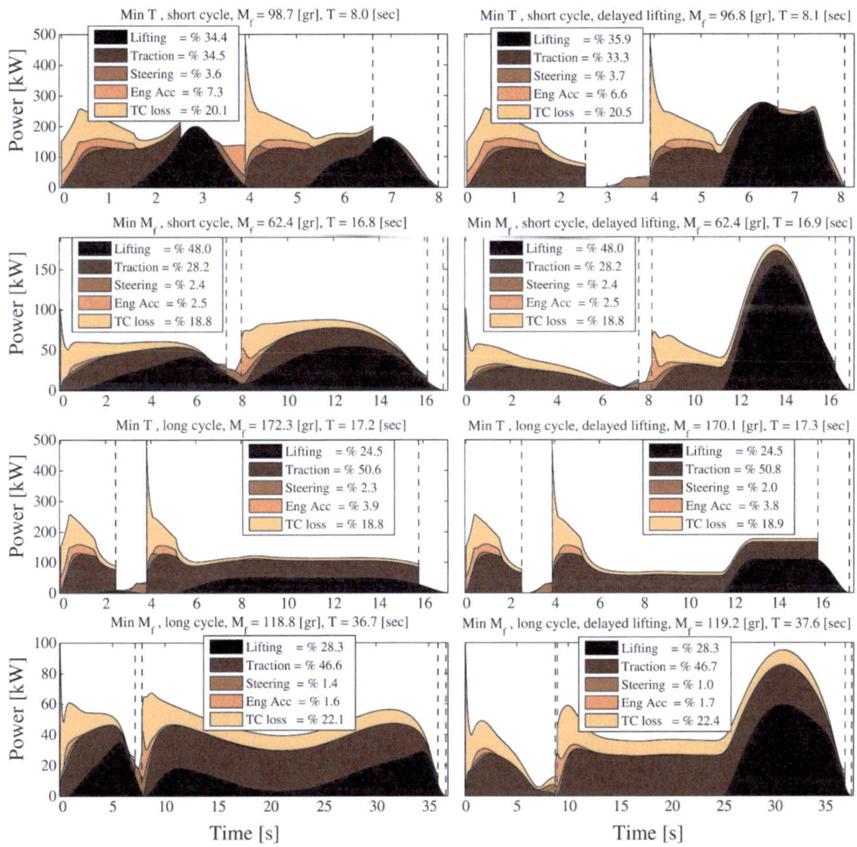

Fig. 18.10 The power brake down for the short and long cycles during the Min M_f and Min T transients for normal operation (*left column*), and delayed lifting operation (*right column*). The *vertical lines* are the phase boundaries

$$P_{loss,TC} = T_{pump}\,\omega_{ice} - T_{gb}\,\gamma\,\frac{V}{r_w} \tag{18.42}$$

A part of the engine power is consumed to overcome the engine inertia $P_{eng,acc}$ during engine acceleration. When the engine decelerates, $P_{eng,acc}$ becomes negative meaning that the kinetic energy of the engine is delivered back to the system.

$$P_{eng,acc} = J_{ice}\,\omega_{ice}\,\frac{d\omega_{ice}}{dt} \tag{18.43}$$

Finally, the power balance in the system is described by the following equation

$$T_{ice}\,\omega_{ice} = P_{lift} + P_{trac} + P_{str} + P_{eng,acc} + P_{loss,TC} + P_{loss,GB} \tag{18.44}$$

The power distribution in the WL system during the Min T and Min M_f transients of the short and long cycles is illustrated on the left column of Fig. 18.10 where $P_{loss,GB}$ is lumped into P_{trac}. The right column in the same figure shows the power distribution for the case where lifting is delayed. It is interesting to note that by changing the lifting pattern, the proportion of energy consumption remains almost unchanged among different consumers and only the course of events gets shifted. The increase in the cycle duration in case of delayed lifting is negligible, and the fuel consumption remains nearly unchanged in the Min M_f transients while it decreases by 1.2 and 1.9 % in case of Min T transients of the short and long cycles respectively. This implies that lifting can be performed closer to the load receiver in the loading cycle without causing major losses in cycle duration or increase of fuel consumption while better stability of the vehicle is achieved during the load carrying operation.

18.5 Conclusion

A wheel loader (WL) is modeled as an integration of three sub systems namely powertrain, lifting and steering where the aim is to describe the main dynamics of the subsystems with highest power consumption during the WL operation. The trajectory generation for the WL from loading point to the load receiver is also included in the model. The diesel engine is modeled including turbocharger limitations while the number of state variables and control inputs are reduced by modeling the intake manifold pressure as a function of engine speed and torque. The torque converter is modeled by static characteristic curves and a constant efficiency gearbox is used. The geometry of the boom is analyzed and the structural constraints during lifting are modeled in the lifting system.

The loading cycle is divided into four phases with constant gearbox gear ratio during each phase in order to avoid facing a mixed integer optimal control problem. The minimum fuel (Min M_f) and minimum time (Min T) dynamics of the system are calculated by solving a multi-phase optimal control problem. The optimal trajectory in the short loading cycle from loading point to the load receiver and the system dynamics are calculated for several load receiver positions. An alternative load lifting pattern where the lifting is delayed until the last 30 % of the cycle duration is studied. The suggested lifting pattern ensures that the WL remains stable during load carrying as the bucket remains on low height during most of the cycle time.

The optimal trajectory to the load receiver is found to be identical for the Min T and Min M_f transients and remains unchanged in case of the new lifting pattern implying that the transport path and longitudinal motion can be solved separately. However there is a coupling between them, as the steering consumes power during the maneuver. The results of the power break down for the system components show that when the vehicle starts from stand still in the minimum time transients, most of the engine power is allocated to vehicle traction in order to enable fast vehicle acceleration whereas lifting starts later when vehicle has reached high speed and less power is required for traction. By delaying the lifting operation, the amount of distributed engine power among various components remains unchanged while the course of events gets shifted. Finally, it is shown that when the load lifting is delayed, the cycle time slightly increases in both Min T and Min M_f solutions, however, the positive side effect is that the fuel consumption decreases in the Min T case.

References

1. Alshaer B, Darabseh T, Alhanouti M (2013) Path planning, modeling and simulation of an autonomous articulated heavy construction machine performing a loading cycle. Appl Math Model 37:5315–5325
2. Egerstedt M, Hu X, Rehbinder H, Stotsky A (1997) Path planning and robust tracking for a car-like robot. In: Proceedings of the 5th symposium on intelligent robotic systems, pp 237–243
3. Eriksson L, Nielsen L (2013) Modeling and control of engines and drivelines. Wiley, New York
4. Fahroo F, Ross IM (2008) Advances in pseudospectral methods for optimal control. In: AIAA guidance, navigation and control conference and exhibit, pp 18–21
5. Filla R (2011) Quantifying operability of working machines. Dissertation, Linköping University, No. 1390
6. Filla R (2013) Optimizing the trajectory of a wheel loader working in short loading cycles. In: The 13th Scandinavian international conference on fluid power, SICFP2013, 3–5 June 2013
7. Frank B, Skogh L, Filla R, Fröberg A, Alaküla M (2012) On increasing fuel efficiency by operator assistant systems in a wheel loader. In: Proceedings of the international conference on advanced vehicle technologies and integration, pp 155–161
8. Ghabcheloo R, Hyvönen M (2009) Modeling and motion control of an articulated-frame-steering hydraulic mobile machine. In: 17th Mediterranean conference on control and automation, 24–26 June 2009
9. Janarthanan B, Padmanabhan C, Sujatha C (2012) Longitudinal dynamics of tracked vehicle: simulation and experiment. J Terramech 49:63–72
10. Kotwicki AJ (2012) Dynamic models for torque converter equipped vehicles. In: Proceedings of the 14th ASME design engineering technical conference, pp 359–368
11. Koyachi N, Sarata S (2009) Unmanned loading operation by autonomous wheel loader. ICROS-SICE international joint conference, Aug 2009, pp 18–21
12. Murray R, Sastry S (1993) Nonholonomic motion planning: steering using sinusoids. IEEE Trans Autom Control 38(5):700–716
13. Nezhadali V, Eriksson L (2013) Modeling and optimal control of a wheel loader in the lift-transport section of the short loading cycle. In: AAC'13—7th IFAC symposium on advances in automotive Control
14. Nezhadali V, Eriksson L (2013) Optimal control of wheel loader operation in the short loading cycle using two braking alternatives. In: IEEE VPPC—the 9th IEEE vehicle power and propulsion conference
15. Nilsson T, Fröberg A, Åslund J (2013) Fuel and time minimization in a CVT wheel loader application. In: AAC'13—7th IFAC symposium on advances in automotive control
16. Product brochure: Volvo L220G wheel loader (2012). http://www.volvoce.com/construction equipment/na/en-us/products/wheelloaders/wheelloaders/L220G/Pages/specifications.aspx
17. Reeds JA, Shepp LA (1990) Optimal paths for a car that goes both forwards and backwards. Pac J Math 145(2):367–393
18. Sarata S, Weeramhaeng Y, Tsubouchi T (2005) Approach path generation to scooping position for wheel loader. In: Proceedings of the 2005 IEEE international conference on robotics and automation, pp 1809–1814
19. Takahashi H, Konishi Y (2001) Path generation for autonomous locomotion of articulated steering wheel loader. Comput Aided Civil Infrastruct Eng 16(3):159–168

20. TOMLAB 7.9: http://www.tomdyn.com/
21. Tsubouchi T, Sarata S, Yuta S (1998) A practical trajectory following of an articulated steering type vehicle. In: Zelinsky A (ed) Field and service robotics. Springer, London, pp 397–404
22. Walström J, Eriksson L (2011) Modeling engines with a variable-geometry turbocharger and exhaust gas recirculation by optimization of model parameters for capturing non-linear system dynamics. J Automobile Eng 225:960–986
23. Zhang Y, Zou Z, Chen X, Zhang X, Tobler W (2003) Simulation and analysis of transmission shift dynamics. Int J Veh Des 32(3/4):273–289

Author Index

H. Waschl et al. (eds.), *Optimization and Optimal Control in Automotive Systems*,
Lecture Notes in Control and Information Sciences 455,
DOI: 10.1007/978-3-319-05371-4, © Springer International Publishing Switzerland 2014

MIX
Papier aus verantwortungsvollen Quellen
Paper from responsible sources
FSC® C105338

FSC
www.fsc.org

If you have any concerns about our products,
you can contact us on
ProductSafety@springernature.com

In case Publisher is established outside the EU,
the EU authorized representative is:
Springer Nature Customer Service Center GmbH
Europaplatz 3, 69115 Heidelberg, Germany

Printed by Libri Plureos GmbH
in Hamburg, Germany